Philosophy of Science

The Link Between Science and Philosophy

PHILIPP FRANK

A SPECTRUM BOOK

Prentice-Hall, Inc.
Englewood Cliffs, N.J.

PRENTICE-HALL PHILOSOPHY SERIES
Dr. Arthur E. Murphy, Editor

Dedicated to
HANIA
in remembrance of our journey
through the Old and the New World

LIBRARY OF CONGRESS
CATALOG CARD NUMBER: 57-6980

Spectrum paperback edition first published 1962

PRINTED IN THE UNITED STATES OF AMERICA
66409-C

Preface

A little learning is a dangerous thing;
Drink deep, or taste not the Pierian spring;
There shallow draughts intoxicate the brain,
And drinking largely sobers us again.

These famous verses of Alexander Pope hit perhaps no field of knowledge as adequately as the philosophy of science. Distances in space and time have shrunk immensely because of the advances of science in the nineteenth and twentieth centuries, and the power put into the hands of men has increased to a hardly imaginable degree; but a great many critics of contemporary civilization would point out that, despite these achievements, mankind has not become happier and is facing today dangers which have their sources in these very achievements of science. The responsibility for this unfortunate situation is ascribed by some authors to the fact that today the social sciences have advanced much more slowly than the physical sciences, and other authors like to point at the paucity of support given to moral and philosophical knowledge compared with knowledge of the material world. If we were to express ourselves in a more abstract and conceptual way, we would say that the rift between science and philosophy has been blamed for the inability of science to make its advances a blessing for men. Quite a few religious, educational, and even political leaders have advised that the advance of science should be retarded administratively in order to accelerate advances in the humanities. Such a promotion of moral values by administrative and financial means would hardly be feasible and cannot be sincerely desired by governments and responsible social groups. It is now a matter of common knowledge that quite a few representative groups have deplored the fact that this country is lagging behind other countries in the training of scientists, countries that are our political and economic rivals. These groups request larger appropriations for the training of natural scientists. How can these conflicting aims be reconciled? A main purpose of the present book is to show that one does not need to diminish research and teaching in science in order to enhance interest in the moral and philosophical aspects of the world.

The fact is stressed that the deeper we dig into actual science the more its links with philosophy become obvious. As the title suggests, this book regards the philosophy of science as the "missing link" for which we have to look. Presentations of this field have very often started from a concept of science that has been half vulgar and half mystical. Other presentations have linked science with a philosophy that has actually been a mere system of logical symbols without contact with the historical systems of philosophy. But these very philosophies have served as support for ways of life and, specifically, for religious and political creeds.

In this book, we attempt to start from the way in which science is understood by the scientist in his most creative and critical moods. However, we shall also attempt to establish links with the historical types of philosophy like idealism and materialism that have actually served as supports of moral, religious, and political creeds.

I have worked about five years on this book and have discussed its approach with audiences of various kinds: with the undergraduates of Harvard College and the Massachusetts Institute of Technology, with the students in the Harvard Graduate School of Physics, and with the Adult Education Classes of the New School of Social Research in New York among others.

In proofreading and editing I had the valuable help of Ralph Burhoe (Executive Secretary of the American Academy of Arts and Sciences). In typing and editing I was supported by Alice Atamian, Jean Brockhurst, Harriet Drell, and Rita Fernald. The figures were drawn by Henry Fernald. The Editorial Department of Prentice-Hall, Inc., has been very helpful in editing my manuscript.

Cambridge, December, 1956   PHILIPP G. FRANK

Contents

v

Introduction: Of What Use Is the Philosophy of Science?

1. The Rift Between Science and Philosophy

When we examine the most creative minds in twentieth-century science, we find that the greatest ones have strongly stressed the point that a close tie between science and philosophy is indispensable. Prince Louis de Broglie, who created the wave theory of matter (de Broglie waves), writes:[1]

> In the nineteenth century there came into being a separation between scientists and philosophers. The scientists looked with a certain suspicion upon the philosophical speculations, which appeared to them too frequently to lack precise formulation and to attack vain insoluble problems. The philosophers, in turn, were no longer interested in the special sciences because their results seemed too narrow. This separation, however, has been harmful to both philosophers and scientists.

Quite frequently we hear from teachers of science that students who are devoted to serious research in science will not bother with idle philosophical problems. However, one of the most creative men in twentieth-century physics, Albert Einstein, writes:[2]

> I can say with certainty that the ablest students whom I met as a teacher were deeply interested in the theory of knowledge. I mean by "ablest students" those who excelled not only in skill but in independence of judgment. They liked to start discussions about the axioms and methods of science and proved by their obstinacy in the defense of their opinions that this issue was one important to them.

This interest in the philosophical aspect of science shown by creative and imaginative minds is understandable if we recall that fundamental changes in science have always been accompanied by deeper digging toward the philosophical foundations. Changes like the transition from the Ptolemaic to the Copernican system, from Euclidean to non-Euclidean geometry, from Newtonian to relativistic mechanics, and to the four-dimensional curved space have brought about a radical change in our common-sense explanation of the world. From all these considerations everyone who is to get a satisfactory understanding of twentieth-century science will have to absorb a good deal of philosophical thought. But he will soon feel that the same thing holds for a thorough understanding of the science which originated in any period of history.

2. The Missing Link Between Science and the Humanities

A great many authors in different walks of life have deplored a great threat to our present civilization: the deep rift between our rapid advance in science and our failure in the understanding of human problems, or, in other words, the rift between science and the humanities that in former periods has been bridged by liberal education.[3]

The fading away of liberal education was sharply dramatized by Robert Hutchins[4] in his remarks on the place of "philosophy" in our universities. At all periods before the nineteenth century, philosophy and theology were the central subjects at every institution of higher learning. All the special fields of knowledge were coordinated by the ideas presented in the courses on philosophy. In the nineteenth and twentieth centuries "philosophy" has become a department among other departments like mineralogy or slavic languages or economics. If the scientists were consulted, the majority of them would regard "philosophy" as one of the least important departments. In the traditional teaching a "link is missing" in the chain that should connect science with philosophy. If it is assumed that man is descended from the animal world, we need, in order to confirm this theory, to discover the "missing link" between ape and man, between nature and mind. Hutchins writes:[5]

The aim of higher education is wisdom. Wisdom is the knowledge of principles and causes. Therefore, metaphysics is the highest wisdom. . . . If we cannot appeal to theology, we must turn to metaphysics. Without theology or metaphysics a university cannot exist.

He states bluntly that a metaphysics that exists independently of science and is of eternal validity is a necessary basis of any meaningful university education. Instead of relegating philosophy to a special department, Hutchins suggests that:

In an ideal university the student would not proceed from the most recent observations back to the first principles, but from the first principles to whatever recent observations we claim significant in understanding them. . . . The natural sciences derive their principles from the philosophy of nature which, in turn, depends on metaphysics. . . . Metaphysics, the study of the first principles, pervades the whole. . . . Dependent on it and subordinate to it are the social and natural sciences.

This program is obviously based upon the belief that there are philosophical principles which are independent of the advances of sciences, but from which instead the statements of science, natural and social, can be derived.

The trouble with such a program is, of course, the problem of finding these principles of permanent validity. As a matter of fact, the permanence of the philosophical principles can be kept up and guaranteed only by spiritual or secular authorities or both. No university education can be based on metaphysics unless the choice of it is decided by an authority that is permanently in control of the teaching.

3. Science as the Balance of Mind

Although the choice of a permanent metaphysics does not seem to be feasible, the main contention of Hutchins, the need for university education based upon principles, is in agreement with the requests of a broadminded philosopher and scientist like Alfred North Whitehead. He writes:[6]

The spirit of generalization should dominate a university. The lectures should be addressed to those to whom details and procedure are familiar. That is to say, familiar at least in the sense of being so congruous to pre-existing training as to be easily acquirable. During the school period the student has been mentally bending over a desk;

at the university he should stand up and look around. . . . The function of the university is to enable you to shed details in favor of principles.

However, what Whitehead calls "principles" are not the statements of a "perennial metaphysics" which Hutchins suggested as the basis of every university. Whitehead says: "The ideal of a university is not so much knowledge as power. Its business is to convert the knowledge of a boy into the power of a man." From our knowledge of facts we proceed to general principles by the method that we learn in science. In his opening address as professor of philosophy and history of science at the University of London, Herbert Dingle spoke about the "Missing Factor in Science" in 1947.[7] He said:

> It is my task to inquire how it has come about that a generation so amazingly proficient in the practice of science can be so amazingly impotent in the understanding of it, and the thesis I wish to propose is that the state of unselfconscious automatism in which science finds itself today is due to the lack throughout its history of a critical school working within the scientific movement itself and performing the function, or at least one of the functions, which criticism has performed for literature from the earliest times.

Science has to do, on the one hand, with hard stubborn facts, and on the other hand, with general ideas. What science teaches us is the correlation between both. The chief thing university educators should give to students is interest in the possibility of coordinating stubborn facts by means of abstract principles. This is the most fascinating topic of university education. Whitehead says about it:[8]

> This balance of mind has now become part of the tradition which infects cultivated thought. It is the salt which keeps life sweet. The main business of universities is to transmit this tradition as a widespread inheritance from generation to generation.

We need a full understanding of the principles of physics or biology, an understanding not only of logical argument but also of psychological and sociological laws; briefly, we need to complement the science of physical nature by the science of man. By pursuing the work of empirical science we shall strive towards the same goal that men like Hutchins have wanted to reach by unvariable metaphysical dogmas. In order to understand not only science itself but also the place of science in our civilization, its relation to ethics, politics, and religion, we need a coherent system of concepts and laws within which the natural sciences, as well as philosophy and the humanities, have their place. Such a system may be called "philosophy of science," it would be the "missing link" between the sciences and the humanities without introducing any perennial philosophy that could only be upheld by authorities.

The need for this "missing link" has been felt strongly among our college students during recent years. The Harvard Student Council set up a Committee on the Curriculum which produced in 1942 a report in which a letter from a Nevada boy to Dartmouth College was quoted:

> We believe that a liberal education should give a picture of the inter-related whole of nature, including man as the observer. . . . We ask that a liberal education give a real philosophy of knowledge based on facts. . . . A good teacher can show the relationship between his course and other courses.

4. Is the Scientist a "Learned Ignoramus"?

About a century ago the rift between science and the humanities in our present world was traced back by Ralph Waldo Emerson[9] to the lack of human appeal in science teaching. He wrote:

> "There is a revenge for this inhumanity. What manner of men does science make? The boy is not attracted. He says: I do not wish to be the kind of man my professor is."

There is scarcely any doubt that the teachers of philosophy, history, or English have a much greater influence upon the intellectual and emotional make-up of the average college student than the teachers of mathematics or chemistry.

Quite a few of our authors have stressed the point that a great danger for our Western culture may emerge from our educational system which trains overspecialized scientists, who are glorified by public opinion. Perhaps no author has characterized this situation as lucidly and aptly as the Spanish philosopher Ortega y Gasset.[10] In his book, *The Revolt of the Masses*, he writes about the scientist of our century that "science itself—the root of our civilization—automatically converts him into mass-man, makes of him a primitive, a modern barbarian." On the other hand, the scientist is the most adequate representative of our twentieth-

century culture, he is "the high-point of European humanity."
Nonetheless, according to Gasset, the scientist who has received
an average training is today

> ignorant of all that does not enter into his specialties and "knows."
> We shall have to say that he is a *learned ignoramus*, which is a very
> serious matter as it implies that he is a person that is ignorant, not
> in the fashion of the ignorant man, but with all the petulance of one
> who is learned.

Our author claims that the organization of scientific research
allows people who are intellectually very commonplace to achieve
important results and to become unduly complacent.

> A fair amount of things that have to be done in physics or biology is
> mechanical work of the kind which can be done by anyone, or almost
> anyone. For the purpose of innumerable investigations it is possible
> to divide science into small sections and to enclose oneself into one
> of them and to leave out of consideration all the rest. . . . In order
> to obtain quite abundant results it is not even necessary to have
> rigorous notions of their meaning and foundation.

The passage quoted from Ortega y Gasset certainly does not
describe the scientific work of men like Newton or Darwin or, for
that matter, of Einstein or Bohr, but it characterizes fairly well the
way in which the "scientific method" is described in textbooks and
classrooms where an attempt is made to "purge science of philosophy" and where a certain routine type of science teaching has been
established. Actually great advances in sciences have consisted
rather in breaking down the dividing walls, and a disregard for
meaning and foundation is only prevalent in periods of stagnation.

If the scientists, who play an immense social role in our present
world, are not to become a class of learned ignoramuses, the education of these men must not restrict itself to the purely technical
approach but must give full attention to the philosophical aspect
and place of science within the general domain of human thought.

5. Technological and Philosophical Interest in Science

The thrill of advance in science has not always arisen from technical innovations that were invented to make human life more
pleasant or unpleasant, like television or atomic energy. The
Copernican system, according to which our earth is moving through
space, brought about a description of the world that could not be
expressed in terms of the common-sense concepts which man had

developed in order to describe the rest and motion in his daily-life experience. Newton's mechanics introduced concepts of "force" and "mass" that were in disagreement with the common-sense meanings of these words. These new theories aroused excitement far beyond the small group of scientists and philosophers; the interest in them surpassed the interest in many purely technical advances.

Such a phenomenon has repeated itself again and again in intellectual history. He who received his education in the first quarter of this century was witness to the thrill produced by the announcement of Einstein's Theory of Relativity, which could not be formulated in terms of the common-sense concepts that had served through ages to describe our experiences about spatial and temporal distances. In a similar way, the theory dealing with the behavior of atomic and subatomic particles (Quantum Theory) could not be formulated by using the common-sense concepts of velocity and position, cause and effect, freedom and determinism. We have seen that at all periods the effect of scientific advance on the common-sense interpretation of nature has been strong and has stimulated interest in science no less than the effect on technical progress has done.

The interest in science that is not due to its technical applications but to its impact upon our common-sense picture of the world we may call briefly the "philosophical" interest. Science teaching in our schools of higher learning has for the most part ignored this philosophical interest, and has even proclaimed it a duty of the teacher to present science as completely isolated from its philosophical implications. As the result of this kind of training the position of the science teachers among their fellow citizens has somehow become unsatisfactory. In the columns of magazines that are devoted to cultural problems and in the pulpits of our churches of all denominations it has been claimed that twentieth-century science has made a great contribution towards the solution of urgent human problems: a reconciliation between science and religion, the refutation of materialism, the re-establishment of the belief in the freedom of will and moral responsibility. In other quarters, however, it has been claimed that modern science has supported materialism or relativism, and contributed to undermining the belief in absolute truth and moral values. To prove these points, contemporary physical theories like the Theory of Relativity and the Quantum Theory have been invoked.

When we ask a trained physicist (not to speak of a graduate in engineering) for his opinion on these questions, we immediately notice that his training in physics has not provided him with any judgment. The graduate in science will, as a matter of fact, often be more helpless than an intelligent reader of popular science magazines. A great many degree holders in physics and engineering will be unable to give any but the most superficial answer, and even this superficial answer will not be the result of their professional training, but the profit they have made from reading some popular articles in newspapers or other periodicals. Moreover, many of them will not even venture to give a superficial answer, but will just say, "This is not my field, and that is all there is to it." If intellectual curiosity is not satisfied by the science teacher, the thirsty student takes his spiritual drink where it is offered to him. In the best case, he gets information from some good popular magazine, but it can be worse, and he can become a victim of people who interpret science in the service of some pet ideology which in many cases has been an antiscientific one. They have claimed that the physical theories of our century have "abandoned rational thinking" in favor of —I do not know exactly what, as I cannot imagine what alternative exists to rational thinking in science.

It may seem paradoxical, but the dodging of philosophical issues has very frequently made science graduates captives of obsolete philosophies. This result of the "isolationist" attitude in science teaching has often been denounced by those scientists who have given intensive thought to philosophy. Every child acquires through his education a common-sense picture of the world, briefly speaking, a "philosophy." He learns how to use words like "rest and motion," "time and space," "matter and mind," "cause and effect," etc. This vocabulary is closely related to the vocabulary in which the "do's" and "don'ts" that are to govern a child's behavior are expressed. This philosophy acquired during childhood and adolescence has only too frequently remained the common-sense belief of the grown up scientist in all fields where he is not a "specialist." On the other hand, within science itself this "common-sense philosophy" has frequently been superseded by a more critical philosophy, by an abandonment of the common-sense language. The most conspicuous example is that of the changes in the conceptual scheme in speaking about "rest and motion," starting with Copernicus and continuing in our period through the work of men like Einstein and Bohr.

6. Obsolete Philosophies in the Writings of Scientists

In this way, the students of science have acquired a "double personality," a kind of schizophrenia due to the contrast between his scientific thought and his childhood philosophy. Perhaps no one has formulated these facts as lucidly as Alfred North Whitehead, equally great in science and philosophy. He starts[11] from the remark that during a period of little change in science some basic principles have not been challenged over a long period and can be accepted without much criticism. He writes:

> It is legitimate (as a practical counsel in the management of a short life) to abstain from the criticism of scientific formulations so long as the superstructure "works." But to neglect philosophy when engaged in the reformation of ideas is to assume the correctness of the chance philosophic prejudices imbibed from a nurse or a schoolmaster or current modes of expression.

Whitehead speaks of "chance philosophy" because it depends upon the chance of our birth what philosophy we imbibe during our childhood. He points out precisely the factors that determine this "philosophy": our preschool education, school including Sunday school, and even the vocabulary and syntax of the language in which we are educated. The behavior of scientists who, without questioning, stick to the chance philosophy of their childhood has, according to Whitehead, an analogy in the field of religion: the behavior of those "who thank Providence that they have been saved from the perplexities of religious inquiry by the happiness of birth in the true faith."

Since their childhood philosophy is often retained by scientists in spite of changes in scientific thought, it very frequently happens that the presentations of science contain as inclusions remnants of obsolete philosophies. This point was very strongly brought out by Ernst Mach who was, like Whitehead, equally penetrating in science and philosophy, although he advocated very different views. Both, however, agreed fully in the assertion that without critical philosophy science will itself become the vehicle of obsolete philosophies. Mach wrote:[12]

> The domain of the transcendent is inaccessible to us. . . . I confess, however, frankly that its inhabitants are not able to stimulate my curiosity. I am no philosopher but only a scientist. . . . However, I don't want to be a scientist who follows blindly the guidance of a single philosopher, as the patient in Moliere's comedy is expected to follow the guidance of his physician. . . . I have not attempted to introduce some new philosophy into science but rather to remove an old and obsolete one. . . . There are some fallacies that have been noticed by philosophers themselves. . . . They have been kept alive longer in science where they have faced less alert criticism, as a species of animal that is incapable of surviving on the mainland can be spared on a remote island where there are no enemies. . . .

However, these relics of obsolete philosophies in science have been denounced by men whose background and aims have been very different from Mach's and Whitehead's. We may quote the example of Friedrich Engels,[13] the most intimate collaborator of Karl Marx[14] in his scientific, philosophical and political endeavor. He wrote:[15]

> Natural scientists believe that they free themselves from philosophy by ignoring it or abusing it. They cannot, however, make any headway without thought, and for thought they need thought determinations. But they take these categories unreflectingly from the common consciousness of so-called educated persons who is dominated by relics of long obsolete philosophies, or from the little bit of philosophy compulsorily listened to at the university (which is not only fragmentary, but also a medley of views of people belonging to the most varied and mostly the worst schools) or from uncritical reading of philosophical writings of all kinds. Hence they are no less in bondage to philosophy, and those who abuse most are slaves to precisely the worst vulgarized relics of the worst philosophies.

Since the philosophy of Karl Marx and Friedrich Engels, Dialectical Materialism, has become the official philosophy of the Soviet Union and its affiliated states, the views expressed in this quotation have had a far-reaching influence upon the attitude of the Soviet Government towards science. Every presentation of science was searched for the hidden philosophy which might be hostile to the philosophy of the ruling party. This argument has served in many cases as a pretext for regimentation of science by the state.

7. Information or "Understanding"?

In our time, the government has to devote a great deal of attention and financial support to scientific research.   In a democracy, no government could embark on such a program unless it were supported by its citizens; but they would not support it unless they understood what it was all about.   The problem arose of how citizens could learn to judge reports of experts, *e.g.*, about appropriations for research projects, without being specialists in science themselves.   James Bryant Conant writes:[16] "Every American citizen in the second hâlf of this century would be well advised to try to understand both science and the scientists as best he can."

Many people have believed that this goal could be achieved by popularizing the results of science, by adult education courses in which intelligent and interested men and women could absorb the "facts" discovered by scientists in a digestive way.   Conant, however, made the point that by absorbing "results" and "facts" laymen could not acquire any judgement about reports of scientists. What the citizen needs rather is an understanding of how the mind of the scientist works in getting results, and along with it, in what sense those results are "valid" or "reliable" and can be used as a basis of judgement.   According to Conant,

> The remedy does not lie in greater dissemination of scientific information among nonscientists.   Being well-informed about science is not the same thing as understanding science. . . . What is needed is methods for imparting some knowledge of the tactics and strategy of science to those who are not scientists.[17]
> What blocks the inexperienced person who attempts to examine critically proposals advanced by scientific experts is his ignorance of the way such experts think and talk.[18]

The systematic way to the understanding of science as well as to the tactics and strategy of science is the main content of any philosophy of science.

8. Footnotes for the Introduction

1. Louis De Broglie, *L'Avenir de la Science* (Paris: Plon, 1941).

2. Albert Einstein, in his obituary on Ernst Mach, *Physikalische Zeitschrift*, Vol. 17 (1916), page 101*ff.*

3. The *Encyclopaedia Brittanica* (Volume 7, article on "education") discusses the antithesis between "vocational" and "liberal" education.   The former stresses immediate needs, the latter "takes rather a long view of the life and the needs of

the community." It attempts to develop the abilities of the student by a "liberal arts" curriculum, including mathematics, basic sciences, grammar literature, history, etc.

4. Robert Maynard Hutchins (1899–   ), American educator, President and Chancellor of the University of Chicago from 1929 to 1951.

5. *Higher Learning in America* (New Haven: Yale University Press, 1936.

6. Alfred North Whitehead (1861–1947), printed in the inexpensive Mentor Book Edition, *Aims of Education*.

7. Herbert Dingle (1890–   ), British scientist and philosopher, Professor of History and Philosophy of Science, University College, London.

8. Alfred North Whitehead, *Science in the Modern World*, Mentor Books.

9. Ralph Waldo Emerson (1803–1882), American essayist and poet; *Essays on Representative Man*, 1849; *Nature*, 1836; *The Conduct of Life*, 1860.

10. José Ortega y Gasset, (1883–   ), Spanish essayist and philosopher. The original of his book, *The Revolt of the Masses*, appeared in 1930. It is now printed in the inexpensive Mentor Books edition.

11. In his book, *The Principle of Relativity* (London: Cambridge University Press, 1922).

12. Ernst Mach (1838–1916), Austrian physicist, psychologist, and philosopher. In his book, *Erkenntnis und Irrtum* (*Knowledge and Error*) (J. A. Barth: Leipzig, 1905).

13. Friedrich Engels (1820–1895), German political philosopher and advocate of socialism.

14. Karl Heinrich Marx (1818–1883), German political philosopher and socialist leader. Marx and Engels produced, in close cooperation, the philosophy of "Dialectical Materialism."

15. In his book, *Dialectics of Nature*. The manuscript of this book was written by Engels between 1873 and 1882, but it was not finished when the author died, and was not published until 1925. It appeared as Volume II of the publications of the Marx and Engels Archive, Moscow, in the German and Russian languages. The English translation, *Dialectics of Nature* (New York: International Publishers, 1940), has a preface by John B. S. Haldane. This book has become the official basis for every presentation of the Philosophy of Science in the Soviet Union.

16. James Bryant Conant (1893–   ), American scientist, educator and diplomat, President of Harvard University 1933–1953, United States High Commissioner in Germany 1953–1955, United States Ambassador to Germany, 1955–   . Published: *On Understanding Science* (New Haven: Yale University Press, 1947); *Science and Common Sense* (New Haven: Yale University Press, 1951).

17. J. B. Conant, *Science and Common Sense, op. cit.*

18. I. B. Cohen and F. G. Watson, *General Education in Science*, (Cambridge, Mass.: Harvard University Press, 1952).

1

The Chain That Links Science With Philosophy

1. Facts and Concepts

In his poem "Sonnet to Science," Edgar Allan Poe[1] indicts science as follows:

> Science! true daughter of Old Time thou art
> Who alterest all things with thy peering eyes.
> Why preyest thou thus upon the poet's heart,
> Vulture, whose wings are dull realities?
>
> . . . .
>
> Hast thou not dragged Diana from her car?
> And driven Hamadryad from her wood?
>
> . . . .

The modern scientist will hardly agree that his science consists of "dull realities." The more we study science, the more we shall notice that science is neither "dull," nor that it speaks of "realities." The "car of Diana" is much nearer to the "dull realities" of our everyday life than the symbols by which modern science describes the orbits of the celestial bodies. "Goddesses" and "nymphs" look much more like people we meet in our everyday life than the electromagnetic field, the energy or the entropy that populates the "unseen universe," which, according to modern science, accounts for the "dull realities" of our direct sense observation.

When we speak of science, we always speak on two levels of dis-

1

course or abstraction.   The first of these is the level of everyday common-sense experience; *e.g.*, we observe some dark spot moving with respect to some other dark spots.   This is the level of direct observation; laboratory reports deal with these simple facts of experience.   One could analyze these simple experiences from the psychological point of view, but we shall not do that here; we shall take it for granted that we all share these experiences.   By this, we do not mean to imply that these simple experiences cannot be discussed in a more profound way, but simply that this discussion does not belong to the philosophy of science.   The second level to which we have referred is that of the general principles of science. This is completely different from the level of common-sense experience.   The latter can be shared by all; the former employs language very far from that of everyday experience.   Science consists essentially of these general principles.   A collection of mere statements about dancing spots is not science.   The central problem in the philosophy of science is how we get from common-sense statements to general scientific principles.   As we have said, these common-sense experiences and statements are understood and accepted by all.   This basis of acceptance is well characterised in the lines of the great American poet, Walt Whitman:[2]

> Logic and sermons never convince,
> The damp of the night drives deeper into my soul,
> Only what proves itself to every man and woman is so,
> Only what nobody denies is so.

Statements of this type are: "In this room stands a round table. Now this table is removed from this room into the adjacent room." Or: "On this scale the pointer coincides with a mark between two and three; now the position of the pointer changes and it covers a mark between three and four."   A general agreement is certainly possible about statements of this type.   We do not claim that such statements describe a "higher reality" than other statements; nor do we pretend that the world described is the "real" world.   We make such statements the basis of all science only because there can be achieved a general agreement among men of average education whether, in a specific case, such statements are "true" or not. We may refer to discourse consisting of such statements as common-sense discourse, or everyday discourse. It "is so," to Walt Whitman, because it "proves itself to every man and woman."

But the situation is completely different if we consider general statements formulated in abstract terms like the "Law of Inertia," or the "Conservation of Energy." Whether we call them principles or premises or hypotheses or generalizations, one thing is certain: We cannot achieve about them a general understanding of the kind we can achieve about common-sense statements. Therefore, naturally, the question arises: Why do we accept some general scientific statements and not others? What are the causes of our acceptance of these general statements? This is partly a psychological and sociological problem. The general statements of physical science are not simply empirical facts. The fact is that people advance and accept these general principles: This fact, however, belongs not to physics but to, say, psychology or anthropology. Thus we see that even the philosophy of physical science is not exhausted by physics itself. In physics, we learn some of the reasons why these general principles are accepted, but by no means all of them. The philosophy of science is part of the science of man, and indeed, we shall not understand it unless we know something of the other sciences of man, such as psychology, sociology, etc. All the reasons for the acceptance of the general principles of science belong to the philosophy of science. What is actually the relationship between common-sense experience and these general principles? Is mere common-sense experience sufficient? Are the general statements of science uniquely determined, or can the same set of common-sense experiences give rise to different general statements? If the latter, how can we choose one of these general statements rather than another? How do we get from the one—common-sense experience—to the other—the general statements of science? This is the central problem of the philosophy of science.

We might describe here, in a preliminary and perfunctory way, what the relationship between science and philosophy is. If we speak in the ordinary way of a chain that connects common-sense experience with the general statements of science, at the end of this chain, as the statements become more and more general, we may place philosophy. We shall see that the more one goes into generalities, the less uniquely are the latter determined by direct observations, and the less certain they are. For the moment we shall not go further into the distinction between science and philosophy. We shall discuss this later.

2. Patterns of Description

By collecting and recording a large stock of common-sense experience in a certain field, we may produce long lists of pointer readings or descriptions of dancing colored spots.   But by mere recording, accurate and comprehensive as it may be, we do not obtain the slightest hint as to how to formulate a theory or hypothesis from which we may derive in a practical way the results of our recording. If we simply set as the problem the finding of an hypothesis which would be in fair agreement with our records, it does not seem possible for us to obtain an unambiguous result.   As early as 1891, C. S. Peirce[3] wrote:

> If hypotheses are to be tried haphazard, or simply because they will suit certain phenomena, it will occupy the mathematical physicists of the world say half a century on the average to bring each theory to the test, and since the number of possible theories may go up into the trillion, only one of which can be true, we have little prospect of making further solid additions to the subject in our time.[4]

If we make an attempt to set up a theory or hypothesis on the basis of recorded observations, we soon notice that without any theory we do not even know what we should observe.   Chance observations usually do not lend themselves to any generalization. It is perhaps instructive at this point to peruse a passage from Auguste Comte's *Course of Positive Philosophy*.[5]   Comte has been regarded as the father of a school of thought known as "Positivism." According to an opinion frequently held by philosophers, he and his school have extolled the value of observations and minimized, or even rejected, the formation of theories by creative imagination. However, he writes:

> If, on the one hand, every positive theory must necessarily be based on observations, it is equally sensible, on the other hand, that in order to carry out observations our minds need some theory.   If, in contemplating the phenomena, we did not attach them to some principles, it would not be possible to combine these isolated observations and to draw from them any conclusions.   Moreover, we would not even be

able to fix them in our minds.   Ordinarily these facts would remain unnoticed beneath our eyes.

Hence, the human mind is, from its origin, squeezed between the necessity to form real theories and the equally urgent necessity to create some theory in order to carry out sensible observations.   Our minds would find themselves locked within a vicious circle, if there were not, fortunately, a natural way out through the spontaneous development of theological concepts.[6]

The theological concepts are very near to common-sense experience.   They interpret the creation of the world by the gods as analogous to the making of a watch by a watchmaker.   We shall see later that this kind of analogy has been the basis of all metaphysical interpretations of science.   At this point, we must be distinctly aware of the fact that a mere recording of observations provides us with nothing but "dancing spots," and that "science" does not begin unless we proceed from these common-sense experiences to simple patterns of description, which we call theories. The relationship between direct observations and the concepts that we use in "scientific description" are the main topics with which any philosophy of science is concerned.

Let us take a relatively simple example, where this relationship is rather direct.   Let us imagine that we launch a body into the air— say, a remnant of cigarette paper—what does it do?   If we do this many times—a hundred, a thousand, hundreds of thousands of times—we shall find simply that the motion is different every time. The accumulation of all these observations is obviously no science. And this is not the way that the physicist works, unless it is in a field that is very little advanced, about which he knows almost nothing.   If we study physics, we learn some rules—for uniform motion, for accelerated motion, for combinations of uniform and accelerated motions.   These are schemes of description.   We must invent them before we can check them, but how are we to invent these schemes?   The human imagination enters here.   We try to imagine some simple scheme.   But what is simple?   We must try out all such different imagined schemes to see if the actual motion of our falling paper is approximately described by any one of them. In textbooks of physics one finds the statement that these schemes are "idealized motion."   This is a very misleading expression; it refers to a metaphysical doctrine which maintains that for every empirical object there is a corresponding idea of it.   The result of "idealization" is entirely arbitrary.   By the word "idealizing" you say nothing except that you compare some empirical object with some "idea" that you have invented.   There is the question of the purpose of your making this invention or "idealization": for example, for some problems it would be more useful to idealize the ordinary atmosphere as a very thick medium, for others as empty space.

Now let us return to the question of the falling cigarette paper. In the mechanics of today, we compare every motion with a scheme that is the motion of a mass point in empty space.   We consider two types of motion as the components of the motion of a launched body, a uniformly accelerated motion downward and a uniform motion horizontally.   The first of these we call gravitational motion and the second inertial motion.   From this scheme we can derive many useful things, but not everything.   This analysis is approximately correct for thin air but not so much for a medium of high viscosity.   We need the invention of another scheme if we want to compute the effect of a dense or viscous medium.

The pattern by which we describe motion in thin air is a motion of constant "acceleration."   The concept of acceleration is very remote from the dancing spots of our direct observations.   If the position of the moving body is described mathematically by an arbitrary function of time, the acceleration is described by the computation of "second derivatives with respect to time" in the sense of differential calculus.   To observe the equivalent of a "second derivative" in the domain of common-sense experience would mean to carry out a very great number of extremely delicate pointer readings; we must not forget that the "second derivative" is defined as the limit of an infinite set of values.

We can, therefore, say that the experimental scientist does not observe at all the quantities that occur in the patterns of scientific description, in the laws of science.   Suzanne Langer[7] in her book *Philosophy in a New Key*, writes:

> The men in the laboratory . . . cannot be said to observe the actual objects of their curiosity at all. . . .   The sense data on which the propositions of modern science rest are, for the most part, little photographic spots and blurs, or inky curved lines on paper. . . .   What is directly observable is only a sign of the "physical fact"; it requires interpretation to yield scientific propositions.[8]

3. Understanding by Analogy

We shall, for the time being, consider motion only in very thin air. Is the human mind then satisfied if it knows this scheme of constant acceleration? No, it asks *why* does it accelerate downward and go with uniform motion horizontally? If you want to explain this to a schoolboy (and in a sense we are all schoolboys of the world), you say that it accelerates downward under the influence of the attraction of the earth. But if you think a little, you realize that this is no explanation at all. What is attraction? In medieval times, explanations were always anthropomorphic, and consisted of a comparison with human actions. It was believed that heavy objects wanted to get as close as possible to the center of the earth. The closer they approached, the more jubilant they became and the faster they went. Although more sophisticated today, we still use the concept of attraction. If we record the positions of the falling cigarette paper, we act on the level of everyday experience. But we try to "understand" the general law of its motion by comparing it directly with attraction, which is a psychological phenomenon of our everyday life. We are not satisfied to introduce everyday experience solely by direct observations of the falling body.

It is harder to explain the uniform motion of the body. We say that it is caused by inertia; we all know what this means because we know from everyday experience that we are inert. Inertia means sluggishness, the lack of a desire to move. For example, there must be some external inducement to get up in the morning—some class that must be attended, or the expectation of a good breakfast. The law of inertia seems very plausible to us on the basis of this comparison. We only wonder why it took so many thousands of years for man to discover it. However, this method of explanation by introducing the experience of our own sluggishness is quite arbitrary. Things are not so simple as they seem.

If we are in bed in a train, we cannot determine simply from our own sluggishness whether without effort we will stay in bed or be thrown out. If the train stops or changes its speed, our "sluggishness" does not help us to stay at rest in bed. What really happens is that "without effort" we keep our velocity with respect to some physical masses. In the example of the train this mass is our earth. But from the example of the Foucault pendulum or the deviation of launched projectiles by the rotation of the earth, we can see that the earth is only a substitute for some larger mass with respect to which we keep our velocity; for instance, the mass of our galaxy. And we shall see later that even this is not completely correct. In any case, the analogy of the everyday experience of sluggishness predicts the observable effects of motion only in a very vague way, which is useful only under very special circumstances. What really matters in physical science is the abstract scheme: Every velocity will remain constant with respect to some specific mass which constitutes what we call an inertial system. Comparison with the phenomena of everyday life will not show any inconsistency with this scheme. Sluggishness has only as vague an analogy to inertia as attraction has to gravitation.

If we find a simple scheme for a group of phenomena—*e. g.*, constant acceleration for a body falling in thin air—we are apt to think as follows: "The motion with precisely constant acceleration is an idealization of the actual fall of a body in thin air." The word "idealization" hints that we omit the accidental deviations of the actual motion, and retain only the "essential part of the motion," the uniformly accelerated motion. To the scientist, the term "essential" means "pertinent to reaching the intended goal." As far as our example is concerned, it means "pertinent to the simplest and most practical description of a fall in thin air."

In this way we can distinguish between the "essential" and the "accidental" components of a certain motion. There has been an urge, however, to put more general questions, such as: What are the "essential properties" of motion in general? or, what is the "essence of motion"? If we wish to use the term "essence" in the same way as in the special case, we would mean by "essential properties" of a thing those properties which are necessary to achieve a certain purpose. Without specifying a purpose, the term "essential" has no distinct meaning, unless there is a purpose which is taken for granted and does not need to be mentioned.

If an object is built by men—*e.g.*, a house—it is clear that the "essential properties" of the house are those which are important to the builder, the properties that make it a good house to live in or a house which can be sold at a large profit. We can, therefore, speak about the essence of a natural object, a stone or an animal or a human being, only if we assume that their maker had a definite purpose in making them.

If we speak of the "essence" of natural objects, we regard those objects as analogous to artificial man-made objects.[9]   This analogy is either implicitly assumed or made explicit by referring to the maker of the physical world.   We shall return to this way of speaking later on, when we shall discuss the metaphysical interpretations of science.

4. Aristotle's Scheme of Natural Science

We noted (in Section 1) that we must speak of science on two levels.   One we called the level of everyday common-sense experience, that is, the level of direct observation.   The other is the level of the general principles of science.   It is not too much to say that most of the misunderstandings in the philosophical interpretation of science have arisen because the distinction between these two levels and the way in which they are connected have not been clearly understood.   In the whole history of philosophy, these two levels of immediate experience and abstract sentences have played a great role.   Professor F. S. C. Northrop[10] is concerned with this distinction in his well-known book, *The Meeting of East and West*. He considers the distinction between Eastern philosophy (Indian, Chinese) and Western philosophy (English, French, German) and concludes:

> The Oriental portion of the world has concentrated its attention upon the nature of all things in their emotional and aesthetic, purely empirical and positivistic immediacy.   It has tended to take as the sum total of the nature of things that totality of immediately apprehended fact which in this text has been termed the differentiated aesthetic continuum.   Whereas the traditional West began with this continuum and still returns to local portions of it to confirm its syntactically formulated, postulationally prescribed theories of structures and objects, of which the items of the complex aesthetic continuum are mere correlates or signs, the East tends to concentrate its attention upon this differentiated aesthetic continuum in and for itself for its own sake.[11]

To speak more simply: The differentiated aesthetic continuum is the central object of Eastern philosophy.   Western philosophy begins with this and sets up theories; if it wishes to test a theory, it goes back to this.   The chief object of Western philosophy is not this differentiated aesthetic continuum but abstract rules like the conservation of mass, energy, etc.   I do not know whether this distinction between Eastern and Western philosophy is correct or not. Whatever may be true about East and West, however, one thing is certain—that there are these two approaches, immediate sense experience and conceptual structures.

In order to give a clear and simple presentation of that "Western" approach to science and philosophy, we may start from Aristotle,[12] whose writings provide the oldest attempt at a systematic approach to science and philosophy. In his book on physics (both physics and the philosophy of physics were included in this ancient work), he describes "the natural path of investigation." He says:

> The natural path of investigation starts from what is more readily knowable and more evident to us, and proceeds to what is more self-evident and intrinsically more intelligible . . . it is one thing to be knowable to us and quite another thing to be intelligible objectively. This, then, is the method prescribed: to advance from what is clearer to us, though intrinsically more obscure, towards what is intrinsically clearer and more intelligible.[13]

To illustrate this path of investigation we may use an example already mentioned: The results of our observations on the falling paper are directly knowable to us because we see them with our eyes, but they are intrinsically obscure because they do not follow a plausible law. On the other hand, the laws of inertia, of causality, and the like are intelligible and plausible because they reflect some analogy with our very familiar experiences. Aristotle wanted to say that it is one of the fundamental characteristics of the scientific method to proceed from what is directly knowable to us to what is intelligible.

5. From "Confused Aggregates" to "Intelligible Principles"

In ancient and medieval science, science and philosophy were part of one chain of thought and were not distinguished from one another. One end of this chain touched ground—directly knowable observations. The chain connected them with the other, more lofty end— the intelligible principles. Aristotle's way of expressing this may rightly be criticized today, but his formulation remains, even today, a practical frame of reference which is useful for all discussions about the relationship between philosophy and science. Aristotle said, "Now what is plain and obvious at first are rather confused aggregates, the elements and principles of which become known to us later by analysis."[14]   Such a confused aggregate was our observation of a falling cigarette paper.   When we have analyzed this confused aggregate, we obtain the principle of inertia, the concept of a mass point, etc.   The latter are intelligible concepts.   This is a description which applies in a certain way to every scientific investigation.   Even the most hard-boiled engineers must recognize that there are two types of statements: on the one hand, statements regarding direct observations and crudely empirical rules which the engineer calls "rules of thumb"; on the other hand, intelligible principles like the law of inertia.   No one can deny that these two levels exist.   One of the most obvious differences between these two levels is this: The engineer will readily change his "rules of thumb" under the impact of new observations, but he will not easily admit that such a general principle as the law of inertia is wrong. If it comes to a choice, he will usually assume that his observations were wrong and not the law of inertia.

The chain will be a useful picture for the understanding of the distinction between science and philosophy.   This distinction has not always existed.   In ancient and medieval times, the whole chain from observed facts to intelligible principles was called science and was also called philosophy.   If we look today at the traditional way of teaching science and philosophy in the universities, we find them taught in different departments.   There is little cooperation between them.   The scientists frequently believe that philosophers are just talkers, and that what they talk is nonsense, at that.   The philosopher says that the scientist is a man with a very narrow mind, who understands only a very small field; whereas the world as a whole is the subject matter of the philosopher.   An explanation frequently proposed is that science has become so specialized that it is no longer possible for a man to know, as Aristotle did, ethics, politics, physics, poetics, rhetoric, etc.   Today, no one, it is argued, can acquire a universal knowledge and understanding.   Everyone is too busy learning to be familiar with a narrow specialized subject. There is a saying: "The scientist knows much about little; the philosopher knows little about much."   Speaking about the increasing specialization in science does not, however, tell us the whole story.   In some ways, science is today less specialized than it was fifty years ago: there are many more cross connections.   Consider, for example, physics and chemistry; fifty years ago they were regarded as very different fields. The students of one of these subjects were discouraged from "wasting time" in classes devoted to the other. Philosophers even gave an "intelligible" reason why physics and chemistry would always be separate from each other: Physics had to do with quantity, chemistry with quality. Then there developed the field of physical chemistry, later the field of chemical physics. Today it would be difficult to say what the difference is between physics and chemistry, and the distinction exists only if the most elementary experiences on the lowest level of abstraction are being described; the higher the level of abstraction, the less distinction. Physicists used to despise chemistry because it was a crudely empirical knowledge, something like "cooking," but now the laws of chemistry are derived from physics, from thermodynamics, electrodynamics, and from quantum mechanics. Therefore, it is now much easier for physicists to learn and understand chemistry and, similarly, for chemists to learn physics. The same condition exists between physics and biology, or between economics and anthropology. Until recently, the latter were considered as completely unconnected. Economists were people who could calculate the trends in the stock exchange; anthropologists studied savage tribes. Today we must understand economics as a tribal custom, and tribal customs from the economic point of view.

Therefore, we cannot say with certainty that today a man cannot acquire an understanding of different fields of science. The disappearance of the old unity between science and philosophy can hardly be ascribed to the increasing specialization in science.

6. "Science" and "Philosophy" as Two Ends of One Chain

We discussed Aristotle's description of the "natural path of investigation," which, "starts from what is more readily knowable and more evident to us and proceeds to what is more self-evident and intrinsically more intelligible . . . " This whole idea is based on the fact that there are such general principles which are clear and intelligible to us although remote from our immediate experience. If we look about us in the world, we observe various kinds of physical phenomena: the motion of planets around the sun, the moving of particles in an electromagnetic field, etc. Why these phenomena take place, and why they follow specific laws is obscure.

The role of the general principles is to make plausible to us why these phenomena take place in this way and not another way. If we consider the chain which connects the statements about our direct experience with the general statements of science, we may ask what the role of this chain is in human life.   We can describe this role by describing both ends of the chain.

We start from the end of the chain which corresponds to directly observed facts, which are described in the language of everyday life. We try to set up principles from which we can derive these observable facts.   From one principle we can in some cases derive an immense number of observable facts.   From Newton's[15] laws we can derive the facts regarding the motions of celestial bodies; from the electromagnetic theory we can derive facts concerning all electric and magnetic phenomena; from the Mendelian[16] laws we can derive heredity patterns, etc.   These principles provide orientation in the world of facts.   They aid us in the practical applications of our observations.   We may call this end of the chain, briefly, the experimental or technical end.   This use of the chain—setting up principles from which we can derive observable facts and applications of observed facts—is what we call "science" today.   "Science" is not much interested in whether these principles are plausible or not.   The latter does not much concern the scientist as a scientist.   In many textbooks we find the statement that it doesn't matter at all whether or not these principles are plausible.   In fact, these textbooks say, the principles of twentieth-century science like relativity or the quantum theory, are not at all plausible, but paradoxical and confusing.   So we may also call this "experimental or technical end" the "scientific end" of the chain as well.

In ancient science, however, men required also that the law of inertia, for example, be able to be derived from plausible or intelligible principles, like the principle of sufficient reason (nothing can happen without a cause) or the law of eternity of substance (all matter is eternal; it cannot be destroyed or created).   This end of the chain in which the laws of physics are derived from intelligible and self-evident principles may be called the "philosophic" end of the chain.   The laws of intermediate generality, the physical laws, are themselves reduced to laws of higher generality which are immediately intelligible.   Everyone will understand why we need the scientific end, but why do we need this philosophic end of the chain? There is no doubt that, for practical purposes, mankind has always needed this philosophic end. It is a fact that it has been so through the centuries, and that it is still so today. When the principles of relativity and of quantum mechanics were developed, some people said: "Maybe you can derive useful results from these principles, but they are obscure, even paradoxical. They serve a certain practical purpose, but they are not 'intelligible.' We do not 'understand' these theories as we understood Newtonian mechanics." There are, of course, very different opinions concerning the precise conditions under which we regard a principle as "intelligible." Some people say that they are "directly intuited." Others emphasize the point that the question of what principles are regarded by man as "intelligible" is a function of historical evolution. In any case, the longing for these "understandable" principles exists; this is a psychological fact. But what need has been actually satisfied by such principles? It cannot be a scientific need, or the principles would be simply scientific principles, like the laws of physics, and justified by their empirical results.

Through the work of scientists, we have learned that the observable phenomena, complex as they may seem to be, can, in many cases, be derived approximately from simple mathematical formulae. The positions of a falling body can be approximately described by the formula: "The acceleration is constant." The positions of a planet relative to the sun can be approximately described by saying that they are "situated" along a conic section called an "ellipse." The scientist would describe these facts as follows: Starting from the observation of positions, the scientist is looking for a simple formula from which one can derive the observed positions. The process by which such a formula is found is called "induction." The finding requires the use of creative imagination on the part of the scientist. If we want to describe this finding of a formula in our everyday language, there are two ways of describing it. We could say that the formula is the "invention" of the scientist, that it did not "exist" before the scientist found it. We compare it with an invention like the telephone, which did not exist before Alexander Graham Bell "invented" it. The hypothesis or formula is a product of human imagination, of the scientist's inventive power. It must be tested by sense experience.

However, the same state of affairs can also be described by a different analogy with common-sense experience. We could say that the formula has always existed within the observable facts. The scientist "discovers" it, as Columbus "discovered" America. The scientist is not an inventor; he "sees" the formula with his "inner eye" by looking at the observable phenomena with his sense organs. The scientist makes use of "intuition" to discover the formula.

This latter way of describing the scientist's activity is in agreement with the "great tradition" of scholastic philosophy,[17] while the description of the scientist's work as "invention" is more in agreement with the line of Positivism[18] and Pragmatism.[19] Hans Reichenbach,[20] in his book, *The Rise of Scientific Philosophy*, points out that it has been characteristic of ancient and medieval philosophy to believe that there is a "seeing with our minds" that is analogous to seeing with our eyes. As we see shapes and colors with our eyes, we see ideas and general laws with our minds. This was the basis especially of Plato's theory of ideas.[21] According to Reichenbach, traditional philosophy argued as follows:

> Since physical things exist they can be seen; since ideas exist they can be seen through the eye of the mind. . . . Mathematical vision is construed by Plato as analogous to sense perception.[22]

The modern scientist says that hypotheses and formulae are the result of imagination, and are tested by trial and error. But the philosopher of the "great tradition" would say that the scientist "sees" the formula through the observable phenomena by the power of his intellect. The analogy between direct sense perception and direct intellectual intuition is strictly stressed by Aristotle, who says that "as the senses are always true as regards their proper sensible objects, so is the intellect as regards what a thing *is*." And St. Thomas Aquinas[23] says: "Hence the intellect is not deceived about the essence of a thing, as neither the sense about its proper object."[24]

The belief in this analogy accounts for the belief that our intellect can "discover" by intuition general laws of nature, and can be certain that they are true.

7. The "Scientific" and the "Philosophical" Criteria of Truth

We may put here the question: On what grounds do we accept some principles and not others? We can distinguish two different criteria for truth, or, to speak a language nearer to common-sense language, two reasons for accepting a principle. It is historically interesting that this distinction is a very old one. It was very well formulated by Thomas Aquinas, the leader of medieval philosophy, in the thirteenth century. The criteria which he developed—and which he described in his *Summa Theologica*—can still be regarded today as the characteristic distinction between the two parts of our chain.[25] One reason for believing a statement is that we can derive results from it which can be checked by observation; in other words, we believe in a statement because of its consequences. For example, we believe in Newton's laws because we can calculate from them the motions of the celestial bodies. The second reason for belief—and medieval philosophy considered this to be the higher one—is that we can believe a statement because it can be derived logically from intelligible principles.

From our modern scientific point of view, we apply only the first of these two reasons. We may call this the "scientific criterion" in the modern sense. As Thomas Aquinas points out, this criterion is never convincing. Judging by it, we find, for instance, that the conclusions drawn from a certain set of principles are in agreement with observation. Then we can only conclude that these principles may be right, but it does not follow that they must be right. It could be that the same observational results could also be derived from a different set of principles. Then our observations cannot decide between two different principles. For example, someone's purse suddenly disappears. We can make the hypothesis that it has been stolen by a boy, and we can draw the conclusion that if a boy has stolen it, the purse will have disappeared. But if the purse has been stolen by a girl, the same result would follow. If we make the hypothesis that some boy has stolen a purse, and then observe that no purse has disappeared, we can conclude that the hypothesis is false. But if the purse is gone, the hypothesis may be true, but not necessarily. Since we can never imagine all possible hypotheses, we cannot say that a certain hypothesis is the right one. No hypothesis can be proved" by experiment. The correct way of speaking is to say that experiment "confirms" a certain hypothesis. If a person doesn't find his purse in his pocket, this confirms the hypothesis that there may be a thief about, but it doesn't prove it. He may have left it at home. So the observed fact confirms the hypothesis that he may have forgotten it.   Any observation confirms many hypotheses.   The problem is what degree of confirmation is required.   Science is like a detective story.   All the facts confirm a certain hypothesis, but in the end the right one may be a completely different one.   Nevertheless, we must say that we have no other criterion of truth in science but this one.

In the second case, the philosophic criterion of truth, an hypothesis is regarded as valid if it is derivable from self-evident, clear, intelligible principles.   These two criteria work at the two ends of our chain.   At the scientific end, we say that principles are proved by their observable consequences.   This holds for the most general principles.   But if we begin with the principles of causality, or of sufficient reason, and try to check them by their consequences by experiment, the sequence is rather hazy and complicated.   In the philosophic view, these principles have the advantage of being self-evident.

This "self-evidence" was originally based on the belief in the analogy between "seeing through our eyes" and "seeing through our intellect."   We shall learn later (Chapter 2, Section 7) why the search for "self-evident and intelligible" principles has survived the belief in the analogy between eyes and intellect.

We have presented Thomas Aquinas' criteria of truth in a "modernized" language.   It is probably useful, however, to know his original formulation. He wrote:

> Reason may be employed in two ways to establish a point: first, for the purpose of furnishing sufficient proof of some principle, as in natural science when sufficient proof can be brought to show that the movement of the heavens is always of uniform velocity.   Reason is employed in another way, not as furnishing a sufficient proof of a principle but as confirming an already established principle by showing the congruity of its results, as in astrology the theory of eccentrics and epicycles is considered as established because thereby the sensible appearances of the heavenly movements can be explained; not, however, as if the proof were sufficient, inasmuch as some other theory may explain them.[26]

8. The Practical Use of "Philosophic Truth"

Before we go into the question of whether these principles are self-evident or not, and why we cherish them, let us ask what the "practical" use of these general principles is. They are supposed to describe the universe as a whole, its ultimate structure. Why do we need to know this? Has it any influence on our lives? What is this influence? We consider that human society is, in a way, a picture of the universe, that we act in a natural way if we act according to the laws of the universe. Man has the belief that as he formulates the general structure of the universe, men in general will imitate this structure in a certain way in their lives. Many people will not realize that they have behaved in this way. If we go to Sunday School, however, we are imbued with one view of the ultimate structure of the universe at a very early age. Traditional religion is one of the theories we may give of the ultimate structure of the universe. One would think, offhand, that physical theories like the theory of motion would have no such influence on the orientation of human actions as we ascribe to traditional religion, but it is worthwhile to examine this theory from that point of view.

The ancient laws were very different from those of today. The laws of motion for terrestrial bodies differed from those for celestial bodies. All terrestrial bodies were considered to have a tendency to move toward a certain goal—stones downward, air and flames upward. This tendency to move toward a certain goal was regarded as the characteristic feature of all terrestrial motion. Celestial bodies were thought to move in permanent circular motion. In other words, the law of motion depended on the substance of the body. Celestial bodies were believed to be made of a completely different matter from terrestrial ones—an immaterial, subtle substance. The universe consisted of the more ordinary substance of terrestrial bodies and the more noble substance of celestial bodies.

Similarly, the world was believed to consist of lower types of beings and of higher types. Thus the theory of motion was of great importance in man's whole life—it supported his belief in the hierarchical structure of society. It encouraged the moral behavior of human beings. Even in antiquity there existed "bad" people who did not believe in this difference between celestial and terrestrial substances—who undermined this belief which people should have. In Plato's *Laws* he said that such people should be in prison.[27]

All those who call themselves educators (and everybody who has been educated wants to be an educator) believe that one way of life is better than another, and that they must support the scientific theories which support their other beliefs.   Thus these general principles influence human behavior.   In a way, these "intelligible" principles are more practical in their effects than the physical principles.   The technical effects of science are more indirect than a blunt command to someone as to what he must do.   Thus the most general principles, the intelligible principles, are also practical but on a different level—in a way, they are more practical.   Bluntly speaking, science proper provides us with the technical means by which we can produce weapons to defeat the enemy, but the philosophic interpretation of science can direct man in such a way that he makes actual use of the weapons.

We can illustrate this situation easily by an example taken from ancient Greece.   Plato discusses in his *Republic*[28] the question of how to educate the future leaders of society by a curriculum which will make them "good" leaders.   One participant in the dialogue raises the question of whether astronomy belongs in that curriculum and how this topic of instruction could be justified.   Socrates, who represents Plato's opinion in the dialogue, rejects emphatically the view that astronomy should be taught for its technical results, for its usefulness in agriculture or navigation.   This kind of knowledge is irrelevant for the future leader.   However, if we look for the "intelligible principles" which account for the movements of the celestial bodies, we find, according to the doctrine of ancient Greece, that the planets are moved by divine beings who move in perfect circles because they are perfect beings.   These philosophic principles of astronomy are not very useful for technical purposes, for the actual computation of the observable positions on the sphere. But the belief in this philosophic interpretation gives support to the belief in the divine beings.   This belief, in turn, is very useful for encouraging the "good" conduct of citizens.   From this angle Plato says astronomy is a very important topic in the curriculum of future leaders.

We get a very clear idea of the chain linking science with philosophy if we consider the example of astronomy as it was conceived by Plato.   The French physicist, philosopher, and historian, Pierre Duhem,[29] pointed out that, briefly speaking, Plato distinguished three degrees of astronomy: observational, geometrical, and theological (or philosophical) astronomy.   They are situated in this sequence along our chain.

Duhem describes Plato's conceptions in a way that is very useful to us if we want to understand the relationship between science and philosophy at a time when both still constituted one coherent system of thought:

> There are three degrees of knowledge.   The lowest degree is knowledge by sense observation.   The supreme degree is knowledge by pure intellect; it contemplates eternal beings and, above all, the sovereign good.

These two degrees of knowledge coincide with what we called previously "confused aggregates" and "intelligible principles," or "things seen with our eyes," and "things seen with our intellect." Then Duhem continues:

> Between the lowest and the supreme degree of knowledge is a kind of mixed and hybrid reasoning which occupies the intermediate degree. The knowledge born of this intermediate reasoning is geometrical knowledge.   To these three degrees of knowledge three degrees of astronomy correspond.

It may seem strange that there is no doubt that what we call "modern science" has developed from this "mixed and hybrid reasoning" that characterizes the intermediate degree of knowledge. Duhem goes on:

> Sense perception is responsible for the astronomy of observation. This kind of astronomy pursues the complicated curves described by the stars. . . . Through geometric reasoning the mind produces an astronomy which is capable of precise figures and constant relationships.   This "true astronomy" replaces the erratic paths which observational astronomy attributed to the stars by simple and constant orbits . . . the complicated and variable appearances are false knowledge. . . . Pure intellect reveals the third and supreme astronomy, theological (philosophical) astronomy. . . . In the constancy of celestial motions it sees a proof of the existence of divine spirits which are united with the celestial bodies.[30]

2

The Rupture of the Chain

1. How the Rupture Occurred

Thomas Aquinas explained the distinction between his two criteria for belief by means of an example which he took from astronomy.[1] If we wish to know the motion of the celestial bodies, we can derive from intelligible principles that they move in permanent circular motion because celestial bodies are perfect, divine beings. Permanent circular motion is obviously more perfect than any non-circular or interrupted motion. Even in antiquity, however, it was known that these laws of motion derived from self-evident principles did not yield precisely the observed positions of the bodies on the sphere. Therefore, astronomy developed the theory of epicycles or superposition of circular motions of different radii from which could be derived the complicated observable motions of the celestial bodies. Thomas Aquinas stressed the point that the theory of epicycles could not be derived from self-evident principles. It was in agreement with observations, but it might be false, as it was not derivable from intelligible principles. The break in the chain connecting science and philosophy developed from the fact that the criterion for the acceptance of a principle was not the same in both parts of the chain science-philosophy, or, in other words, through the whole axis science-philosophy.

We have spoken several times of this chain which connects science and philosophy, direct observations and intelligible principles.

This point may be illustrated by a rough drawing:

direct   general and
observations   scientific   philosophic   intelligible
                  end   end   principles

Figure 1

This chain is what one calls science *plus* philosophy. We have along the chain statements of various degrees of generality. On the one hand, statements of fact; on the other, general principles that are clear and intelligible in themselves. Between these, we have statements of intermediate generality—Ohm's law,[2] Newton's law of gravitation,[3] the laws of electrodynamics,[4] Mendel's laws of heredity[5]—not intelligible by themselves, but useful in theories.

This distinction is obviously connected with the double criterion for belief. If we have statements of intermediate generality—laws of physics, for instance—why do we believe that they are true? In science we use the criterion of truth, which requires that we can derive from these laws facts which are in agreement with experience. We say that the law is confirmed by experience. As we have mentioned, it is false to say that these laws of intermediate generality are ever "proved" by experiment, or worse, that they can be "derived from the facts." One can derive a statement only from a more general statement, never from one which is less general. For example, from the statement, "All men are mortal," we can derive the fact that a particular man is mortal, but from the fact that all particular men we know of have been mortal, we cannot derive the statement that "All men are mortal." Among the Greeks there was a man who said that no one could prove to him that he was mortal. As long as he was alive he would refuse to believe that he was mortal, and when he was dead no one could prove anything to him. A general statement is always the product of an ability of the human mind; this process may be called induction, inductive guessing, imagination. In any case, it is not logical derivation.

Thus, in the words of Saint Thomas Aquinas, we may believe in a statement for its consequences. The more consequences there are which check it, the more we shall believe in it. But, as he also said, we can never prove any statement this way. The Ptolemaic system or the Copernican system, the wave or the corpuscular theory of light[6]—a great many facts can be derived from both theories. It is practical to set up these statements, which are then called principles or hypotheses. (There is no distinction between a principle and an hypothesis. When we begin to take an hypothesis seriously, we call it a principle.) The scientific point of view is that general statements are only proved or confirmed by their consequences, that what they mean "intrinsically" does not matter. From this "purely scientific" point of view, one should have no particular predilection. We connect this point of view with the scientific end of our chain.

The other end of the chain comes from the longing to know "why." Science does not tell us "why"; it only answers questions concerning what happens, not "why" it happens. This longing to find out "why" is nothing more than the longing to derive scientific statements from general principles that are plausible and intelligible. Such a longing stems from the belief that there are such principles. There have been, of course, a great many opinions about the criteria for what is plausible and intelligible.

2. Organismic and Mechanistic Philosophy

Before we discuss the meaning of the term "intelligible," let us give an historical example of some changes in what are called "intelligible principles." We shall discuss the change from an organismic to a mechanistic philosophy. This provides an example of "intelligible" principles from which the derivation of the principles of intermediate generality was attempted.

In ancient and medieval science, what were the "intelligible" principles from which the laws of mechanics were derived? It was believed that everything had a certain nature, and acted according to this nature, which was meant for a certain purpose—the nature of a bird was to fly, of a frog to jump, of a doctor to cure (optimistically speaking), of a stone to fall down, of smoke to go up, of celestial bodies to move in permanent circular motion. Everything acted according to its nature.[7] In a general way, without details, one could derive from this statement how a stone would behave, etc. Of course, one would never believe in principles from which one could derive anything that was in flagrant disagreement with experi-

ment.   The fact that a principle was in agreement with experiment
would not, however, be the only reason for believing in it.   This
view may be called the organismic view because it pictured everything acting as an organism would do.   The general idea was that
the way in which an organism acted was intelligible.   Aristotle
said that the motion of an animal was easier to understand than that
of a stone.   Nowadays we are astonished at this statement, since
our view is just the opposite.   This statement is characteristic of
the organismic view.

About the year 1600 (we usually date the birth of modern science
from Galileo and Newton), the idea developed that we must base
the laws of motion on new principles.   The most characteristic
is the law of inertia, which pictures a body "by its nature" going to
infinity, where it has no business to go, exactly contrary to the
organismic view.   After men became accustomed to them, however, at the beginning of the nineteenth century, the Newtonian
laws were considered as intelligible, plausible principles in themselves.   The "organismic" world view was replaced by the "mechanistic" one.   From this point of view, the Newtonian laws are
considered the most intelligible and plausible laws.[8]   Now the
motion of animals is difficult to explain.   The acceleration of a
man leaving the classroom is very easy to understand, according to
the organismic principles, by describing the man's purpose—for
instance, that of going to lunch—but it is very difficult to under-
stant this acceleration from the mechanistic point of view.

Years ago in Vienna, the advent of the first streetcar was a great
event.   There is a story that the engineer explained the streetcar to
an Archduke, who listened very attentively, and, when the engineer
had finished, the Archduke said that there was only one thing he
did not understand—where was the horse?   In the organismic tradition, he could not understand that anything but an organism could
produce force.   On the other hand, in the twentieth century we
have the story of the boy from New York City who had never seen
a horse—we must assume that for some reason he had never been
to a horse race, for even in this mechanized age there seems to be
this one use for the horse.   You can imagine his astonishment then
when he went to the country for the first time and saw a horse pulling
a load.   In the mechanistic tradition, he immediately asked—where
is the motor?

3. How Science in the Modern Sense Was Born

One of the greatest twentieth-century philosophers, A. N. White-head, wrote:

> All the world over and at all times there have been practical men, absorbed in irreducible and stubborn facts; all the world over and at all times there have been men of philosophic temperament, who have been absorbed in the weaving of general principles.[9]

In antiquity and the Middle Ages, there was very little coopera-tion between these two types of men.   Whitehead emphasizes the point that science in the modern sense was born when such coopera-tion started, and when both interests, in facts and in ideas, were combined in one and the same person.   "The union of passionate interest in the detailed facts with equal devotion to abstract gen-eralization forms the novelty in our present society."[10]

William James[11] described these two personality types in his lec-tures on "Pragmatism."   He called them the "tender-minded" and the "tough-minded" types;[12] an exclusive interest in hard facts seemed to him to hint a "toughness" of character.

Whitehead assumed that cooperation between these types could not have occurred before our "present society" was born.   In the society of ancient Greece the "philosophers" or "scientists" who were interested in general principles belonged to a higher social class than those who were interested in the "hard facts" of tech-nical application, the artisans and craftsmen.   The latter belonged to a low class and had no understanding of general ideas.   We know, however, that the ancient Greeks and Romans displayed a marvelous art and skill in building and even in some fields of mechanical engineering, but the knowledge of these ancient builders and engi-neers was not "philosophic" or "scientific"; it was purely tech-nological.   Their methods were not derived from Aristotle's organ-ismic physics.

The contrast between the ancient and modern approaches to technical knowledge is described by a professor of applied engineer-ing in contemporary Rome:

> What modern science and industry accomplishes by laboratory research tests, by theoretical hypotheses expressed in formulae . . . was accomplished for the science and industry of ancient times by the

transmission of technical knowledge . . . and by empirical formulae, jealously guarded and handed down in mysterious symbolic form.[13]

We might say that the "lower" strata collected facts, while the "higher-ups" advanced principles. Contact between the two types of knowledge was discouraged by social custom. If a man of high social status attempted to apply his "philosophy" or "science" to technical problems, he was severely criticized. Experimental testing of general principles requires manual labor, which was regarded by the ancient Greeks as the appropriate occupation of slaves but not of free men.

We can understand this attitude if we read in Aristotle's book on politics his defense of the institution of slavery. He compared the master's rule over the slave with the rule of man's intellect over his body. He said:

> Nor can we doubt that it is natural and expedient for the body to be ruled by the soul and for the emotional part of the soul to be ruled by the intellect or the part in which reason resides, and that if the two are put on an equality the consequence is injurious to both.[14]

He derived from this remark the adequate relationship between man and animal, between male and female. "The same law of subordination," he continued, "must hold good in respect of human beings generally." According to him:

> There are two classes of persons and the one is as far inferior to the other as the body to the soul or a beast to a man . . . these persons are natural slaves and for them a life of slavish subjection is advantageous. . . . A natural slave is only so far a rational being as to understand reason without himself possessing it. And herein the slave is different from other animals, as they neither understand reason nor obey it.[15]

The slave was regarded as a being who was not able to conceive general ideas, but only to understand orders as to how to act in special cases. This is the exact difference between the "philosopher-scientist" and the craftsman. The latter type of person included, according to the view in ancient Greece, not only artisans, but also what we call "artists"—painters, sculptors, musicians.

How deeply embedded the contempt for manual work was in the Greek mind can be seen in Plutarch's[16] biography of the great Athenian statesman, Pericles.[17] Today we regard the flourishing of the arts as the great glamor of the "Periclean Age," but Plutarch wrote:

> Admiration does not always lead us to imitate what we admire, but, on the contrary, while we are charmed with the work, we often despise the workman. Thus we are pleased with perfumes and purple, while dyers and perfumers appear to us in the light of mean mechanics. . . . If a man applies himself to servile or mechanical employments, his industry in those things is a proof of his inattention to nobler studies. No young man of high birth or liberal sentiments would, upon seeing the statue of Jupiter at Pisa, desire to be Phidias (the sculptor) . . . or wish to be Anacreon or Philetas, though delighted with their poems. For, though a work may be agreeable, esteem for the author is not the necessary consequence.

We see that the artists who produced the perennial glory of Greece, men like Phidias[18] and Anacreon,[19] were "despised" by their contemporaries because they did not devote themselves exclusively to "nobler studies," meaning politics and philosophy.

A similar evaluation was made in the field of science. While pure mathematics as an intellectual endeavor belonged to the "noble" or "liberal" studies, the illustration of geometry by mechanical models was regarded as "despicable." Plutarch, in a biography of the Roman general Marcellus,[20] reported that the Greek scientist Archimedes[21] contributed by his mechanical devices to the defense of his home town, Syracuse, against the Roman invaders,[22] but, Plutarch wrote, Archimedes "did not think the invention of engines for military purposes an object worthy of his serious studies."

The great philosopher, Plato, severely criticized those scientists who confirmed theorems of pure mechanics or mathematics by individual tests. According to Plutarch, "Plato inveighed against them with great indignation, as corrupting and debasing the excellency of geometry, by making her descend from incorporeal and intellectual to corporeal and sensible things." Whoever applied mechanical instruments in geometry had to "make use of matter, which requires much manual labor and is the object of servile trade."[23]

From this statement we see clearly that experimental research in mechanics and physics was regarded by the ancient Greeks as an occupation that would degrade a free man and prevent him from pursuing the "noble studies" of philosophy and politics. We can now understand that:

> [tight] union between the search for general ideas and the recording of hard facts could not have taken place before the prestige of artisanship and technical achievement had been substantially increased. This happened after 1600 when everywhere in Europe, in Italy as well as in France and Germany, the craftsmen and artisans in the large cities became a social class who regarded themselves as the equals of the landowners and their staffs of lawyers and clerics.[24]

The "new science" or "new philosophy" consisted in the combination of general ideas, logical conclusions, and experimental investigation. "This balance of mind," Whitehead says, "has now become part of the tradition which infects cultivated thought." He points out that this new way of thought has become the basis of Western education and culture:

> It is the salt which keeps life sweet. The main business of universities is to transmit this tradition as a widespread inheritance from generation to generation. . . . Since a babe was born in a manger, it may be doubted whether so great a thing has happened with so little stir.[25]

4. Science as a Fragment of Philosophy

We shall now try to understand why the chain science-philosophy broke. In antiquity and the Middle Ages, the requirements for checking general principles against observed facts were not very strict. Usually only very vague results were derived from the "intelligible principles." However, as we have seen, the old Romans and Greeks built very interesting structures on the basis of a tradition of craftsmanship passed on from one generation to another without much theory. They used what we call today "know-how." From what we call science and philosophy, they could derive no technical "know-how" at all. The practical application of science was completely provided for by the tradition of craftsmanship. There was no demand for it from science.

From about the year 1600, however, science became more pretentious; it wanted to derive practical mechanics from theoretical mechanics. Then the chain split in the middle. From the principles of intermediate generality, the physical laws, observed facts could be derived. "Scientists" were no longer interested in whether the physical laws could be derived from principles of higher generality. The great example in history is the failure of the theory of concentric circles to explain the positions of the planets in the sky, which led to the introduction of the "ugly" theory of epicycles,[26] and the theory of epicycles could not be derived from intelligible principles. The break in the chain produced science in its modern sense as one fragment of the ancient chain "science-philosophy." Man had become aware that statements derived from intelligible and beautiful principles could account only in a very vague way for observed facts. The union between science and philosophy was possible only during a period of separation between science and technology. Modern science was born when technology became scientific. The union of science and technology was responsible for the separation between science and philosophy.

It would be a great exaggeration to say that the scholars of antiquity and the Middle Ages believed only in deductions from general principles and not at all in agreement with experience. If we want to be sincere, we must admit that everybody has believed in both. In the late Middle Ages, there arose a philosophical movement which was to represent the transition from medieval to modern thought. This movement emphasized the decisive role of experience in science, and depreciated to a certain degree the role of logical argument. It advocated a shift in emphasis with respect to Thomas Aquinas' two criteria of truth. The new movement stressed the importance of the "scientific criterion." As a predecessor of this movement we may quote Roger Bacon, a writer of the thirteenth century.[27]

> There are two ways of acquiring knowledge; namely by argumentation and by experience. . . . Argumentation arrives at a conclusion and makes us agree with it. But argumentation does not banish doubt so effectively that the mind rests in intuition of the truth until the truth is discovered by way of experience.[28]

In modern science, supposedly very hard boiled, no theory checks with all the facts. We accept some general principles which seem to be plausible and try to derive the facts as well as possible. It sounds very nice to say that we reject a theory upon one disagreement with the facts, but no one will do this before a new theory is found. A good example was the failure of eighteenth-century scientists to abandon the hypothesis of "phlogiston" when a fact was discovered that was in disagreement with the conclusions drawn from it. When a pure metal, like tin, is heated in air, the metal becomes an earthy stuff that we call a "calx," and the process itself is called "calcination." This phenomenon was explained by the hypothesis that when a pure, shiny metal was heated in air it emitted a stuff called phlogiston (a Greek word meaning heat-stuff). By losing this stuff, the shiny metal became a dull calx. Since calcination consisted in the separation of phlogiston from the metal, it seems that the resulting calx should have weighed less than the metal, but the opposite was the case. "That a calx weighed more than a metal was known throughout the eighteenth century, but this fact was not recognized as being fatal to the phlogiston theory." After having stated this as a matter of fact, James Bryant Conant[29] wrote:

> Here is an important point. Does it argue for the stupidity of the experimental philosophers of that day? Not at all, it merely demonstrates that in complex affairs of science one is concerned with trying to account for a variety of facts and with welding them into a conceptual scheme; one fact is not by itself sufficient to wreck the scheme A conceptual scheme is never discarded merely because of a few stubborn facts with which it cannot be reconciled; a conceptual scheme is either modified or replaced by a better one, never abandoned with nothing left to take its place.[30]

If a specific fact is discovered that is contradictory to some conclusion drawn from a theory or conceptual scheme, the only thing that we can learn with certainty from this contradiction is that "there is something wrong" with this theory, but we do not know precisely what is wrong. A theory consists of a great number of statements which may be interlocked in a complex way. The newly discovered fact does not tell us which of these statements is false. In the common parlance of the scientist, we would say that the "theory is refuted" by the facts if statements which are "essential" to the theory have to be dropped. Then by what criterion do we distinguish between essential and accidental parts of a theory? We know from Chapter 1, Section 3, that "essential part of a theory" actually means "essential to a certain purpose of the theory." Therefore, we cannot say that a certain fact refutes a particular theory, but only that it is incompatible with a certain purpose of that theory.   We are at liberty to modify the statements that are not essential to this purpose and thus achieve agreement with the new fact.

We shall discuss examples of this situation later on, but we can easily understand it by comparing a theory with the blueprint of an airplane.   If the plane starts losing altitude, we can only conclude that there must be "something wrong."   This may be in any part of the blueprint, or it may be in the quality of the fuel, or something else.   We cannot conclude that the "blueprint is wrong"; perhaps with only a small alteration we would have a blueprint of a plane with excellent flying qualities.   We can question whether the original blueprint would then be proved to be wrong.   This would depend on whether the necessary modifications were "essential" or not.   We have learned, however, that "essential" always refers to a certain purpose.   The failure of an airplane to perform an expected motion would not "prove" that the blueprint had to be discarded.

Much has been said about the "crucial experiment" that can decide whether a certain theory must be rejected or not.   A single experiment can only refute a "theory" if we mean by "theory" a system of specific statements with no allowance for modification. But what is actually called a "theory" in science is never such a system.   If we speak of the "ether theory" or the "corpuscular theory" of light, or of the "theory of evolution" in biology, each of these names covers a great variety of possible systems.   Therefore, no crucial experiment can refute any such theories.   A famous example was the "crucial experiment" which Arago[31] proposed in 1850 to test the corpuscular theory of light.   This theory was refuted in 1855, but in 1905 Einstein[32] again made use of this theory in a greatly modified form known as the hypothesis of "light quanta" or "photons."

Pierre Duhem, in his book, *La Theorie Physique, son Objet et sa Structure*, said bluntly: "In physics the crucial experiment is impossible."[33]   As a matter of fact, Duhem discussed as an example Arago's experiment, which was intended to bring about an irrefutable decision between the corpuscular and wave theories of light.   Duhem pointed out that it was not possible to prove that there was no third possibility besides these two.   In precisely the same year, 1905, when Duhem wrote this remark, Einstein had actually discovered (or maybe "invented") this third possibility, the theory of light quanta.[34]

A new theory, on the other hand, has never been accepted if it did not possess a certain degree of simplicity and beauty. These criteria have a definite connection with the philosophic end of our chain. The fact that these two criteria do not always fit very well together has led to the idea that science and philosophy are two entirely different fields of knowledge. Some people believe that they will never challenge one another, that they are to be two autonomous fields of discourse. This state of separation has been the prevalent relationship between science and philosophy in university curricula during the nineteenth century and the first half of the twentieth. Today, it is still the typical attitude in our institutions of higher learning. On the other hand, we shall see later that there have been vigorous attempts to restore unity by a more catholic concept of science.

5. How "Science" Can Become "Philosophy"

We have learned that the axis science-philosophy broke because the plausible and intelligible principles—which somehow described the ultimate structure of the universe—did not yield practical results in the field of observable facts and technical applications. When the mechanistic science of Galileo and Newton was built up, one didn't consider whether these laws were "intelligible." Later, when these laws were seen to serve their technical purpose very well, they came more and more to be regarded as "intelligible" or "philosophic" principles. We can see this mechanistic science going through three stages. In the first stage, the laws were accepted because of their agreement with observable facts, but they were regarded as purely descriptive because they could not be derived from intelligible principles which were at that time the organismic principles. In the second stage, the mechanistic laws gained a reputation of being themselves self-evident and intelligible. In the twentieth century, however, new physical theories have been advanced which are regarded as yielding the observable facts better than the mechanistic principles do. In this third stage, the mechanistic principles are still regarded as intelligible but no longer practical. Today people say that the new theories—quantum mechanics, relativity—are accepted because they are practical (in other words, we can build new devices such as the atomic bomb which we could not build before), but that they are not intelligible.

Thus, from an historical study, it would seem safe to say that there is no essential difference between intelligible principles and statements of science from which observable facts can be deduced. After a hundred years, Einstein's formula, $E = mc^2$, will probably be regarded as a self-evident statement. Nevertheless, it is upon this distinction—between intelligible and merely practical statements—that the separation between science and philosophy depends. From permanent motion in concentric circles to the Ptolemaic system of epicycles, to the Copernican system, to the abandonment of circular motion altogether, and the conception of the elliptical orbits of the planets, men have had to accept these succeeding theories because they yielded practical results, even though it meant the breakdown of their intelligible principles. We shall now illustrate this general remark by some examples.

When Copernicus advanced his heliocentric theory, he was opposed not only by the advocates of traditional theology and philosophy, but also by authors who believed strongly in empiricism in science. Francis Bacon[35] called Copernicus a man "who thinks nothing of introducing fiction of any kind into nature provided his calculations turn out well."[36] This means, in other words, that Copernicus applied only the "scientific" criterion of truth, and disregarded the philosophic criterion (Chapter 1, Section 7). Bacon called the Copernican system a "fiction," while he considered the geocentric system as an hypothesis or theory. The distinction between "fiction" and "theory" is still made today by quite a few scientists and philosophers.[37] Very often Einstein's theory of relativity is called a "fiction" while Newtonian mechanics is regarded as a "theory." What is the difference? If we follow Bacon's way of speaking, a "fiction" is a system of statements from which the observed facts can be derived by mathematical reasoning, but the statements which constitute the "fiction" are not intelligible or plausible by themselves. They cannot be understood by means of analogies with the experiences of everyday life. Those authors who pin the tag of "fiction" on the physical theories of our twentieth century mean this word in exactly the same way as Bacon did. We shall learn, in Chapter 4, the specific reasons which led Bacon and his contemporaries to tag the Copernican theory as "not plausible or not intelligible."

The technical superiority of the Copernican theory over the Ptolemaic has never been denied; it has even been recognized by the Church. The more astronomical experience and theories advanced, the greater recognition was accorded this superiority. In Newtonian mechanics the sun became the system of reference with respect to which the laws of motion were valid; this was not the case with respect to the earth. The superiority of the sun as a system of reference was then established without any doubt, but when this role of the sun was recognized, it was regarded as very "plausible" and "intelligible" that the sun should be "at rest." It was now regarded as "improbable" that the big sun with all the fixed stars, the fundamental system of reference, should all together rotate around our small unimportant earth. From being "technically useful," the Copernican system had developed into a theory which was "intelligible" or "philosophically true."

However, a theory which was intelligible by itself would be of eternal validity. If it were not true because of its observable consequences, but "by its own light," no further experience could bring about any change in our belief in its validity. In the twentieth century, when Einstein advanced his general theory of relativity, it turned out that every system of reference is equally admissible in mechanics, and that the superiority of the sun exists only within a very restricted part of the universe. The belief that the Copernican theory was intelligible by itself again turned out to be an illusion.

In a very similar way, the attitude toward Newton's laws of motion has undergone radical changes. His theory of planetary motion was based on two pillars, the law of inertia and the law of gravitation. Neither of these hypotheses seemed to be "intelligible," or even "plausible," to Newton's contemporaries. However, the mathematical conclusions drawn from them were in excellent agreement with all known observations of planetary motion, including even the mutual perturbations. Newton's theory was accepted because of its technical excellence as a scientific "truth," but originally it was not recognized as a "philosophical truth." The greatest scientists of his time, men like Huyghens and Leibniz, were reluctant to accept principles which were not "intelligible."[38]

For Leibniz, the laws of inertia and gravitation were "fictions," as the Copernican system was for Bacon. Newton, like Copernicus, was regarded as a man who would accept any fiction provided that it led by correct mathematical reasoning to results which agreed with experience.

Newton himself explained his views in a letter addressed to Leibniz in a journal:

> To understand the motions of the planets under the influence of gravity without knowing the cause of gravity is as good a progress in philosophy as to understand the frame of a clock and the dependence of the wheels upon one another without knowing the cause of the gravity of the weight.[39]

Newton regarded his theory of gravitation as analogous to the description of a clockwork that keeps the planets moving. He agreed that if his laws of gravitation and inertia could be derived from an intelligible principle, this would make for progress in understanding, but he preferred to restrict himself to what we have called the "purely scientific" aspect, and to abandon the search for intelligible principles. He started from principles of "immediate generality." His famous statement, "hypotheses non fingo" (I do not make up hypotheses) means in other words: "I restrict myself to fictions and do not care for intelligible principles." His goal was definitely "scientific truth" and not "philosophical truth."

However, after the great technical successes of the Newtonian laws, since the beginning of the nineteenth century there has been a steady growth of the belief that the Newtonian laws are themselves intelligible. Analogies were drawn between the law of inertia and the personal experience of sluggishness and, eventually, Newton's laws were regarded as "intelligible principles." When they reached this status, they were no longer dependent upon further experimental research. They were declared to be self-evident statements which would be valid in any future system of physics.

In this way Newton's scientific theory became a "philosophical system." Henceforth, every attempt to modify Newton's laws would be regarded as contradicting self-evident principles. Mechanistic physics became a mechanistic philosophy. Every new physical theory that contradicted Newtonian physics was now "absurd." We shall see the consequences of this attitude in the reluctance to accept such twentieth-century concepts as the theory of relativity and the theory of quanta.

6. Speculative Science and Metaphysics

The principles and the observations of science are not formulated in the same language. We have described (in a perfunctory way) the language of observation as being statements about some dancing spots, while the general principles of science use such terms as "force," "potential," "energy," etc. From statements about abstract terms, we can never derive anything about observable facts. In mechanics we may learn what functions the variables x, y, and z are of the variable t. But this does not tell us anything about the observable world. How shall we observe the variations of x, y, z? The logic of science must include, besides the principles and observations, the link between the abstract concepts of science and the observational terms. These links are called "operational definitions,"[40] or sometimes "semantical rules."[41] The discussion of these rules belongs also to the logic of science. Here we are not interested in whether the principles are intelligible or not. The only requirement is that the results must be in agreement with experience. This is their full justification, and there is no other justification, from the point of view of science.

By taking this general scientific point of view, we have disregarded a large part of our chain. The scientist can say that the rest of the chain does not interest him at all, we shouldn't speak of it nor think of it. This is one way of looking at it, but a great many people do not agree that the other end of the chain should be completely disregarded. Since the intelligible principles cannot be checked directly by the methods of science, we must ask how we can check to see whether or not a principle is intelligible. Some believe that there is another type of thinking, in addition to scientific thinking, which is called philosophical thinking. Still others say that man cannot know of this at all; that we need help that transcends reason, whose source is religion. It is clear, however, that people are eager to extend knowledge beyond "science" in the modern sense to the realm of those intelligible principles. We find also those who combine both points of view, who do not want to extend human reason beyond what we have called the logic of science, but who believe that, since men are concerned about general principles, these belong to religion, which goes beyond human reason and appeals to the supernatural. This combination of the hard-boiled scientist with a belief in the supernatural is not rare.[42]

Philosophy is also thought to be concerned with hypotheses of a more speculative nature than those found in science. I do not think that this is true, since all hypotheses are speculative. No distinction can be made between scientific and speculative hypotheses. One says that Newton's laws, the laws of electricity, etc., are scientific but considers the hypothesis that all men survive after death as speculative. Many have tried to check it by experiment. If it is taken seriously, it can be a scientific hypothesis. Of course, it can be formulated in such a way that it cannot be checked in principle. We can say that after death men become spirits with their own language and laws, and with no means of communication with human beings. This is not a scientific hypothesis, since there is no means of checking it. What kind of hypothesis is this? It may be called a metaphysical hypothesis. Its nonscientific character comes from the fact that essentially it cannot be checked by experience, not from its fantastic nature, since a scientific hypothesis can also be fantastic. It may be said that all things are material, that there is no spirit. If this statement is formulated in such a way that it cannot be checked, it is a metaphysical statement. If it means that all facts about the world can be deduced from the laws of matter, e.g., of electrodynamics, dynamics, etc., then it is a scientific hypothesis. It may be a fantastic statement, but not a metaphysical statement. The thesis of materialism may also have a different meaning. We may say that everything in the world is matter, but that nevertheless we cannot derive everything from the laws of mechanics, etc. Such a statement cannot in principle be checked, and we should, therefore, call it a metaphysical statement.

Thus we distinguish between metaphysical and scientific statements. What do these metaphysical statements mean, and why are we eager to make them? To say that there is only matter, or that there is only spirit is to make a direct statement about the nature of the universe. On what grounds is such a statement accepted? What is its practical function? Such statements have just as practical results as the scientific ones; they have a direct effect on human behavior.

7. The Belief in Intelligible Principles

The unity of science and philosophy in the old classical sense was perhaps best described by the famous tree of Descartes:[43] The roots of this tree corresponded to metaphysics (the intelligible principles), the trunk to physics (statements of intermediate generality), and the branches and fruit to what we would call applied science. He regarded the whole system of science and philosophy as we today regard science alone; he felt that the metaphysical principles were ultimately justified by their "fruits," not merely by their self-evidence. What we today call applied science consisted for him not only in mechanics (engineering) but also in medicine and ethics; even today we speak of social engineering. The difficulty was that from the general principles of Cartesian or Aristotelian science-philosophy no results could be derived which were precisely in agreement with observation, but these principles seemed to be intelligible and plausible. So the tree was cut in the middle. For the derivation of technical results, it was necessary to start from the physical principles in the trunk, the middle of the tree. Science in the new sense was to think only of how the fruits would develop from the trunk without regard to the roots from which they sprang.

Later, in the nineteenth century, it was felt that the mechanical laws of Newton had restored the old unity. The philosophy of materialism developed the idea that the laws of mechanics played the same role as the old organic laws in the Aristotelian philosophy, and that everything could be derived from them. At the beginning of the twentieth century, it became evident that the laws of mechanics were not entirely satisfactory either. The attempt was then made to keep these mechanical laws as metaphysical laws,[44] and the facts of nuclear physics, etc., were thought to be derived from principles of intermediate generality. If we look through the history of science, we see a whole strange phenomenon. After Newton, his laws were respected for their practical use. They proved to be so practical that after a while they acquired a certain dignity and, in turn, were then regarded as "intelligible" principles. Later it was found that their practical use had been exaggerated— that the phenomena of nuclear physics, the conversion of mass into energy for example, could not be derived from them. It was said, then, that the Newtonian laws must be kept for their "dignity," because they were intelligible.[45] We come now to the last point in this chapter. What is really the criterion by which we judge whether or not these principles are "intelligible?"

This "dignity" referred to above was ascribed at one period to the Aristotelian laws, and at another to the Newtonian laws. The belief in this quality survived when the belief in the scientific validity was gone. We can explain this in many ways. A perfunctory way is to say that the human mind is sluggish, that people adapt very slowly to advances in science; just when we come to understand some general laws of science, they are proved to be wrong. There is something to be said for this explanation, but it is probably not the whole truth. It is certainly true that the laws of science which we use are of different degrees of stability. Some we do drop easily—like the rules of thumb used by the engineer in his daily work—but the laws of Newton have been alive a long time. We may say that such laws are intrinsically plausible.

Why are some laws of greater plausibility than others? To answer this, we must consider some examples, such as the law of sufficient reason, or the law of conservation of substance. Why do they seem plausible to us? No one would say that Ohm's law or the law of electromagnetic induction is "plausible" or "intelligible," let alone "self-evident." If we analyze this fact psychologically, we see that the plausibility of these general laws lies in their apparent analogy to observations that are very familiar to us. Conservation means to the physicist that a function of certain mechanical, caloric, and electrical quantities remains constant. The sum of these quantities, which are very different in different fields, remains constant. From this law, the physicist can calculate such consequences as the speed of a falling weight or the cost of electricity. Then he says that "energy" is a substance that cannot be destroyed. We see many things in the world of our immediate, everyday experience which apparently cannot be destroyed. We do not expect the houses we live in, for example, to disappear before our eyes; and if they are destroyed, we comfort ourselves with the thought that they are only subdivided into atoms and molecules. Of course, today we know that atoms can be destroyed, but we still comfort ourselves with the thought that electrons cannot be. Finally, we know that everything can be destroyed but energy. Comparing something very complicated with something simple and familiar is not a very profound way of thinking—we replace the complicated statement of the conservation of energy used by physicists by the observation of immediate experience that objects do not disappear—but it is satisfying. Principles become obscure when they lose this close analogy to everyday experience, as is the case today with the principles of quantum mechanics and relativity.

We may give a simple example of this point. One who studies mechanics often starts with simple machines, among which is the lever. If he asks what the condition for equilibrium is for a simple, frictionless lever, he is told that the condition is that $gl = g'l'$. (See Figure 2.) How can this be derived? It is not "plausible"

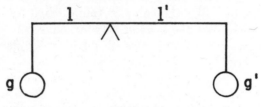

Figure 2

by itself because one must learn physics to know it. The argument used by Archimedes[46] was that if the weights and lengths were equal, the lever would not move at all because it wouldn't know which way to move. If this principle is accepted, the general principle can be mathematically derived. Archimedes had the impression that he had derived everything from an intelligible principle. Ernst Mach[47] said that this argument is illusory. It involves the presupposition that the motion depends only on the length of the arms and the magnitude of the weights. It could depend on the color of the arms or of the weights, the material they were made of, the weather, the magnetic field of the earth, etc. In other words, Archimedes presupposed everything he wished to prove. He did not derive it from the principle of sufficient reason. In order to do this, he would have had to know what reasons exist in the world. There are many other possible asymmetries in the world that we do not know. It is a completely vicious circle. The lever cannot move because it has no reason to move. Why has it no reason to move? Because it depends only on the length of the arms and the size of the weights—but then we already know what we are trying to prove. We have a vague experience or idea of symmetry. If we analyze it, we must decide what factors really matter, but when we have done this we have done everything— we have no need for the principle of sufficient reason.

The reason why we believe in these intelligible principles is certainly very superficial. Many do not like to admit to themselves that this strong belief comes from vague analogies with everyday experience. This dislike has been put in words as follows: These principles are of a nature which is very difficult to describe—we know them from "intuition," from a kind of ability which is different from the ability used in ordinary science and which yields more certain results. These principles may be very plausible, but they are not applicable; they lead to circular arguments, just as the principle of symmetry is plausible but not applicable unless we describe what qualities are important. Analogy between general statements and everyday experience can only be superficial. The "dignity" of these intelligible statements—other than that which comes from their agreement with observable facts—comes from this vague analogy to everyday experience. Thus, if we cut off the roots of Descartes' tree, there remains a longing for these vague analogies, to give us back the feeling that we can understand the general scientific principles other than and better than by their observable results.

8. "Science Proper"

If we want to use the language in which we have been brought up by parents and teachers, we can recognize a double purpose of science: to provide technical knowledge, and to promote an "understanding" of the universe. This double purpose became especially obvious when the split between science and philosophy occurred. Then it appeared to be impossible to fulfill both purposes by a single system of thought. Many have held and do hold that science can give only technical knowledge, that it has only a certain technical value. For "real understanding," we need philosophy, which sets up principles that are intelligible and plausible, but does not yield precise practical knowledge. This is the way that science and philosophy have separated. There is, however, no doubt that philosophy also serves a practical purpose. Whereas science gives methods of devising physical and chemical devices, philosophy gives methods of directing the behavior of men. Thus the philosophical side reaches its practical purpose in an even more direct way than science proper.

What I mean by "science proper" is science in its stage of separation from philosophy as it is taught in our usual science classes. From this "scientific aspect," science should contain as little philosophy as possible. The instructor starts from observed facts and sets up principles from which these facts can be derived. "Science proper" is not interested in whether these principles are "intelligible." He is interested, however, in the fact that from a small number of such principles "of intermediate generality" there can be derived a great number of observable facts. This is called the principle of economy in science. Setting up a small number of principles from which one can derive as many facts as possible is a kind of minimum problem. The dream of science is to derive all facts from one principle. This probably cannot be achieved. If this cannot be achieved within science, it may be imagined that the principles of science could be derived from one master principle in philosophy, where no precise agreement with observable facts is required. Deriving everything from water, from fire, from spirit, as some of the ancient Greeks[48] attempted to do, is an extreme case of economy.

It is very important always to keep in mind that science is not a collection of facts. No science is built up in this way. A collection of statements indicating on what days snow has fallen on Los Angeles is no science. We have science only when we can set up principles from which we can derive on which days snow will fall on Los Angeles. Furthermore, if the principles we set up are as complicated as experience itself, this will be no economy and no "science proper." A great many principles or one very complicated principle amount to the same thing. If the principles are as complicated as the facts themselves, they do not constitute a science. The mere observation of the positions of the planets in the sky is no science. Ancient scientists attempted to set up curves which would describe this motion. These were once thought to be circles; they were later thought to be ellipses, but this is only true if perturbations are neglected. Considering perturbations, the equations of those curves are very complicated—there are so many terms that they could fill a volume of a hundred pages. This is just as complicated as recording all the positions of the planets. It gives us no advantage; nor is there any science in it.

If there is not a small number of principles, if there is no simplicity, there is no science. If a man says that he doesn't want speculation, that he just wants to be given all the facts—he is asking for only the preliminary step to science, not for science itself. The scientist is often accused of oversimplifying. This is true; there is no science without oversimplification. The work of the scientist consists in finding simple formulae. Some say that the scientist doesn't help us to understand anything because he oversimplifies everything. Who knows another way of "understanding" complicated things than by oversimplifying them?

After the scientist has set up a simple formula, he must derive from it observable facts. Then he must check these consequences, to see if they really are in agreement with observation. Thus the work of the scientist consists of three parts:

1. Setting up principles.
2. Making logical conclusions from these principles in order to derive observable facts about them.
3. Experimental checking of these observable facts.

These three parts make use of three different abilities of the human mind. Experimental checking makes use of the ability to observe, to record sense impressions; the second part requires logical thinking, but how do we get the principles in the first part? This is an extremely controversial point. Many authors say "by induction from observed facts"—the inverse of deduction.[49] If the scientist observes that the same sequence appears frequently, he will conclude that it will always do so. This reminds us of the story about a man who bought a horse, and wished to accustom the horse to live without eating. For thirty days he succeeded in preventing the horse from eating, and so concluded that the horse was then trained to live without eating; but on the thirty-first day the horse died. "Induction" is not as simple. We can hardly set up any method of finding a general principle like gravitation by it. We all know the story of how Sir Isaac Newton was supposed to have hit upon the theory of universal gravitation when an apple fell on his head. Whether this story be true or not, the point is that we cannot set up a system for induction upon such a basis.

For the analysis of science, however, the way in which we obtain general principles is not so relevant. The general principles may come to one in a dream. The way in which we obtain them would play a role if we were to make a sociological or psychological analysis of science. In the "logic of science" what matters with regard to the general principles is not the way in which we get them by induction, but the way in which we derive the rest of the body of science from them by deduction. The ability we need in order to obtain the general principles of science we may call imagination.[50] We immediately encounter the difficulties of induction in the simplest

Figure 3

case. Suppose we plot the results of a series of measurements by a series of points on coordinate paper, and we wish to represent these results by a function. We imagine that the arc should be as smooth as possible. If we have no idea what this curve should be, we will not find it. The points do not determine the curve in any case; we have to imagine a criterion of "smoothness." (See Figure 3.) The problem of induction will be discussed more elaborately in Chapter 13.

9. Science, Common Sense, and Philosophy

We shall now describe the relationship between science and philosophy after the split had occurred in a way that seems a little paradoxical and is a definite oversimplification. It will, however, direct our attention to the central characteristics of both domains of human enterprise. The principles of science can be formulated in such a way that they are very far from common sense, but the checking of them by experiment is always done on the level of common-sense experience. The paradoxical situation arises that, in a way, philosophy is nearer to common sense than science. Philosophy has always required a close correspondence between the general principles themselves and common-sense experience. The more science has advanced into the theoretical field, the more remote from common sense its general principles have become.

The results of observations and experiments which form the factual basis of science can be described in the language of everyday life or, in other words, by common-sense statements.   In Aristotelian and medieval physics, a distinction was made between "heavy" bodies, like rocks, that fall to the ground, and "light" bodies, like smoke, that ascend to the heavens,   This is the language of the "man in the street."   Before the rise of modern physics, about 1600, this common-sense language was used, not only in the description of observations, but also in the formulation of the general principles of science: "If a body is heavy, it falls."   Herbert Dingle[51] has written: "The undying glory of Galileo's contribution to thought is that, though only half-consciously, he discarded the everyday common-sense world as a philosophical necessity."[52]   In his theoretical system, all bodies fall with equal acceleration to the ground.   He paved the way for the Newtonian system, in which the planets move according to the same laws as a falling stone, although our common-sense experience seems to show a fundamental diversity between these two types of motion.   It is a matter of fact that the advance in science has consisted to a large extent in the replacement of the common-sense world by a world of abstract symbols.

If we want to formulate general principles from which a wide range of observable facts can be derived, we must discard the language of common sense, and make use of a more abstract terminology.   Herbert Dingle has remarked that on the common-sense level there is a clear-cut distinction between physics and chemistry. If we speak on the level of modern atomic and nuclear physics, however, there is no longer any such distinction.   Dingle wrote: "The truth is that chemistry indeed has no place in the strict scientific scheme. . . . The part played by chemistry in the growth of science has been a pragmatical, heuristic one."[53]   To speak briefly, chemistry is today a common-sense term, but not a scientific term.

These remarks are of great importance for an understanding of contemporary science.   Many terms that have previously been used in scientific language can no longer be used because the general principles of contemporary science now employ terms which are much more remote from common-sense language.   Expressions like "matter," "mind," "cause and effect," and similar ones are today merely common-sense terms, and have no place in strictly scientific discourse.   In order to become aware of this evolution, we must compare twentieth-century physics with its predecessors in the eighteenth and nineteenth centuries.   Newtonian mechanics used terms like "mass," "force," "position," "velocity" in a sense that seemed to be near to their common-sense usages.   In Einstein's theory of gravitation, the "coordinates of an event" or the "tensor potentials" are terms that are connected with expressions in our common-sense language by a long chain of explanations.   This is even more true for the terms of quantum theory like "wave function," "position matrix," etc.   Einstein spoke, in a lecture given at Oxford in 1933, about the "ever widening gap between the basic concepts and laws on the one side, and the consequences to be correlated with our experience on the other, a gap which widens progressively with the developing unification of the logical structure, that is, with the reduction of the logically independent elements required for the basis of the whole system."[54]

Our observations and experiments, however, have invariably been described in common-sense language, notwithstanding all changes in the principles.   Hence, science has become more and more accustomed to the use of different languages in the same picture of the universe, and it has become an important task of the scientist to fit these different languages into one coherent system.   Herbert Dingle said rightly: "If I emphasize the necessity of freeing scientific philosophy from the intrusion of common-sense conceptions, it is not in order to depreciate common sense but because the great danger today lies in this confusion."[55]

Because of this confusion, it is often the case that if a philosopher and a scientist discuss general principles, the philosopher will object that the scientist's principles are abstruse.   There lies the chief distinction between the two ends of our chain.   At the scientific end, the agreement with common sense is reached on the level of direct observations, whereas at the philosophic end, agreement with common sense is found on the level of the abstract principles themselves. The French philosopher, Edouard le Roy,[56] described this in a very graphic way.   Science starts from common sense, and from generalization by induction or imagination one derives science; but the derived principles themselves may be very far from common sense.   To connect these principles directly with common sense—

this is the work done by philosophers.   We may draw a diagram:

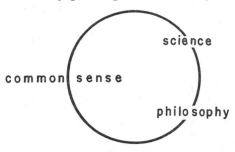

Figure 4

The diagram indicates that from science to common sense there are
two ways to go.   The scientific way (via mathematical derivation
and experimental verification) is often a very long one.   Therefore,
man requires a way by which these principles become directly
plausible; this means a way by which they can be connected with
common sense by a "short circuit."   By philosophical interpretations, the scientific principles are directly related to common sense.[57]
I would not say that this diagram is very exact, but it does give one
idea of the structure of the human mind.   Philosophy introduces
into science something in which the scientist "as a scientist" has
no interest.   As a matter of fact, the scientist is also human and
has his weaknesses, if one may call this requiring that the general
principles of science be plausible in themselves a weakness.   The
teacher of physics always finds the students grateful for any hint
which makes the laws more plausible.   So we may say that everyone is interested in this.   The scientist "as such" is not much concerned with it, but it shows us the way in which people in general
rationalize science, how they envision science.

3

Geometry: An Example of a Science

1. Geometry as the Ideal of Philosophy

"Metaphysics has always been the ape of mathematics," wrote C. S. Peirce in 1891,[1] and it is well known that Plato did not admit a student of philosophy to the Academy unless he had had training in geometry.[2] Peirce explained this requirement by continuing: "Geometry suggested the idea of a demonstrative system of absolutely certain philosophical principles, and the ideas of the metaphysicians have at all times been in large part drawn from mathematics."[3] When it was demonstrated by the example of non-Euclidean geometry that even the axioms of geometry were not self-evident and not of "eternal validity," the belief in the self-evidence of metaphysical principles was severely shaken. Peirce wrote: "The metaphysical axioms are imitations of the geometrical axioms; and now the latter have been thrown overboard, without doubt the former will be sent after them."[4]

There is no doubt that the high degree of certainty which has been reached in geometry has encouraged the hope that a similar certainty could be achieved in other fields of knowledge and, above all, in the synthesis of all knowledge, in philosophy. René Descartes,[5] in his famous *Discourse on Method*, a guiding beacon at the start of modern philosophy (after 1600), described precisely the role that he ascribed to geometry as a guide to philosophy, thus:

> The long chain of simple and easy reasoning by means of which geom-
> eters are accustomed to reach the conclusions of their most difficult

demonstrations, has led me to imagine that all things, to the knowledge of which man is competent, are mutually connected in the same way, and there is nothing so far removed from us as to be beyond our reach or so hidden that we cannot discover it, provided only we abstain from accepting the false for true, and always preserve in our thoughts the order necessary for the deduction of one truth from another.[6]

Since the procedure in geometry had led to more satisfactory results than that in any other field of science, Descartes drew generalizations from it and advanced four "precepts of logic" which would guide him in finding truth. He described these precepts as follows:

The *first* was never to accept anything as true which I did not clearly know to be such; that is to say, carefully to avoid precipitancy and prejudice, and to comprise nothing more in my judgment than what was presented to my mind so clearly and distinctly as to exclude all possible grounds for doubt.[7]

To know something "clearly and distinctly" has been called the "Cartesian criterion of truth." In substance, it is not much different from Aristotle's requirement that the general principles of science should be "intelligible" or "intrinsically knowable," in contrast to the vague sense impressions which are "knowable to us" but "intrinsically obscure" (see Chapter 1).

Descartes went on: "The *second*, to divide each of the difficulties under examination into as many parts as possible, and as might be necessary for its adequate solution."[8] This "second precept" of Descartes' is obviously also a generalization of the actual method used by the geometrician. If the latter is to prove from the axioms of geometry the theorem that the sum of the angles of a triangle is 180°, he proceeds by small steps, each of which is a simple, logical conclusion that seems valid to the most untrained mind. This proceeding by small steps is just what Descartes requires in his "second precept."

The characteristic of geometry that has made it an example for all sciences, and, moreover, for philosophy, can be simply formulated as follows: There are two types of statements in geometry, axioms and theorems. Only the latter can be proved by reasoning; the truth of the axioms must be recognized not by reasoning but by

direct intuition, by the eyes of the mind or whatever one may call
this ability.   This conception of geometry is the one which has set
the example for philosophers of all times.   At the beginning of
modern philosophy we have the words of Blaise Pascal:[9]

> Our knowledge of the first principles, such as *space, time, motion,
> number,* is as certain as any knowledge we obtain by reasoning.   As
> a matter of fact, this knowledge provided by our hearts and instinct
> is necessarily the basis upon which our reason has to build its conclu-
> sions. . . . If our reason denied its consent to the first principles unless
> our heart had provided a demonstration, this requirement would be as
> ridiculous as if our heart would deny its consent to all demonstrations
> unless they were enforced by added sentiment.[10]

No matter how wide the gaps between different philosophical systems, all have two beliefs in common.   First, there are propositions
about observable facts which we know with certainty although (or
perhaps because) they are not based on induction from sense observations.   Second, the existence of such propositions is "proved"
by the example of mathematical propositions.   For these propositions are known with certainty, and this certainty is not based on
empirical facts.   There is a great variance between the German
idealistic[11] philosopher, Immanuel Kant, and the French rationalist,[12] Descartes.   Kant, however, stressed even more emphatically
than Descartes or Pascal the point that the belief in the possibility
of "philosophy proper," of "metaphysics," was ultimately based
upon the example of geometry, which proved by its mere existence
the possibility of "intelligible principles."   To understand Kant's
statement we have only to note that he meant by a "synthetical
a priori judgment"[13] what we have called a statement about
observable facts which we perceive by the eyes of the mind without
actual sense observation, but which can and should be scientifically
checked by actual sense observations.   Kant wrote in his *Prolegomena to Any Future Metaphysics*:[14]

> It happens fortunately, that though we cannot assume metaphysics
> to be an actual science, we can say with confidence that certain pure
> *a priori* synthetical cognitions, pure mathematics and pure physics,
> are actual and given; for both contain propositions which are thor-
> oughly recognized as absolutely certain . . . and yet as independent
> of experience.   We have therefore some at least uncontested syn-

thetical knowledge *a priori*, and need not ask whether it be possible for it is actual. . . .

If we consider this common opinion of leading philosophical schools, it seems advisable to investigate geometry from the purely scientific point of view and to find out whether geometry actually consists—on the one hand, of axioms which are determined by "internal intuition," and, on the other, of theorems which are logically derived from them. As a matter of fact, during the whole nineteenth century this was the common opinion among mathematicians. We can observe this by looking into any average textbook of geometry. We may choose, for example, W. W. Bemann and D. E. Smith's *New Plane and Solid Geometry* of 1899.[15] We read: "There are a few geometrical statements so obvious that the truth of them may be taken for granted." The authors distinguish, as does Euclid, two types of such "obvious statements," axioms and postulates. All the profound philosophical terminology of Aristotle and Kant, the predicates of "intrinsically intelligible" and "synthetically *a priori*," appear in this textbook under the very harmless designation of "obvious" and "may be taken for granted."

Around 1900 a new conception of geometry developed which deprived "philosophy in its isolated state" ("metaphysics") of its favorite example, and made a reunion of science and philosophy possible. This change in the conception of geometry was, in fact, a decisive one in the relationship between science and philosophy. It is not an accident that at about the same time great changes in physics occurred, the establishment of the new theories of relativity and quanta, which required a fundamental adjustment in our traditional ideas on science and philosophy.

2. "Intelligible Principles" and "Observable Facts" in Geometry

We are now going to discuss the transition from the traditional nineteenth-century conception, in which science was topped by a piece of "separated philosophy," to the twentieth-century conception, the transition from the role of axioms as "intelligible principles" to their twentieth-century role. The different aspects of science can be characterized by the different places they assign to sense observation, to logical reasoning, and to creative imagination.

If we want to understand this, it is best to attempt a thorough understanding of one specific science. We shall take as our example *plane geometry*. There is an old saying, "If you understand one leaf of grass, you understand the whole universe." Thus, if we understand the structure of science in plane geometry, we have gained much toward understanding it in other sciences.

It is well to start from a field in which you can apparently "prove" a great deal. In geometry no one will doubt that logical argument plays a large role. If we understand what role logical argument plays in geometry, we shall understand the whole role it plays in science in general. The question is: in geometry how do we "prove" facts which can be checked by sense observation? We start from certain "axioms" which are ordinarily said to be statements that are self-evident. Then we try to derive other statements called "theorems" from the axioms by logical conclusions. In geometry, the most elementary student is presented with the distinction between "intelligible principles" (axioms) and observable facts—he does not need to read Aristotle for this. The impression is given in the ordinary teaching of geometry that there is a certain harmony between what can be proved and what can be observed in experiment. For example, consider the triangle:

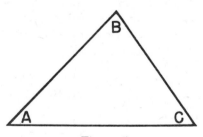

Figure 5

We have $A + B + C = 180°$ in every triangle. The student learns how to "prove" this. Then he takes a protractor and measures the sum of the angles and, if he is lucky, the sum will be about 180°. He gains the impression that there is a certain established harmony between logical thinking and nature. This idea is really produced by the traditional method of teaching geometry. If the student has once acquired this impression in geometry, he goes further with it in physics. In the latter, he learns some proofs in which logical conclusions and results of experiments are so mixed up that even intelligent students will hardly understand them. One theorem is assumed and another derived from it, but the first is just as uncertain as the other. If this is explained correctly, no confusion results. In geometry it is easier to stress from the beginning what can be proved and what cannot be proved. It is easy to distinguish between what is observed and what is proved. And what is an "intelligible principle"? We can learn all this from geometry.

Very often a statement does not seem by itself to be "intelligible" or self-evident, but some corollary of it looks very plausible and even self-evident. At first glance, the statement that the sum of the angles of a triangle equals 180° does not look very convincing, but this can be expressed in another form which makes it look much more plausible. If the quadrangle *ABCD* is divided by the diagonal *BC* into two triangles, and if in each of these triangles the sum of the angles is 180°, the sum of the four angles of the quadrangle is 360°. Then we can set the problem in which *A*,

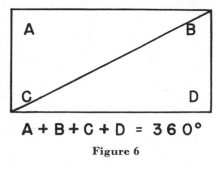

$$A + B + C + D = 360°$$

Figure 6

B, *C*, and *D* are all equal. In other words, $A = B = C = D$. Then each one is a right angle, and we have a rectangle.

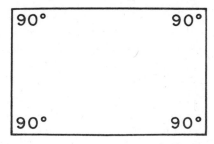

Figure 7

Hence, from the statement that "the sum of the angles of a triangle equals 180°," it follows that we can construct rectangles. However, the existence of rectangles is very plausible to us. We would be reluctant to believe that there could be no rectangles or squares. The existence of rectangles makes it possible to build a wall with bricks without gaps. Without rectangles we could not build in our usual way—our whole way of life would be different. We can see that the theorem about the sum of the angles in a triangle is very closely connected with our technical civilization.

On the one hand, we have the impression that these laws of geometry are logically derived; on the other hand, they seem to be laws of technical "know-how." This lends force to the idea that human beings must act in a way that can be derived from intelligible principles. The belief that we can derive empirical facts from intelligible principles is an important part of the web of our ideas. It is very important to examine to what extent this is true or is not true in geometry. Without axioms, there is no geometry; everything in geometry has to start from axioms. Very little space in geometry textbooks is devoted to the question: How do we know that the axioms are true? This question does not belong to mathematics and is not studied in any of the other recognized fields of science. Many teachers of mathematics have volunteered the answer that the question has no meaning. From the purely mathematical point of view this is true, since there is no mathematical method for discussing it. But, as we shall see later, it can be discussed in another way.

3. Descartes, Mill, and Kant

We shall discuss three opinions concerning the foundations of geometry. One goes back to Plato and Aristotle—to the idea of intelligible principles. In other words, we can see quasi-intuitively, with the "eyes of the mind," that the axioms are true. This has perhaps been best described by Descartes, the French mathematician and philosopher.[16] According to him, the statement that certain principles are self-evident means that if you understand them well, you also understand that they are true. He argued: "I can demonstrate properties (by imagining a triangle) which turn out to be really true (by observation); it follows that they come from the essence of the triangle. My mind must be able to grasp this essence. Otherwise I could not demonstrate these properties." This is a school of thought which is called "rationalism"—it holds that by the power of one's mind, one can penetrate to, for example, the essence of the triangle. Descartes wrote:

> I discover innumerable particulars respecting figures, numbers, motion, and the like, which are so evidently true, and so in accord with my nature, that when I now discover them I do not so much appear to learn anything new as to call to memory what was before in my mind, but to which I had not hitherto directed my attention. . . .

In contrast to Descartes' "rationalism," the school of "empiricism" has claimed that there are no principles the validity of which can be confirmed by the power of reason alone. According to the empiricist philosopher, John Stuart Mill, axioms are empirical statements just like any others—they only differ from others in being simpler and in having a wider basis. The rationalist refers to a triangle as an object of our imagination, while the empiricist refers to a triangle as a physical object. Both aspects of a triangle are legitimate in some way, or we could never check principles against facts. Two hundred years after Descartes, John Stuart Mill wrote in his book *A System of Logic* in 1843:[17]

> The peculiar accuracy, supposed to be characteristic of the first principles of geometry, appears to be fictitious. When it is affirmed that the conclusions of geometry are necessary truths, the necessity exists in reality only in this, that they correctly follow from the suppositions from which they are deduced. These suppositions are so far from being necessary that they are not even true; they purposely depart, more or less widely, from the truth. . . . It remains to inquire what is the ground of our belief in axioms—what is the evidence on which they rest. I answer: they are experimental truths, generalizations from observations. The proposition: Two straight lines cannot enclose a space (Euclid's formulation of the axiom that "two points determine one and only one straight line") is an induction from the evidence of our senses.[18]

We see that the axioms of geometry which have been regarded as the most conspicuous instance of Aristotle's "intelligible principles" are, according to the empiricist, Mill, results of sense observations. The conclusions drawn from the principles, on the other hand, are products of our reason. It seems that the rationalist idea and the empiricist idea refer to completely different things, both of which exist. What is the connection between an imaginary triangle and the physical object? The rationalist thinks that he can find the properties of a triangle by "looking with the eyes of his mind upon the triangle." But his mind can obviously look only upon an imagined triangle, and not upon a physical triangle which belongs to the world of material objects. On the other hand, the empiricist thinks that he obtains the properties of a triangle by looking at a physical triangle with his sense organs. How can we then understand that the geometrical propositions are more certain than any result of sense observations?

Immanuel Kant[19] found a way out of this dilemma which must certainly be called ingenious. He asserted that our sense organs, our eyes, do not see the real triangle which exists in the external world. This real triangle, the "thing in itself" as Kant called it, is inaccessible to our sense organs. If we look at a triangle, we see it in a way which is determined by the properties of our minds. What we call in our ordinary language the "seen triangle" is a result of the cooperation of the real triangle and our minds. Our minds are responsible for a "frame" through which we see every external object. This means that what the empiricists call the "real triangle" seen by our senses is actually an "imagined triangle." Therefore, there is no wonder that the eyes of our minds can see its properties. The geometrical properties are actually properties of the imagined triangle, while the properties of the real triangle are unknown, or perhaps even nonexistent. According to Kant, the knowing of properties by our minds is only possible if we assume that these properties are not properties of the real triangle. He said, "By sensuous intuition we can know objects as they *appear* to us (to our senses), not as they are in themselves, and this assumption is absolutely necessary if synthetical propositions *a priori* be granted as possible."

This new idea has been given by Kant and his school the name of "critical idealism." The word "idealism" denotes a world view according to which the results of our sense observations are not pictures of real objects. These objects may not exist at all, or may be very different from the way they appear to us. The first view which denies the reality of the world of our experience is called plain "idealism." The Kantian view asserts that the external world exists in itself, but appears to us in a way that is determined by the nature of our minds. This view is called "critical idealism."

The twentieth century has, in close connection with "science proper," developed a new conception of the place of geometrical axioms which has absorbed some elements of rationalism, empiricism, and critical idealism, but has attempted to eliminate, so far as possible, superfluous concepts.

4. "Axioms" and "Theorems"

We shall consider how the theorems of geometry have been proved traditionally; for example, the theorem that the sum of the angles of a triangle is equal to two right angles. We shall see that this is closely connected with the statement that there are similar triangles—in other words, triangles with the same angles but with sides of different lengths. The existence of similar triangles is one of the fundamental ideas with which we approach the external world. It accounts for the possibility of figures which have the same shape but different sizes. On it is based our belief that what is true on a small scale can be repeated on a large scale, and vice versa. We believe this in a very naive way. No student would find it easy to doubt that what his teacher demonstrates as true for triangles on the blackboard is also true for triangles too big to put on the blackboard.

Before proceeding with this discussion, however, let us first "polish up" our knowledge of plane geometry. We may start with the axiom: if we have two points, A and B, and we have two straight lines which connect these two points, "between these two lines there is no space," in the language of Euclid.[20]

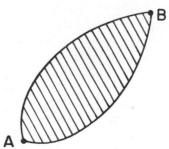

Figure 8

In other words, if there are two points, A and B, there is only one straight line which connects them. This is one of the first axioms in Euclidean geometry.

What is a point? And what is a straight line? In ordinary geometry these are usually defined in a vague way. A point is that which has no part. Intuitively this has a certain meaning, but it is difficult to use. We shall return to these questions later. For the moment we have only a vague idea of points and lines.

However, we can at once ask the question: is this axiom which we have just described self-evident, or is it not self-evident? It is not as self-evident as it looks because it means this: through a point A, two straight lines which diverge can never meet again.

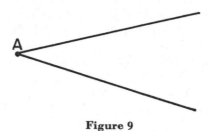

Figure 9

If we look into ourselves, it seems at first that this is intuitively clear, but how far does our imagination go toward imagining straight lines? I would say hardly ten feet. Intuitive imagination certainly does not go very far. What we use is actually an inference from the increasing distance between the segments of the lines. We imagine that the distance between them will continue to increase. But this is just a vicious circle; it amounts to saying the same thing —that lines which are diverging will never meet. If we pursue two "straight lines," along the surface of the earth, they will certainly meet on the other side of the earth. When the earth was thought to be a plane, the situation was clear; but we know now that this is an illusion—that there is no way to distinguish a small section of a sphere of large radius from a plane. So what happens farther and farther along on these two diverging "straight lines" we really do not know; there is no intuitive evidence that they never meet again. Our axiom, then, is an hypothesis about the behavior of straight lines.

There is a more sophisticated difficulty involved: someone may say that a line which comes back on itself is not a straight line, but if we define a straight line as a line which never meets itself, this is a tautological statement—a straight line is a straight line. Are axioms only definitions? If they are only definitions we can never derive physical facts from them. Thus there are two aspects of geometrical axioms—"pure definition" and "hypothesis about physical objects." In this first axiom we see already all the difficulties involved.

Note that from this first axiom we can deduce that two straight lines can have either one point or no points in common.

We shall now proceed to the conception of "congruence." Consider a straight line g, which contains two points A and B, and also another straight line g' containing points A' and B'. What do we mean by saying that the two distances AB and $A'B'$ are "congruent?" The two segments are said to be congruent if they can be brought into coincidence. This presupposes that we know what we mean by transposition—that during motion, roughly speaking, the two distances do not change their size, which means that they remain congruent. Again we have a vicious circle. We have, however, a definite idea of a rigid body; we can define it by its physical properties: elasticity, hardness, etc. Then we can define congruence by the transposition of a rigid body. Two segments are "congruent" if they can be brought into coincidence by being moved as "rigid" rods.

We also have the concept of congruent angles. Two straight lines define an angle in the following figure:

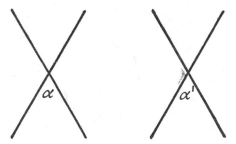

Figure 10

Angles are congruent if the straight lines which define them can be brought into coincidence. Two triangles are defined as congruent if all sides and angles are congruent: then the triangles can be brought into coincidence.

Thus we can state the first theorem of congruence. Let ABC and $A'B'C'$ be two triangles:

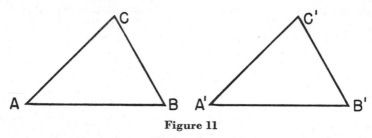

Figure 11

Let $AB \equiv A'B'$, angle $CAB(\alpha) \equiv$ angle $C'A'B'$, and $CBA(\beta) \equiv$ angle $C'B'A'$, where the sign $\equiv$ stands for "is congruent with." This means that we could transpose triangle ABC on to triangle $A'B'C'$ so that the line segment AB would coincide with the line segment $A'B'$, the AC with the line $A'C'$, and the line BC with the line $B'C'$. But we know that the lines AC and BC meet at C. The lines $A'C'$ and $B'C'$ meet at C'. But C must coincide with C', since our first axiom states that two straight lines can have only one point (or no point) in common. Thus triangle $ABC \equiv$ triangle $A'B'C'$. This means that we could transport triangle ABC so as to bring it into coincidence with triangle $A'B'C'$. Hence, if $AB \equiv A'B'$ and $\alpha \equiv \alpha'$, $\beta \equiv \beta'$, the triangles are congruent. This is the first theorem of congruence.

5. The Euclidean Axiom of Parallels

We are now much closer to being able to prove that the sum of the angles of a triangle is equal to two right angles, but first we must prove an important theorem about the conditions under which two straight lines do not intersect. Consider a straight line h which is cut at a point A by another straight line g under the angle α. Then consider that h is also cut an another point B by a straight line g' also under the angle α. (See Figure 12.) We wish to prove that the two straight lines g and g', thus drawn, can never meet. How can this be proved? Let us assume that they do meet to the right of h at C. Then we have triangle ABC. Now using the theorem that vertical angles are congruent (which we have not proved), and following the same argument as we have made above, we see that to the left of h there must also be a triangle ABC' congruent to triangle ABC.   Thus, if the two straight lines g and g' meet on one side of h, they must also meet on the other side; but this is impossible, because then there would be two points C and C' connected by two straight lines g and g'; hence the straight lines g and g' can never meet.

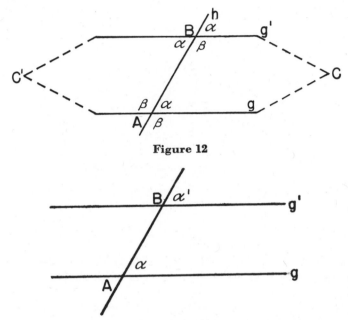

Figure 12

Figure 13

Now how can we prove that the sum of the angles of a triangle is 180°?   Note that the above proof does not show that it is *only* when two lines g and g' cut the line h under the same angle α that they will never meet.   The line g' may also cut the line h under some angle α' (not equal to the angle α under which the line g cuts the line h), and yet g and g' may never meet.   In order to derive the theorem about the sum of the angles of a triangle, however, we must make use of the assumption that it is *only* when the lines g and g' cut the line h under the same angle α that they will never meet.   (See Figure 13.)   Then we say that g' is "parallel" to g. Our assumption is called "Euclid's Axiom," or the "Axiom of Parallels."   This states that through a point B outside a straight line g there is one and only one straight line g' which is "parallel" to g. Both g and g' cut h under the same angle α.

We are now ready to prove the theorem about the sum of the angles of any triangle. Consider a triangle ABC and through the vertex C draw the parallel to the base AB. Since g is parallel to g', AC cuts g and g' under congruent angles α (see Figure 14).

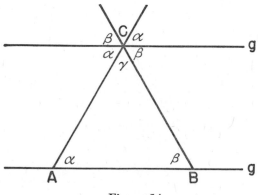

Figure 14

BC cuts g and g' under congruent angles β. Since g' is a straight line, angle α plus angle γ plus angle $\beta = 180°$, or two right angles. But these are the same angles as are in the triangle, so that whatever triangle we may be considering, the sum of its angles, $\alpha + \beta + \gamma = 180°$ of two right angles. To reach this conclusion we obviously need the Euclidean axiom. We have proved first that if two straight lines g and g' intersect a transversal h under the same angle α, they can never meet. But in our present proof we have used the theorem that if two straight lines g and g' meet, they must intersect a transversal under the same angle. Otherwise there may be a line g'' which intersects h under a different angle and still will never meet g. To exclude this possibility we must use the "Euclidean Axiom:" the line g' which intersects a transversal under the same angle as the line g is the *only* line which will never meet g.

This is a very important point. The theorem that the sum of the angles of a triangle is equal to $180°$ presupposes the Euclidean axiom. It plays a particular role for different reasons. We shall see that if this axiom is not accepted, not only the law regarding the sum of the angles of a triangle breaks down, but that much more of our view of the universe breaks down also. I have already mentioned the concept that for every figure there exist similar figures; if we assume the Euclidean axiom, this can be easily proved. Consider a triangle *ABC*. Through a point *D* on the side *AC* draw a parallel to the base *AB*. If we assume that the parallel axiom is true, then the base angles of the small triangle *CDE* are equal, respectively, to the base angles of the large triangle *ABC*, as indicated in Figure 15. Since the sum of the angles of any triangle is

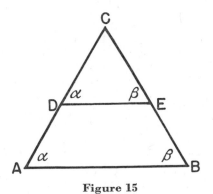

Figure 15

equal to 180°, we get a small triangle *CDE* with the same angles as the big triangle *ABC*. These two triangles have the same shape because they have the same angles, but they differ in size. If we do not know that the parallel axiom is true, or if we know that it is not true, we cannot give this proof. Thus, if the parallel axiom is not true, we cannot prove the existence of similar figures. These two things exclude each other.

Can we prove that the parallel axiom is true? Apparently not, or it would not be an axiom. Logically then, we can set up a different axiom. If the angles of a triangle do not equal 180°, what follows from this? One thing we can easily see without much calculation. Consider an isosceles triangle *ABC* (Figure 16). If the sum of the angles is equal to 180°, angle *ABC* = 60°. Now divide the triangle into two equal parts, by dropping a median from the vertex *C*. The question is: Is the sum of the angles in each of the small triangles the same as in the big triangle? The answer is yes, since 60° + 30° + 90° = 180°. If we assume now that the sum of the angles in a triangle is not 180°, we shall see that the sum of the angles in a small triangle is completely different from the sum in a big triangle. Again consider an isosceles triangle where the base angles equal 60° (Figure 17). Let us assume that the sum of the angles in this triangle is 160°. Then angle ACB equals 40°. Divide this triangle into two equal parts, by dropping a median to AB. We observe then that the sum of the angles in each of the two small

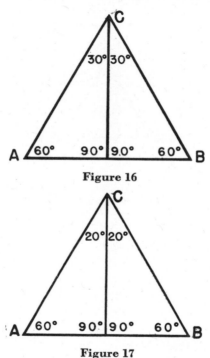

Figure 16

Figure 17

triangles is equal to only 170°. If we pursued this, we would see that the smaller a triangle becomes, the nearer the sum of its angles approaches to 180°.

In the case where the sum of the angles of a triangle is less than 180°, let us not describe the sum of the angles itself, but the difference between this sum and 180°—what is called the "defect." In other words, the defect is equal to $[180° - (\alpha + \beta + \gamma)]$. In the big triangle above, the defect is equal to 20°. In the small triangle, the defect is equal to only 10°. Between the area of a triangle and its defect there is a very simple relation. The area of each of the

small triangles above is half the area of the big triangle, and the
defect of each of the small triangles is half the defect of the big triangle. For very small triangles the defect approaches zero. A
very small triangle behaves as if the Euclidean axiom were true.
This can be proved in a very general way; here we have only given
some examples to illustrate it.   If the Euclidean axiom is not true,

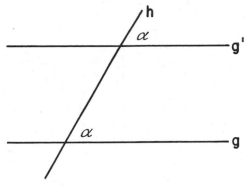

Figure 18

there are no similar triangles, but small triangles behave in a different way from large ones.   For this reason it is very difficult to
check by measurements whether the sum of the angles of a triangle
is generally 180°.

6. Non-Euclidean Geometry[21]

We shall now consider the possibility of dropping the Euclidean
axiom. By what is it to be replaced?   If we accept the Euclidean
axiom, the geometry built on this axiom is called Euclidean geometry.   If we reject the Euclidean axiom, and replace it, the geometry
built on its replacement is called a non-Euclidean geometry.

If the Euclidean axiom is rejected, there are two possibilities.
This axiom asserts that a straight line which made only the slightest
departure from g' (Figure 18), on either side, would intersect
with g.   One possibility is that there may be *no* line g' which
will never intersect g.   In other words, all lines which exist will
intersect.   There is also the possibility that if a straight line is
tilted from g', on either side, by a sufficiently small angle ϵ, it
will not intersect g.   In other words, there may be a "bundle"

of lines—symmetric about g' and bounded by the lines which are tilted from g' by the angle ϵ on either side—which will not intersect (Figure 19). This is the second type of non-Euclidean geometry, and it is the only type of non-Euclidean geometry which we shall discuss here. We shall investigate how the world would look if this assertion replaced the Euclidean axiom. One thing is certain— the sum of the angles of a triangle would not be equal to 180°.

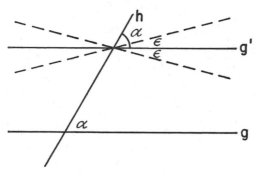

Figure 19

The first type of non-Euclidean geometry we have mentioned, which asserts that there are no parallel lines, is known as Riemannian geometry; it also drops the axiom that only one straight line connects two points. The second type, which asserts that there is an infinite number of straight lines which do not intersect g, enclosed in a certain angle about g', was built up by Lobatchevski,[22] a Russian mathematician, and too, at about the same time, by the Hungarian, Bolyai.[23] Euclid's axiom is replaced by "Lobatchevski's axiom." The conclusions drawn from this axiom may be further described as follows: Let us draw a straight line g and a point A outside g (Figure 20). Then we draw through A a transversal h normal to g and a straight line g' under an angle of 90° to h. Then g' will not meet g. If we draw straight lines from the point A, always making larger and larger angles with h, at first these lines will intersect the line g. Eventually, however, (if we exclude the case that all straight lines meet each other—the "Riemannian axiom") we will come to a line which makes a certain limiting angle α^* with the transversal h and is the first line that will not intersect g. In Euclidean geometry $\alpha^* = 90°$. According to

Lobatchevski's axiom, the angle α^* is less than 90°. The straight line that includes this angle with g is called "parallel" to g. In Lobatchevski's geometry, we must distinguish between parallel lines and nonintersecting lines. All the lines which are contained in the "bundle" around g' are called nonintersecting lines. Only the boundary lines of this "bundle" are called parallel lines. In Lobatchevski's geometry there is a parallel line to the left of the transversal as well as to the right of it. Thus, it is characteristic

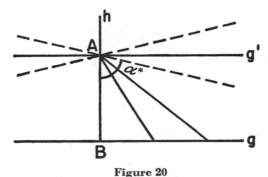

Figure 20

of Lobatchevski's geometry that there is not one parallel line, but two. All lines between them are nonintersecting, *i.e.*, they do not intersect g.

Let us consider what the sum of the angles of a triangle would be in Lobatchevski's geometry. If we consider a line from A which makes an angle with AB which is smaller than α^*, then it will intersect g. We may, however, take a line which makes an angle $\alpha = \alpha^* - \eta$ just a little smaller than α^*. Such a line will intersect g under an angle ϵ which is very small, as small as we like to make it. Thus we see that there are triangles in which the sum of the angles is less than 180°, since $\alpha^* < 90°$ and ϵ may be as small as we wish it to be. This is clear from the beginning if we speak in terms of the defect. In the above triangle, the defect = $180° - 90° - (\alpha + \epsilon) = 90° - (\alpha + \epsilon)$. Considering that we have made the line from A almost parallel to g, ϵ and η will be infinitesimal and we may consider the defect to be $(90° - \alpha^*)$, a positive number.

How do we know how big $(90° - \alpha^*)$ is? Lobatchevski's axiom obviously does not determine the angle α^*, which a "parallel line" makes with a given line.   Under the name "Lobatchevski's axiom" are covered, in fact, an infinite number of axioms.   The parallel line to g (called g' in Figure 21) can be almost normal to AB, or can make any angle with AB.   We are free to require that for any one specific distance AB of A from g the angle α^* may have a value that can be arbitrarily prescribed.   To every choice of the angle

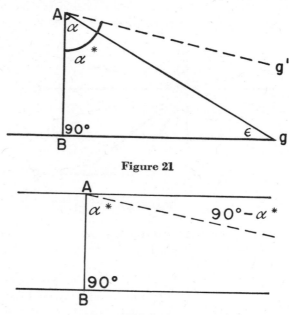

Figure 21

Figure 22

α^* there corresponds, actually, a special form of the Lobatchevski axiom.   If we want a geometry that differs little from Euclidean geometry, we choose that the defect $(90° - \alpha^*)$ be small; if we want a very different geometry, we choose that this defect be large (Figure 22).   We can ask the question: At what size does a triangle begin to be noticeably non-Euclidean?   When we examine physical triangles made of rigid materials like steel, we have never noticed that the sum of the angles is less than 180°, by measurement, but this does not prove anything.   The triangles measured may all have been too small for the defect to be noticeable.   We can only say that all the triangles we have measured have been "small" triangles.   We may define the concept of a "small triangle" as being small compared with a certain "unit" triangle; we may define a "unit triangle" as a triangle in which the defect is one degree.   Instead of giving the value of α^* we can also define a special type of Lobatchevski's geometry by giving the areas of the "unit triangle."   If the size of the "unit triangle" were comparable to the dimensions of a galaxy, all physical triangles would be "small," and we would not notice the defect in any of the triangles that we measure.   The larger the area of a triangle compared with the "unit triangle," the larger will be the "defect."   Therefore, the sum of the angles in a large triangle is less than in a small triangle.   A triangle with an area that is very small compared with the area of the "unit triangle" has nearly a zero defect, and the sum of the angles is nearly 180°; with increasing defect, the sum decreases.   If we have a small triangle that is similar to a large triangle, both have the same angle-sum of 180°.   Since, in Lobatchevski's geometry, large triangles have greater defects than small ones, a large and a small one can never be similar.   Therefore, there are no triangles with the same shape but of different sizes.   From this, one can easily conclude that there are no geometrical figures of any kind which have the same shape but are of different sizes.   By the shape the size is determined.   A triangle with the angle-sum of nearly 180° is only possible if it is of very small size.

7. "Validity" of Propositions in Geometry

We shall now leave the purely mathematical argument for the time being, and ask what the relationship is between geometry and experience.   What we have proved up to now has had nothing to do with experience.   We have simply shown that if triangles fulfill the Euclidean axioms, there are similar triangles.   If, however, triangles fulfill the axioms of Lobatchevski's geometry, there are no similar triangles.   These are merely conditional statements; we cannot derive anything from them about physical triangles made of wood or iron.   If some axioms are true, certain results are true. All that we know to be true in geometry are these conditional statements.   Whatever may happen in the world, these statements remain true.   Purely logical statements are true independently of the physical occurrences in the world.   The same is true of geometry if we take it in the purely mathematical sense.   We can characterize "logical statements" by saying that they are true because of their form, without regard to the meaning of their terms. We can replace all the terms by other terms, and the statements remain nonetheless true. The most familiar example is the logical syllogism: If Socrates is a man and all men are mortal, Socrates is mortal. This statement remains true even if we replace "Socrates," "man," and "mortal" by other terms. For example, if the fox is a mammal and all mammals are vertebrates, the fox is a vertebrate. All statements of geometry are ultimately of this kind.

In the presentation of the elementary textbooks, the statements of geometry are not purely logical statements. They are a mixture of logical and empirical statements. The concept of congruence, for example, is defined by reference to a physical operation, the transfer of rigid bodies. However, we can convert Euclidean geometry by reformulating the axioms into a system of purely logical statements. We shall discuss this reformulation in Section 8. At this point, we shall take it for granted that such a "formalization of geometry" is possible and ask bluntly: Which is true, Euclidean or non-Euclidean geometry? We cannot answer this question from the point of view of mathematics. By mathematical argument, as we have learned, we can only prove that if we assume the Euclidean axiom, it follows that there are similar triangles, and if we reject that axiom, it follows that there are no similar triangles. However, we cannot decide whether it is "true" that there are similar triangles, i.e., whether "Euclidean geometry is true." On the other hand, we are accustomed to applying geometry to physical objects. It needs careful examination to understand how it can be done. Nowhere in the whole system of geometry can there be found a definition of a straight line or a point. Since, however, logical conclusions are independent of the meanings of the terms involved, without defining straight lines and points we can say that if these objects have the properties assumed in the axioms, they also have the properties developed in the theorems. Whatever straight lines and points may be, if we adopt the Euclidean axioms, it follows that there are similar triangles—if Lobatchevski's axioms, no similar triangles—so how can geometry be applied to triangles made of wood or steel? For this purpose, we need obviously, a "geometry" that is entirely different in its structure from the mathematical, formalized geometry of which we have spoken up to now.

We have noted that all the results obtained so far are valid without considering the "meaning" of the geometrical terms. In order to get at the application to physical triangles, we must build up another kind of geometry that considers the meanings of terms like "point," and "straight line." Rudolph Carnap[24] has described, in the introduction to his book, *Formalization of Logic*, "two tendencies in modern logic."

> The one tendency emphasizes form, the logical structure of sentences and deductions, relations between signs and abstractions from their meaning. The other emphasizes just the factors excluded by the first: meaning, interpretation, relations, . . . compatibility and incompatibility as based on meaning, the distinction between necessary and contingent truth, etc. The two tendencies are as old as logic itself and have appeared under many names. Using contemporary terms, we may call them the syntactical and the semantical tendency respectively.

There have been frequent attempts to build up geometry from scratch, not as a logical discipline, but as a science that deals with physical bodies, *e.g.*, wooden and iron triangles. A remarkable attempt of this kind was made by the prominent British mathematician, William Kingdon Clifford, who worked more than most mathematicians have done for the integration of mathematics into our general system of knowledge. Clifford[25] wrote in 1875:

> Geometry is a physical science. It deals with the sizes and shapes and distances of things. . . . We shall study the science of the shapes and distances of things by making one or two very simple and obvious observations. . . . The observations that we make are:
> First, that a thing may be moved about from one place to another without altering its size or shape. Secondly, that it is possible to have things of the same shape but of different sizes.

The "things" about which Clifford speaks here are obviously what are called in physics "rigid bodies." He assumes that the criterion used to make sure a "thing" is rigid is the criterion that is generally used in experimental physics. The "size" and "shape" of a thing are measured by the standard meter in Paris or the standard foot in Washington, using the methods of correction as they are legally prescribed. Then the two observations described by Clifford can be performed using these standards. Clifford continued:

> Applying these [two] observations to triangles we can prove: (a) Two straight lines cannot intersect in more points than one. (b) If two straight lines are drawn in the plane so as not to intersect at all, the angles they make with any third line which meets them will be equal.

We have learned in previous sections (4, 5, and 6) that from the existence of similar triangles we can derive Euclid's axiom of parallels and from this axiom the theorem that a line drawn so as to intersect two parallels will make equal angles with both. In our mathematical argument, we drew these conclusions from "axioms" without using the meaning of the geometrical terms. Clifford started with generalized observations which were, of course, statements about physical facts and conclusions drawn from them which were also statements about physical triangles.

We shall now, in the following sections, discuss the precise relationship between conclusions drawn from axioms without using the meaning of the terms and conclusions drawn from statements about physical facts, where every term denotes a physical object. In twentieth-century science, the "axioms" have been formulated in such a way that in drawing conclusions one actually does not use any information about the meaning of the terms; with the establishment of this completely formalized system of axioms, one draws conclusions about physical triangles by using a peculiar method of coordinating the purely formal, purely logical conclusions with the statements about physical objects.

8. "Formalization" of the Axioms

Geometry, as it is treated in elementary textbooks and courses, is not a purely logical system. The meanings of some of the terms, *e.g.*, of "congruence," are defined by means of physical operations, such as the displacement of a rigid body. However, the Euclidean system can be modified so that it becomes purely logical. We shall illustrate this by a very simple example. This means that we must formulate it in such a way that the validity of its statements are dependent only on their form and not on the meaning of the geometrical terms: "straight line," "point," "intersecting," and "connecting."

Previously we laid down the axiom that two points A and B can only be connected by one straight line (Axiom I). Then with the help of the following diagrams, we derived the conclusion that two straight lines can only intersect in one or no points. (See Figure 23.) If, in Figure 24, the lines p and p' intersect not only at A

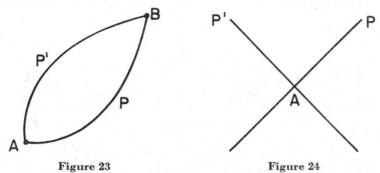

Figure 23   Figure 24

but at a second point B, we would have Figure 23, which is, according to Axiom I, impossible if p and p' are different lines. This was an intuitive proof, which depended on visualizing in a certain way straight lines and points and straight lines intersecting. The meanings of straight line and point and intersecting in a physical way, however, are not necessarily involved. We can formulate this proof in such a way that it becomes completely logical; in other words, so that the physical meanings of these terms is not involved.

We shall demonstrate that the geometrical proofs remain valid even if we replace "straight lines" and "points" by "apples" and "oranges." Let us recall "Axiom I": If there are two points, A and B, there is only one straight line which connects them. (See Figure 25.) From this we can derive Corollary I:

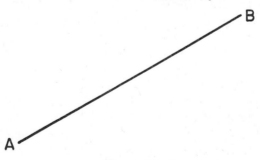

Figure 25

Two straight lines, P and P', can never intersect in more than one point.

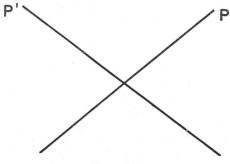

Figure 26

Now we must "formalize" these propositions (Axiom I and Corollary I). This means bringing them into such a form that one can clearly see why the meaning of the geometrical terms has no bearing upon the validity of the proofs.

First, we shall eliminate the terms "connect" and "intersect." If a straight line passes through a point, we shall say the "straight line" "coincides with a point." If a point is on a line, we shall say that the "point coincides with a line." Axiom I now becomes: If a straight line P coincides with two points, A and B, and a straight line P' also coincides with the same two points, A and B, the P and P' are not different from each other. If the corollary were not valid, we could assume that a straight line P coincides with two points A and B, and that a *different* straight line P' coincides with the same two points, A and B. But the first axiom says that P is not different from P'. Thus, from the assumption that P and P' are different, we conclude that P is not different from P'. This would mean: From a proposition "S is true" it follows that "non-S (the negation of S) is true." Therefore, the corollary must be true: Two straight lines, if not identical, can coincide with only one point; or, in other words, cannot intersect in more than one point.

Now we can replace the terms in the above argument as follows: "point" by "apple," "straight line" by "orange," "coincident" with "are on the same plate." Then we shall see that the physical properties of "straight lines," "points," and "coincidence" have nothing to do with the validity of the proof. The first axiom would become: There cannot be on the same plate two apples and more than one orange. And the first corollary: If there are two oranges and one apple on the same plate, then it is impossible that there is still another apple on the plate. If there were a second apple, there would be two apples and two oranges on the same plate. This would contradict Axiom I. Hence, from Axiom I we can obviously conclude the corollary. From this we see that the meanings of "straight line," etc., are not important, that we can draw the same conclusions by changing the meanings of the geometrical terms. In the above argument, we used only the meaning of the term "not," but we could also formalize the logical system so that the meanings of the logical terms would not enter either.

9. Formalization of "Congruence"

The traditional teaching of geometry is still almost the same as it was written by Euclid. In the strictest sense, this is only partly a logical system; it makes use of some empirical notions. Among the basic concepts, the most "physical" seems to be the notion of "congruence." The traditional definition says: Two figures are "congruent" if we can bring them into coincidence. This definition clearly refers to the idea that these figures are "rigid bodies" and can be moved without altering their shape or size. It clearly refers to a physical operation, the displacement of rigid bodies.

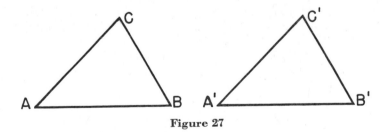

Figure 27

Let us consider, for example, the first theorem of congruence. We shall use the symbol $\equiv$ for "congruent." The first theorem of congruence says that if: $AC \equiv A'C'$, $AB \equiv A'B'$, and angle $CAB \equiv C'A'B'$, then $BC \equiv B'C'$. How do we prove this? According to

Euclid, two sides are congruent if they can be brought into coincidence. Therefore, according to the conditions given above, AC can be brought into coincidence with $A'C'$ and AB can be brought into coincidence with $A'B'$, simultaneously. Therefore, BC must coincide with $B'C'$ and so $BC \equiv B'C'$. We have used here, intermixed, both logical and empirical arguments.

Now let us use a purely logical proposition. We denote by p, q, and r simple propositions. Then a typical example of a purely logical proposition is this: Assume that if p is true, q follows, and that if q is true, r follows. We can conclude that if p is true, r follows. It doesn't matter whether p, q, and r are statements about points, congruence, coincidence, or whatever.

Toward the end of the nineteenth century, attempts were made to make of geometry such a purely logical system. These attempts were summarized and perfected through the work of David Hilbert[26] in 1898–1899, and published in his book, *The Foundations of Geometry*. The idea was that geometry should be developed in such a way that one could go from the axioms to the theorems without depending on what the concepts in the axioms meant. If we formulate the properties of straight lines in the axioms, and use in the deductions only these properties of straight lines, we do not need to know what a straight line is. We have seen that in traditional geometry, the physical meaning of the predicate "coincidence" is used in order to prove the theorems of congruence. So now let us see how the first theorem of congruence which we have discussed above is proved by Hilbert without using the physical meanings of terms. We can no longer say that two systems are "congruent" if they can be brought into "coincidence." This would refer to a physical operation. We must instead refer to the properties of congruence by means of axioms. What properties do two segments have if they are congruent? Consider the following diagram.

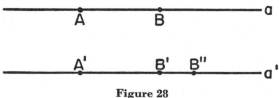

Figure 28

We consider a straight line a and two points A and B in this line. Then the first axiom of congruence says: On any straight line a', starting from any one of its points A', and on each side of it respectively, one can always find a point B' such that $A'B' \equiv AB$. This is one of the properties of congruence. A characteristic point is that this axiom does not exclude the possibility that there may also be a point B'', different from B', such that $AB \equiv A'B''$. The axiom does not say that there is only *one* point B' which has the described property. If we define "congruent" by "bringing into coincidence," this would be the case—in Hilbert's geometry the fact that there is only one such point has to be proved. Two more properties of congruence are given in the following axioms: Two segments

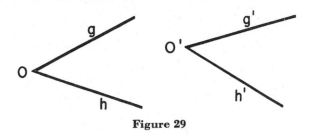

Figure 29

congruent with a third are congruent with one another; and two angles congruent with a third are congruent with one another. An "angle" is defined as two straight lines, g and h, that meet at a point O. Then we can give the axiom: Given a line g', issuing from a point O', in any plane which contains it and on each of the two sides of it, there can be found one, and only one, line h' issuing from O' so that the angle $(g'h')$ is congruent with a given angle (gh). Next we introduce an axiom which states, in a certain way, the relationship between the congruence of segments and the congruence of angles. In order to prove the first theorem of congruence, we shall need such a relationship. In order to avoid the physical meaning of congruence, Hilbert advanced this axiom from which we can derive what Euclid derived from the physical meaning (Figure 30). Hilbert's new axiom is: If $AB \equiv A'B'$ and $AC \equiv A'C'$ and angle $CAB \equiv$ angle $C'A'B'$, then angle $ACB \equiv$ angle $A'C'B'$. These axioms as we have just stated them do not bring in any physical operations; they just give properties of congruence.

Now let us proceed to use them to prove the first theorem of congruence. Consider the following diagram:

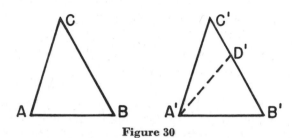

Figure 30

It is given that $AB \equiv A'B'$, $AC \equiv A'C'$, and angle $CAB \equiv$ angle $C'A'B'$. According to the "first theorem of congruence," the relation $B'C' \equiv BC$ should hold. If we define "congruence" by "coincidence," the proof is obvious. Hilbert, however, proved it from his axioms as follows: If $B'C'$ were not congruent with BC, then by the first axiom above (Hilbert's) we can always find a point D' such that $B'D' \equiv BC$. (If the point lay beyond C', the argument would be similar.) Now compare triangle ABC with triangle $A'B'D'$. $AB \equiv A'B'$, $BC \equiv B'D'$, and, by Hilbert's axiom, angle ABC is congruent to angle $A'B'D'$. Therefore, also by Hilbert's axiom: Angle $D'A'B'$ is congruent to angle CAB, and angle $A'D'B' \equiv$ angle ACB. Therefore, angle $D'A'B' \equiv C'A'B'$, and $A'C'$ must coincide with $A'D'$. We can see, in this way, how, by introducing somewhat more complicated axioms, we can prove the theorems of congruence in a completely logical way without referring to the physical idea of "bringing into coincidence." We did not give any physical interpretation of the terms; we simply stated certain axioms in which the terms occur. From this, it is clear, we cannot tell anything about the external world. It follows only that if some congruence exists, other congruences exist also, but we do not know "what congruence is."

On the other hand, we learn that geometry provides us with laws about the properties of physical bodies. If we make a triangle from rigid steel, we check, by real measurements, that the sum of the angles is approximately 180°. Now the problem arises: How can the logical system of mathematical geometry, *e.g.*, Hilbert's system of axioms, help us to obtain the physical laws about

triangles of steel or wood?   This connection will be discussed in the
following section.

10. Operational Definitions in Geometry

We have seen that the system of mathematical geometry, if
properly formalized, becomes independent of the meanings of terms
such as straight lines and points.   The whole system can then be
considered as a definition of these terms, inasmuch as it gives all
the properties of them.   Axiom I, for example, can be formulated
as follows: "Points" and "straight lines" are such objects, and
"coincidence" is such a property that one and only one straight line
can coincide with two given points.   This is an "implicit definition"
of geometric terms.   Axiom I (Section 8) expresses the same thing
in a different form.   We call it an "implicit definition" of points
and straight lines.   These definitions are, as every definition is,
arbitrary.   Whatever happens in the world of experience, no one
can prevent us from formulating these definitions.   They are
neither true nor false; they stipulate rules according to which the
geometrical terms, "point," "straight line," "coincidence," etc.,
are to be *connected with one another;* but they do not stipulate any
rules for connecting these terms with physical objects like triangles
of wood or steel.   If we wish to proceed now to the problem of how
to make use of formalized geometry for our orientation in the physical world, we must ask the following question: Are there objects
in the physical world which have the properties formulated in the
axioms?   If so, then they also have the properties formulated in
the theorems.   We look for a "physical interpretation" of the
axioms of geometry.

We could, for example, say that we interpret a straight line in the
physical world as the edge of an iron cube.   In physics such a body
can only be defined by the technological procedures by which such
a cube is produced.   We have to consider corrections with regard
to changes of size and shape caused by changes in temperature and
pressure.   These procedures include the establishing of the standard
meter in Paris and the measurement of our prospective cube by
comparison with this standard.   Eventually, we obtain the "edge
of a rigid body."   The ends of the edge are a physical interpretation
of "points."   In this way, we obtain what P. W. Bridgman has
called "operational definitions" of the terms "straight line,"

"point," etc. These definitions are, obviously, different in kind from the definition of these terms by the axioms of formalized geometry which we may call "axiomatic definitions." We can also give other "operational definitions" of a straight line—we may say that it is the path of a light ray, or of a stretched cord. We may say that it is the shortest distance between two points—in this case, we need an "operational definition" of what is shortest, *i.e.*, a way of measuring length.

If we give a physical interpretation of "point," "straight line," and "intersection," then the axioms and theorems of geometry acquire a completely different character. The axiomatic definitions of "straight line," "point," etc., are arbitrary, but if we substitute for these terms their operational definitions, they become statements about physical things, and have to be checked by experiment: they may be confirmed or refuted. If, then, we make a triangle of wood or steel, measure the sum of its angles, and find it to be about 180°, is this a confirmation of Euclidean geometry? Strictly speaking, no; it is a confirmation of a "special physical interpretation of Euclidean geometry." If we find "simple" objects with a certain importance for us which fulfill the axioms of Euclidean geometry, we say that Euclidean geometry is "true" in the sense that it has a certain application for us.

It is impossible to check Euclidean geometry against Lobatchevski's geometry directly by comparing physical interpretations of the axioms. How can we determine whether there are one or more edges of bodies which do not intersect if prolonged far enough? Practically, this is impossible to determine by direct experiment. Perhaps, as in physics, we can check some consequences of them rather than the axioms themselves. For example, from the Euclidean axiom follows the theorem that the sum of the angles in a triangle equals 180°. If we measure it, we may find a small "defect"; the observed sum may be a little less than 180°. We can ascribe the difference to experimental errors. We could also assume, however, that Lobatchevski's geometry is valid for the same physical interpretation. If we admit this possibility, we do not know whether or not the "defect" is really due to errors. It depends upon which specific Lobatchevski geometry is valid (Section 7). We must make a specific assumption with regard to how large the unit triangle is. If our measured triangle is much smaller than the unit triangle, the defect will be very small. If we were to make the base of the measured triangle a million miles long, then we would find a much larger defect. Thus we can see two possibilities of accounting for a small measured difference from 180°: either Euclidean geometry applies and we interpret these differences as "errors," or a specific Lobatchevski geometry applies in such a way that triangles here on earth are very small compared with the unit triangle. If there are two possibilities, we will choose the "simpler"—if we can find an obvious criterion of simplicity.

These remarks have dealt with one and the same physical interpretation of two systems of axioms. We should also consider two different physical interpretations. Then, it might happen that one of these two physical interpretations—say, straight lines shown by light rays in one case, and by the edges of rigid bodies in the other—may confirm Euclidean geometry and the other Lobatchevski's geometry. Thus, experiment can never confirm a system of geometrical axioms, but only a "geometry" *plus* a physical interpretation of it. The problem is always this: If the enlarged system, consisting of the axioms of geometry plus their physical interpretation, fails to check with experiment, we can drop either one or the other part of the enlarged system. We have repeatedly stressed the point that the formalized system of geometry does not tell anything about the world of physical experiments, but consists of "arbitrary" definitions. This fact was formulated by the great French mathematician and philosopher, Henri Poincaré.[27] He claimed that the laws of geometry were not statements about reality at all, but were arbitrary conventions about how to use such terms as "straight line" and "point." This doctrine of Poincaré has become known as "conventionalism." It has angered many scientists because it claims that the statements of geometry, which they have regarded as "truth," are only "conventions." They have emphasized that geometry has been of great practical use to man. This fact was not denied by Poincaré. There are useful conventions and useless conventions. If there were nothing in the physical world which fulfilled the axioms of geometry (*e.g.*, no rigid bodies), that system of conventions would be of no practical interest because it could not be applied to anything. Nevertheless, because of its "if-then" character, geometry would remain true. Thus we can say that such logical structures as geometry are true by themselves, independent of what happens in the world and independent of the meaning of their terms. The meaning of their terms is irrelevant. We might say that geometry is an instrument we have constructed to deal with rigid bodies. If we give the laws of geometry a physical interpretation, they are physical laws just like any other physical laws. We must consider that geometry has a double aspect. As a logical structure, it has no connection with reality, but it has the characteristic of certainty. Once we give it a physical interpretation, it no longer has this characteristic of certainty. Einstein has expressed this as follows: "In so far as geometry is certain, it says nothing about the physical world; and in so far as it says something about our physical experience, it is uncertain." Frequently the question is asked whether our "real space" is Euclidean or non-Euclidean. Some wish to prove that our "real space" is actually Euclidean, and that non-Euclidean space is only a fiction, a product of our imagination or construction. This alternative is not correctly formulated.

We must distinguish between geometry as a logical structure and geometry as a physical interpretation. We must understand the degree to which geometry is conventional. From the purely logical aspect, Euclidean and non-Euclidean geometry are two logical structures which are equally consistent and, therefore, equally "true." The question whether or not our "actual space is Euclidean" means: Are there simple physical interpretations of "point," "straight line," etc., which fulfill the axioms and, therefore, also the theorems of Euclidean geometry?

11. The Twentieth-Century Conception of Geometry

When modern science developed, around 1600, there was a certain amount of suspicion with regard to presentations of science which stressed logical systems of terms. Long before the concept of "operational meaning" was thought of, logical systems, as exemplified in medieval scholasticism, had been applied to the world of experience in a rather loose way. It was believed that by formulating a logical system, man had also advanced a theory about the world of experience. In a rather perfunctory way, this belief was not wrong. Some operational definitions were taken for granted, no necessity was felt to formulate them explicitly. Even men like Lobatchevski or Hilbert spoke about straight lines as things of the physical world, as if there were not several different ways of giving an "operational definition" of a straight line. The "edge of a rigid body" has usually been taken for granted as the natural physical interpretation. As we have mentioned, however, the lack of a well-defined link between logical systems and the world of experience was noticed and attacked by the earliest advocates of experimentation as the basis of science. Francis Bacon directed his *Novum Organum* against Aristotle's *Organum*, *Metaphysics*, and *Physics*, in which the philosopher of ancient science stressed the role of logical systems, without paying sufficient attention to the role of "operational definitions," even though he did pay more attention to the latter than his great predecessor Plato. Francis Bacon wrote:[28]

> The syllogism consists of propositions, propositions of words; words are the signs of notions. If, therefore, the notions be confused and carelessly abstracted from things, there is no solidity in the superstructure. . . . The present system of logic . . . rather assists in confirming and rendering inveterate the errors founded on vulgar notions, than in searching for truth. It is, therefore, more hurtful than useful.

Certainly, the urge to present geometry as a purely logical system has been particularly strong. Geometry has been proud of its "absolute certainty," and this claim could not be based on experimental research. Louis Rougier has said[29] in his book *The Geometrical Philosophy of Henri Poincaré*:

> Geometrical theorems seemed to enjoy a double certainty; the apodictical necessity which arose from demonstration, and the sensual evidence that originated in spatial intuition. They seem to have a double truth: formal truth that originates in the coherent logic of discourse, and material truth originating in the agreement of things with their objects.

According to the usual nineteenth-century presentations of geometry, the theorems were based on formal logical deductions from the axioms, but the truth of the axioms was based upon "spatial intuition." The logical derivation of theorems has not been regarded as debatable, but the "spatial intuition of axioms," which is about the same notion as "seeing with the eyes of the mind," has been strongly criticized. Particularly since the middle of the nineteenth century, voices have been raised by those who regarded the axioms as results of experience, even though this would not be compatible with the alleged "certainty" of geometrical theorems.

Two great scientists of the middle of the nineteenth century, Bernhard Riemann[30] and Hermann Helmholtz,[31] made the point that the axioms of geometry were results of physical observation, and that, therefore, the theorems were of no greater certainty than any statements of physics. In his paper *On the Hypotheses of Geometry* Riemann wrote in 1854: "The properties by which space is distinguished from other thinkable three-dimensional continua can only be provided by experience." Around 1900 this view penetrated the vanguard of textbooks, while the average ones were still sticking to the "self-evidence" of axioms. After 1900, some prominent mathematicians published textbooks of geometry in order to bring them into line with the modern view.

The French mathematician, Emile Borel,[32] for example, wrote in his textbook in 1908:

> The goal of geometry is to study those properties of bodies which can be considered independent of their matter, but only with respect to their dimensions and their forms. Geometry measures the surface of a field without bothering to find out whether the soil is good or bad.

The Italian mathematician, Giuseppe Veronese,[33] stated explicitly in his *Elements of Geometry* in 1909: "An axiom is a proposition the content of which is verified experimentally and which is neither contradictory to any other proposition nor deducible from it." The axioms began now to play the role of physical hypotheses. The theorems were, then, also statements about physical facts. The question then arose: What was the difference between an axiom and a theorem? The Italian mathematicians, Frederigo Enriques[34] and Umberto Amaldi, answered thus in their *Elements of Geometry* in 1908:

> The first geometrical properties of figures are *evident;* they are suggested to us by immediate sense *observations* of real bodies which have been the sources of our concepts of these figures. From these first intuitive properties, we can derive by logical conclusions other properties, without recourse to further observations. These properties are, in general, less evident.

Thus the authors stressed the very important point that actually no general proposition can be derived from sense observations, but only suggested and verified by observations. Enriques made the point that it is difficult to see that theorems, such as the sum of the angles in a triangle equals 180°, could be suggested by direct observations, while the proposition that there is only one straight line through two points was strongly suggested by observations. The first is a theorem, the latter an axiom. This situation at the turn of the century was described by Einstein in 1921 in the following way:[35]

> Here arises a puzzle that has disturbed scientists of all periods. How is it possible that mathematics, a product of human thought that is independent of experience, fits so excellently the objects of physical reality? Can human reason without experience discover by pure thinking properties of real things?

The twentieth century has favored a solution to this puzzle that is due neither to the pure mathematicians nor to the "pure" philosophers, but to the mathematical physicists. This solution was given in two steps. The first step was the "axiomatic method" or, in other words, the "formalization of the axioms," which was, following the work of some predecessors like Moritz Pasch,[36] finally accomplished by the German mathematician, David Hilbert. He constructed a system of axioms which were actually "axiomatic definitions" of the geometrical terms with all definitions by physical operations excluded. Hilbert recognized, however, that "this formal system was also a logical analysis of our ability of intuition." He refused to discuss "whether our spatial intuition is *a priori* [seeing with the eyes of the mind] or empirical." We note that in the year 1899 the connection of the formal system of axioms with the properties of physical bodies was still only described by vague and ambiguous terms like "spatial intuition."

The second step was due, above all, to the French mathematician, physicist, and philosopher, Henri Poincaré.[37] Toward the end of the nineteenth century, he attempted to build up a geometry which would embrace the formal-logical as well as the empirical-physical aspect. Hilbert defined the "geometrical terms" by "axiomatic definitions" and referred to their physical interpretations only in such vague terms as "spatial intuition." According to Poincaré, the terms which are defined by a system like Hilbert's are physical things. The claim of the axioms is that there are physical objects in our world, or that physical objects can be manufactured, that fulfill these axioms. If we say, for example, that "light rays" can be substituted for "straight lines," the axioms become "statements of physics." If we want to check whether a triangle of light rays in empty space actually has an angle-sum of two right angles, we face a particular difficulty. If we find that the sum in question is different from two right angles, we can also interpret the result by saying that the "defect" is not due to the nonvalidity of Euclidean geometry, but to the fact that the rays have been deflected by some hitherto unsuspected law of physics. From considerations of this type, Poincaré concluded that we can check whether or not light rays fulfill the Euclidean axioms only if we know all the physical laws about light rays. Otherwise, we can never find out by experiment whether or not Euclidean geometry is valid. We could maintain the validity of the axioms under all circumstances if we assumed the validity of physical laws that compensated any "defect" ascribed to a departure from "Euclidicity."

If we formulate "checking the validity of Euclidean geometry" in this way, it follows, certainly, that there is no experimental method by which it can be decided whether Euclidean or non-Euclidean geometry is true. Einstein wrote: "According to my opinion, Poincaré is right *sub specie aeternitatis* [under the aspect of eternity]." But Einstein[38] thought it would be advisable to give the expression "test the validity of Euclidean geometry" a narrower meaning. He said:

> But it is my conviction that today one still must use measuring yardsticks and clocks which are defined by Euclidean geometry. This means, one has to start from the hypothesis that the yardsticks and clocks which are manufactured in the traditional way obey within small regions of space and time the laws of Euclidean geometry and Newtonian physics. Then (G) and (P) are fixed within certain spatial and temporal limits. Then, you can put the question: If one assumes the validity of the known laws of physics in the whole world, is it possible to uphold Euclidean geometry in the whole world? If the answer is yes, one would say that the validity of the Euclidean geometry is confirmed; if it is no, it is refuted.

One could, for example, make small cubes of rigid steel. According to Einstein's assumption, one could use them to build up small walls without gaps. Then, one could try to see whether he could also build up walls of dimensions of millions of miles without gaps. In this way, the validity of Euclidean geometry could be tested.

The solution of the "puzzle" reached by Poincaré and Einstein has been precisely outlined as follows by Einstein:

> The progress attained by axiomatic geometry consists in the clear separation of the logical form from the factual and intuitive content. According to axiomatic geometry, only the logical-formal is the object of mathematics; but not the intuitive content that is connected with the logical-formal. . . . The statements about physical objects are obtained by coordinating with the empty concepts of axiomatic geometry observable objects of physical reality. In particular: solid bodies behave according to the theorems of three-dimensional Euclidean geometry.

The appropriate and general scheme by which the relation between axiomatic geometry and the behavior of physical objects can be treated was developed by P. W. Bridgman.[39] He introduced the concept of "operational definitions" which must be added to the "axiomatic definitions" in order to fulfill the entire task of geometry. Bridgman stressed the point that any term, *e.g.*, "straight line," which occurs in axiomatic geometry must be coordinated with a technical procedure for manufacturing the object described by this term. Every procedure of this kind can be described in terms of our everyday language, hence the name "operational definition." The core of the definition is the reduction to "physical operations." We may quote, as a simple example, the definition of "length." Bridgman wrote:

> What do we mean by the length of an object? We evidently know what we mean by length if we can tell what the length of any and every object is. To find the length of an object we have to perform certain physical operations. The concept of length is, therefore, fixed when the operations by which length is measured are fixed. That is, the concept of length involves as much as and nothing more than the set of operations by which length is determined.

The evolution of geometry before and after 1800 gave rise to an advance in the philosophy of science that can hardly be overrated.

L. Rougier,[40] who was one of the first among French philosophers
to anticipate the trend of twentieth-century ideas on science, wrote
in his book *The Geometrical Philosophy of Henri Poincaré*:

> It will turn out that the discovery of non-Euclidean geometry has been
> the origin of a considerable revolution in the theory of knowledge and,
> hence, in our metaphysical conceptions about man and the Universe.
> One can say, briefly, that this discovery has succeeded in breaking up
> the dilemma within which epistemology has been locked by the claims
> of traditional logic: the principles of science are either *apodictic truth*
> [logical conclusions synthetic *a priori*] or *assertoric truth* [facts of sense
> observation].   Poincaré, taking his inspiration from the work of
> Lobatchevski and Riemann, pointed out in the particularly significant
> case of geometry that another solution is possible: the principles may
> be simple arbitrary conventions. . . . However, far from being inde-
> pendent of our minds and nature, they exist only by a tacit agreement
> of all minds and depend strictly upon the factual external conditions
> in the environment in which we happen to live.

If we consider this evolution of thought in geometry, we can solve
two questions which have puzzled scientists and philosophers alike
since the birth of non-Euclidean geometry.   First, there is the question of whether our "real space" is Euclidean or non-Euclidean.
The second is whether or not non-Euclidean geometry is as mentally picturable or as intuitive as Euclidean.   The question of
"space" can probably be formulated as follows: Is it possible to
find physical objects that fulfill the axioms of Euclidean geometry?
Since this question, strictly speaking, could never be answered with
certainty in the negative, we should ask rather: Do some specific
simple objects which we, in the language of our daily life, connect
with "straight lines" fulfill the Euclidean axioms, such as light
rays or the edges of solid cubes?   As for the other question, we
must consider that there is an ambiguity in the use of the term
"intuitive."   It can mean "perceivable by sense-observation," but
it can also mean "perceivable by the eyes of the mind or by inner
intuition."   In the first sense, our knowledge of the table on which
we write is "intuitive"; in the second sense, the axioms of geometry
are "intuitive" to those who believe that their validity is self-
evident.   This ambiguity plays an even greater role in German
philosophy, in which the word "anschaulich," corresponding to
"intuitive," is frequently used in the philosophy of science; it has

been a favored word and has brought about a great deal of confusion. If we keep to the first meaning of the word "intuitive," which is the only one admissible in science, we shall be adhering to the way in which the great German physiologist, mathematician, physicist, and philosopher, Hermann Helmholtz,[41] defined an "intuitive" presentation of geometry or of any other science.   He described this as follows:

> It means to imagine completely the sense impressions which the object [*i.e.*, the physical objects defined by the axioms and operational definitions] would produce in ourselves, according to the known laws of our sense organs, under all conceivable conditions of observations. . . . If the series of sense impressions can be given completely and unambiguously, one must recognize that the object can be *intuitively represented*.

In this sense, non-Euclidean geometry is certainly as intuitive as Euclidean.   If we accept, for example, Lobatchevski's axioms, and wish to measure a triangle of light rays in empty space, we can predict the sense impressions we shall have if we measure the angles by means of a protractor or by any of the methods of measurement accepted in physics.

The decisive steps toward a clear understanding of non-Euclidean geometry were taken by Riemann, Helmholtz, and Poincaré, who recognized the essential unity of geometry and physics.   However, this understanding did not come into its own until Einstein showed that such a combination of geometry and physics was really necessary for the derivation of phenomena which had actually been observed.

4

The Laws of Motion

1. Before Galileo and Newton

We are now going to examine a science in which practical confirmation by sense-observations plays a greater role than in geometry. We shall turn to the theory of motion, which is basic to all the sciences. The development of an adequate theory of motion was perhaps the greatest step in the history of science. We shall find here the same distinction between the formal system (the axioms) and its physical interpretation that we found in geometry, but the connection will be much more complicated. It took a long time for the present-day theory of motion to develop, and we do not know now whether or not it is the right scheme for the future. We must study the background of this theory to see how it has developed, and realize that the theories of motion we hold today have not always existed; previous theories were very different, and not what we would consider practical today. However, it is very important to appreciate correctly that the old theories had their own logical consistency and beauty. We cannot say simply that the people who believed in them were "wrong," but it is certain that they used a different symbolic scheme.

Geometry has been, as far as recorded history goes, a fairly stable science. Even the difference between Euclidean and non-Euclidean geometries does not much affect the application of geometry to technological problems. The history of mechanics, however, reveals great differences between modern mechanics (since the seventeenth century) and earlier theories.   This history is very well described in Herbert Butterfield's[1] book, *The Origins of Modern Science*.   Today, Newtonian mechanics is applied in all domains of technology, with the exception of nuclear technology.

When we use the term "modern mechanics," we shall mean "Newtonian mechanics," the basis of almost all applications. The first and fundamental axiom in Newtonian mechanics is the first law, the "law of inertia."   We learn today in school that it states: "A body at rest, if left to itself, will remain at rest, or if moving, will continue to move along a straight line with its initial speed."   Furthermore, most textbooks try to convince us that this is self-evident.   We shall discuss this law later, but now we may just point out that it is certainly not self-evident.   In order to understand the law of inertia at all, we should have to discuss what a "rectilinear motion" is and the meaning of the statement that a "body is left to itself."   Motion parallel to the edge of a rectangular table in a room is not a rectilinear motion with respect to the fixed stars; the earth is rotating with respect to the so-called "fixed" stars, and actually they are not fixed.   It is very difficult to define a rectilinear motion.   The law of inertia is a very complicated law, and not at all plausible to common sense, or self-evident.

What did people believe before Newtonian mechanics was accepted?   How was the theory of mechanics formulated, say, by Aristotle, or by the medieval Aristotelians, of whom the greatest was St. Thomas Aquinas?[2]   A medieval poem, *The Divine Comedy*, by Dante[3] described this early theory of motion in a very pictorial way.   Dante is led through the two lower realms of the next world, Hell and Purgatory, by Virgil, and when he comes to enter Paradise, he is astonished at being able to rise up to it, since he is so heavy. But his "old flame," Beatrice, who guides him through Paradise, gives him instruction in the elements of mechanics.   She says that the fundamental law of mechanics is not that all heavy bodies fall to the ground, but that all bodies move to the place where they belong.   Since they are now spirits, she explains, they do not contradict the law of mechanics, but follow it, by going up.

Today, we believe that all bodies are made from about one hundred elements, but the ancient and medieval idea was that there were two different kinds of bodies, terrestrial and celestial.   Terrestrial bodies were thought to be composed of four different elements—

earth, water, fire, and air—and these four elements were thought to
have a certain natural distribution in the world, as depicted in the
following diagram:

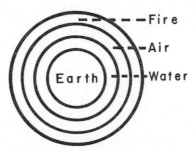

Figure 31

If all these elements had stayed in their places, there would have
been no motion at all, but since there were some disturbances on
earth, it was necessary to distinguish between natural and unnatural
motions.   It was unnatural to take a piece of earth up into the air,
but natural for it to fall back again.   It was natural motion for all
things to go back where they belonged; all bodies tried to fall back
to their natural places as quickly as possible, but there might be
some obstacle.   The greater the obstacle, the more slowly they went
back.   Generally, motion was considered to be a combination of
natural and unnatural motions.

At first glance, this theory does not look so bad; it accounts for
the most obvious phenomena.   There is an important difficulty,
however—if we launch a projectile, we can give it a violent motion
while it is in hand, but we observe that afterward it persists in this
unnatural motion for a while.   Today, we attribute this to "inertia,"
but in ancient mechanics there were only two ways of accounting for
motion.   First, a body could strive toward "its place" (natural
motion); obviously there was no "place" toward which a launched
body could move in a horizontal direction.   Secondly, a body could
be driven by "violence," *e.g.*, by the direct push of one's hand.
Since one's hand touched a launched body only in the first moment,
however, the continuous motion in the horizontal direction was
explained by a continuous push of air.   Another difficulty arose
from the fact that according to the above theory a stone should fall
down with constant speed because the velocity of a body was deter-

mined only by its drive toward its "natural place" and the resistance of the air; but it was known that a falling stone accelerates. There were two explanations for this: the greater column of air pushing down on it as it fell, and the theory that the nearer it came to its place, the more jubilant it became and the faster it went.

2. The Ancient Laws of Motion Were "Organismic"

This ancient and medieval theory can be called an "organismic" theory; it treated the motion of inanimate bodies (such as rocks) by analogy with the motions of animals. Just as we say of a dog that it performs a certain motion in order to obtain a certain piece of meat, medieval mechanics assumed that a stone fell in order to reach its "natural place." What about celestial bodies? Beyond the ring of fire depicted in Figure 31 began the realm of celestial bodies or of planetary spheres, regarded as rotating.

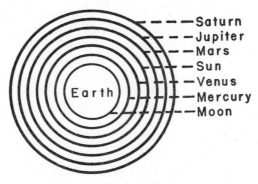

Figure 32

These celestial bodies consisted of a much more subtle matter than terrestrial bodies. To each of these seven spheres belonged a spirit, a kind of god. There were different ideas about these spheres, but the general idea was that they consisted of a nonterrestrial material. They were described by some authors as crystalline spheres. Their fundamental characteristic was that they moved in a most perfect way—in permanent, uniform, circular motion. What, then, was outside the seventh sphere? The eighth sphere of the fixed stars— the stars were supposed to be fixed on the eighth sphere and to move around with it. Beyond the eighth sphere was heaven. Beatrice received Dante at the entrance to the lunar sphere, and the poet describes all seven heavens.

The aim of this whole theory is in the question: What makes these heavens move all the time? One idea was that they could move themselves because of the spirit in each sphere. In his book, *Aristotle*, Werner Jaeger[4] describes how Aristotle himself gradually modified his ideas. Eventually the idea of a "prime mover" or "primum mobile" developed—that beyond the eighth sphere was a ninth which did not move itself, but which moved everything. The common-sense experience which later developed into the law of inertia was formulated by some ancient and medieval scientists in a way that had a vague similarity to that law, but led practically to very different conclusions. These conclusions concerned not only mechanics in the modern sense, but also the domain of human conduct that is directed by metaphysics and religion. The formulation found in Aristotle and Thomas Aquinas reads: "Whatever is in motion is moved by another." This statement obviously describes, in a general way, experiences of our daily life. According to Aristotle and St. Thomas, it was a "philosophical truth," derived logically from "intelligible principles." Several kinds of derivations were proposed by philosophers. I shall mention only one argument which was advanced by St. Thomas: He attempted to disprove the assertion, "a body can be moved by itself without action of another." Let us assume that a certain solid body could be moved as a whole "by itself." Then, when a part was at rest, the whole remained at rest; and when a part was moved, the whole was moved. Then the motion of one part depended on the motion of another part. Therefore, the body as a whole was not moved by itself, in contradiction to the assumption. Thus it was demonstrated that "whatever is in motion is moved by another."

From this statement, ancient and medieval philosophers drew far-reaching conclusions. St. Thomas wrote:[5]

> Whatever is in motion is moved by another: and it is clear to the sense that something, the sun for instance, is in motion. Therefore, it is set in motion by something else moving it. Now that which moves it is itself either moved or not. If it be not moved, then the point is proved that we must needs postulate an immovable mover; and this we call God. If, however, it be moved, it is moved by another mover. Either, therefore, we must proceed to infinity, or we must come to an

immovable mover. But it is not possible to proceed to infinity.
Therefore, it is necessary to postulate an immovable mover.

In this way, from intelligible principles, was derived the existence
of a "prime mover" which was identified, as we learned above, with
the ninth heavenly sphere embracing the sphere of the fixed stars.
The sphere of fixed stars, in turn, was believed to move all the others
by moving the one next to itself and that the one next to itself, and
so on. Thus all motion derived from the sphere of fixed stars—the
basis of astrology. As we have seen, the idea of a prime mover was
regarded by Thomas Aquinas as proof of the existence of God.
Thus it was an essential point of the world picture, very definitely
related to religion. It had very deep roots in medieval thought or,
as we like to say today, in the behavior pattern of medieval man.

This organismic theory could apply as well to animals as to inanimate nature; in fact, it was easier to apply to animals. In spite of
modifications which formed a gradual transition to the modern
theory, it dominated the thoughts of scientists up to about 1600.
As a matter of fact, the organismic theory of motion has never completely disappeared. Although its role in science has faded away, it
has been preserved in the philosophical interpretations of science
until today.

If we use the metaphor of the chain, science-philosophy, we would
say: This theory satisfied the need for an intelligible and plausible
world system, but, going from the philosophical to the scientific end
of the chain, there were difficulties in the realm of celestial as well as
of terrestrial bodies. The difficulties concerning celestial bodies
had already attracted the attention of the Greeks. The laws governing the visible motions of the planets (including the sun and
moon) had been known for centuries—for example, how to predict
eclipses was known. It soon became apparent that the general laws
of the organismic theory were not in very close agreement with these
old empirical laws. This disagreement was one of the sources of
the split between philosophy and science—man became accustomed
to having two world systems, one in astronomy or physics for technical purposes, and one in philosophy for "intelligible truth." (See
Chapters 1 and 2.)

There were also difficulties in accounting for the motion of terrestrial bodies—*e.g.*, the motion of a launched projectile. The hori-

zontal component of velocity remains constant.   As we learned in
Section 1, this could be neither a "natural" nor an "unnatural or
violent" motion.   These difficulties served as focal points, around
which modifications of the organismic theory developed, but there
was no satisfactory theory from which mathematical conclusions
could be drawn concerning the motion of terrestrial bodies before
the period around 1600.   For this reason, they did not have so great
a bearing on philosophical discourse as the difficulties in planetary
motion.   Around 1600, the bridging of the gap between celestial and
terrestrial phenomena started by Galileo and completed by Newton
was one of the most important turning points in the history of the
philosophy of science.

3. The Universe as an Organism

In order to understand the "organismic" laws of motion precisely,
we must realize that they were part of a more general conception:
the analogy between the human organism and the Universe.   The
following is a characteristic statement from the Jewish medieval
philosopher, Moses Maimonides.[6]   In 1194 he wrote:

> Know that the Universe, in its entirety, is nothing else but one indi-
> vidual being. . . . As one individual consists of various solid sub-
> stances, such as flesh, bones, sinews, of various humours, and of vari-
> ous spiritual elements, . . . in like manner the Universe in its totality
> is composed of the celestial orbs, the four elements and their combina-
> tions. . . . The centre is occupied by the earthy earth, this is sur-
> rounded by water, air encompasses the water, fire envelops the air,
> and this is again enveloped by the fifth substance (quintessence). . . .

In a more specific way, Maimonides elaborated upon the analogy
between the human body and the physical universe as follows:

> The principal part in the human body, the heart, is in constant motion,
> and is the source of every motion in the body. . . . The outermost
> sphere by its motion rules in a similar way over all parts of the uni-
> verse. . . .

In the following passage, Maimonides brought out the core of the
organismic theory of motion precisely, by stressing the analogy
between the human body and the universe:

> When for an instant the beating of the heart is interrupted, man dies,
> and all his motion and power come to an end.   In like manner would

the whole Universe perish, and everything cease to exist if the spheres were to come to a standstill.

This analogy between the movements of inanimate bodies, on the one hand, and of human and animal bodies, on the other, becomes very clear to us if we look into Aristotle's book, *On the Movements of the Animals*.[7] He investigated how movement originated in the world. He observed that if one disregarded the mutual impact of inanimate bodies, one found that "things with life" were the cause of movement. A simple example would be the throwing of a ball by a player. Aristotle stressed the point that:

> . . . all living things both move and are moved with some object, so that this is the term of all their movement, the end that is in view. We see that the living creature is moved by intellect, imagination, purpose, wish and appetite. And all these are reducible to mind and desires.

Aristotle compared the role of desires in the movements of men or animals with the role of the immovable mover (prime mover) in the motion of the heavenly bodies thus:

> In one regard that which is eternally moved by the eternal mover is moved in the same way as every living creature, in other regard differently. While celestial bodies are moved externally, the movement of living creatures has a termination.

In his book on *Physics*, Aristotle discussed the movements of animals much as he did those of ships. He applied the same general rules to both. We learned in Section 2 that, according to the organismic laws of motion, no body can be moved by itself; it must always be moved by another. However, it would appear that a human being or an animal can be moved "by himself." In his *Physics*,[8] Aristotle removed this apparent disagreement by stressing "that in animals, just as in ships and things not naturally organized, that which causes motion is separate from that which suffers motion, and that it is only in this sense that the animal as a whole causes its own motion." In a more general way, Aristotle explained the same situation as follows:

> When a thing moves itself it is one part of it that is the mover and another part that is the moved. . . . Therefore, in the whole thing we may distinguish that which imparts motion without itself being

> moved and which is moved. . . . While we say that AB is moved by
> itself, we may also say that it is moved by A.

In other words, the mind or desire was thought to move the body, as, to use a modern example, the engine moves a steamship. For Aristotle or Thomas Aquinas, the effect of human desire upon the movement of the human body was the same sort of thing as the effect of the mechanical "force" exerted by a wind blowing upon the sails of a ship. According to the "organismic" laws of motion, the movement of a man under the impact of his will was the same kind of "physical" event as the motion of a ship under the impact of a wind.

After the time of Galileo and Newton, when the "organismic" laws of motion had been dropped as physical laws, the effect of the human will upon the movement of the limbs and the effect of wind upon the movement of a ship became two entirely different kinds of events. In the first case, one invoked an "effect of mind upon body"; in the second case, an effect of matter upon another piece of matter. A "mental or spiritual force" was distinguished from a "physical force." Frequently, the conviction was voiced that it is "immediately understandable" from our "internal experience" how our will manages to lift our legs, but that we do *not* understand by direct internal experience how a wind can move a ship. Therefore, it has become popular among scientists to say: If a ship is pushed by the impact of the wind on its sails, this is a fact that cannot be "understood" philosophically; it is a stubborn fact that has to be accepted. We can, however, improve the development of our understanding by stressing the analogy between ship and man. We express this analogy by saying that the wind exerts a "force" on the sail, where the word "force" reminds us of our will, of force in the mental sense. In this way, the mental concept of "force" entered technical mechanics. The most hard-boiled engineer would say that a current of air exerts a "force" on an airplane, and would believe that by stressing this analogy with the effect of our "will" on our limbs, he gained an "understanding" of the motion of an airplane.

In this way, a certain infiltration of the "organismic laws of motion" into modern mechanics has taken place. This is noticeable in every classroom and workshop where the laws of motion are taught or used. These leftovers from "organismic" science were justified in the period of Newtonian mechanics and later as a means for "understanding" the motion of bodies. The feeling was that without this analogy of mechanical "force" with "will power," we could calculate the action of mechanical devices but not "understand" them. The expression "understand" in contrast with "merely calculate" denotes a peculiar psychological satisfaction, the release of a certain painful tension. Obviously, the introduction of the concept of "force" into mechanics provides this satisfaction only if we take it for granted that we "understand" the action of our "wills" on our legs better than the action of a mechanical machine upon a tow rope.

The infiltration of the ancient doctrine of organismic science into modern science could not be stopped until it became clear that the action of our will power upon our muscles is neither more nor less "understandable" than the pull of a rope on a ship. We may remember how Newton, in his controversy with Leibniz[9] presented the effect of a falling weight upon a clock and the effect of the human will upon the muscles as two phenomena of the same kind. In both cases, a regular sequence could be observed but no "causation" could be "understood." It was the great achievement of David Hume,[10] in the middle of the eighteenth century, to clarify this point and to purge mechanics from the remnants of the "organismic" laws:

> The influence of volition over the organs of the body . . . is a fact which, like all other natural events, can be known only by experience, and can never be forseen from any apparent energy or power in the cause which connects it with the effect and renders the one the infallible consequence of the other. The motion of our body follows upon the command of our will. Of this we are every moment conscious. But the means . . . by which the will performs so extraordinary an operation: of this we are so far from being immediately conscious, that it must forever escape our most diligent inquiry.

Thus Hume asserted emphatically that the action of our will upon our body is as mysterious as what we could call today "telepathy." He wrote: "Were we empowered, by a secret wish, to remove mountains or control the planets in their orbits; this extensive authority would not be more extraordinary, nor more beyond comprehension" than the action of our soul upon our body. Hume rejected, on all these grounds, the idea that by analogy with our will power we can "understand" the action of mechanical force. He continued:

> We may, therefore, conclude from the whole, I hope without any temerity, though with assurance; that our idea of power is not copied from any sentiment or consciousness of power within ourselves, when we give rise to animal motion or apply our limbs to their proper use and office.

If we accept Hume's analysis of our mental experience, the remnants of the organismic theory disappear completely from scientific mechanics; but they retain their role if we keep to the analogy between mechanical force and will power, not because of the feeling of satisfaction it provides, but because of other benefits which we get from the organismic analogy. We remember the direct bearing of the organismic world picture upon our attitude toward religious, moral, and social traditions. Today we note that the belief in the organismic laws of motion is not based upon science proper, but upon a "metaphysical interpretation of science" which frequently serves practical goals. It has even been said that the belief in the truth of organismic mechanics is the core of all metaphysical interpretations of science. Auguste Comte, in his *Positive Philosophy*,[11] wrote: "The spirit of all theological and metaphysical philosophy consists in conceiving all phenomena as analogous to the only one which is known through immediate consciousness—Life."

4. The Copernican System and the "Organismic" Laws of Motion

We have seen in the example of geometry that to attempt to set up in a direct way a description of observed facts is not practical, but that it is preferable to set up a formal system (of axioms) which describes observed facts in an indirect way. A tour around the formal system and its physical interpretations is the best description of observed facts. We have learned about the Aristotelian theory of motion, in which every motion was compared to that of an organism. In the motion of terrestrial bodies, the essential concepts were the beginning and end of motion; little attention was paid to what happened in between. The emphasis was laid on where the body should go; the end of the motion inspired much more interest that the motion itself—this is typical of an "organismic" view.

The general idea was that the motions in themselves were very complicated, accidental, or whimsical, that they did not follow any clear laws. Biologists and physiologists have always described movements of living beings in this way. We describe the movement of a human hand by describing the initial and final position but pay no attention to the precise orbits that are traversed by muscles and bones. We speak vaguely of a "contraction" or "relaxation" of muscles without indicating all the intermediate states.

Some theories in modern physics remind us of these views. For example, in Bohr's[12] original model of a hydrogen atom, an electron was conceived of as moving around a nucleus in specific stable orbits—if excited, it was to jump from one of these orbits to another. The theory was interested only in these stable states; what the electron did while jumping was not investigated. Thus we see that the emphasis is still sometimes placed on the goal or end of a motion rather than on how it happens. In ancient times, the idea that everything happens according to very simple laws, that nothing is the result of chance, was reserved for celestial bodies—there was always an accidental character to the motion of terrestrial bodies. The belief that terrestrial beings follow not very precise and not very beautiful laws characterizes a certain attitude toward the universe.

If we look into Copernicus'[13] writings, we find that even he was still deeply imbued with the ideas of organismic physics. He investigated such questions as which is more dignified for a body, to move or to be at rest. He thought it was more dignified for a body like the sun to be at rest and give light in the center of the universe. Copernicus represents a transition between medieval and modern physics which is said, traditionally, to have started with Galileo. When Copernicus advanced the hypothesis that the sun is at rest in the center of the universe and the earth is performing a circular orbit around the sun, as well as a daily rotation from west to east about its own center, his hypothesis was obviously in flagrant contradiction to the generally accepted laws of organismic physics. According to these laws, the earth consisted of terrestrial stuff and followed laws of motion that were not valid for celestial bodies that consisted of much nobler material. It was therefore impossible that the earth was performing a circular orbit around the center of the universe. The earth, a heavy body, should move in a straight line toward its "natural place," near the center of the universe. The idea that the earth could remain at a constant distance from the center seemed to contradict man's most certain experiences about the behavior of "heavy bodies" made of terrestrial material. We could invoke some very obvious observational facts against the Copernican hypothesis: If we observe clouds floating in the sky, we should expect that they would have a common velocity westward, if the earth were rotating eastward; but no such westward trend is observed. Copernicus suggested that the eastward rotation of the earth might, by a kind of friction, carry along the atmosphere and clouds; therefore, no westward motion of the clouds would take place.

If we observe a falling stone, it should come to rest west of the place where it was first dropped for the same reason; actually, a falling stone moves vertically downward and shows no westward deviation. Copernicus suggested that here again the air which is carried eastward along with the earth might carry along the stone, but since it does not seem very plausible that the air could carry along a heavy stone, Copernicus suggested an alternative explanation. He assumed that the "natural place" of a piece of earth, like a stone, was near the center of the earth and not near the center of the universe, which was, according to him, the sun. Therefore, a falling stone would be carried along eastward, not by the friction of the air, but because it was "attracted" to its natural place, the surface of the earth. By this assumption, Copernicus anticipated to a certain degree Newton's hypothesis of universal gravitation.

In any case, in order to make the heliocentric system compatible with physical science, the old "organismic" theory of motion had to be "softened up." As a matter of fact, Copernicus had no consistent theory of motion. A new theory which would replace the "organismic" laws of motion did not come into being before the common-sense experience about the sluggishness of bodies was formulated in a way that was much more remote from common-sense laws than the old formulation that "a body cannot move unless it is moved by another."

Galileo[14] advanced a new generalization of our common-sense experience. According to him, a moving body did not stop by itself; it continued moving without being pushed at every instant; as a matter of fact, a body in motion would continue to move with undiminished speed in a straight line. Galileo arrived at this generalization by examining the motion of a smooth body along a smooth inclined plane. (See Figure 33.)

If the ball in (a) rolls downward, its speed increases; if the ball in (b) moves upward, its speed decreases. Therefore, Galileo argued, there must be a boundary case (c). If the plane is parallel to the surface of the earth (perpendicular to the direction of gravity), a pushed ball will move parallel to the surface of the earth in a constant direction with constant speed. This is the original formulation of the law of inertia. It says, to speak precisely, that a body moved in a horizontal direction keeps its speed and direction with respect to the surface of the earth. Its path is, obviously, only a straight line as far as we can regard the surface of the earth as plane. This means that the law is valid only within certain limits. It does not say anything about motion in the world space.

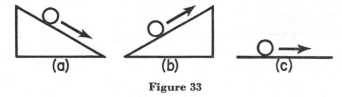

Figure 33

We shall not go into the historical development of this generalization here. We shall jump to the final formulation of the law of inertia as we find it in Newton's *Mathematical Principles of Natural Philosophy*.[15] He assumed that a body that is not moved by another will move with constant speed along a straight line into the infinite if no obstacles intervene. This conception destroyed the whole structure of the Aristotelian universe, where the "natural motion" of a stone was toward its "natural place" in the universe. The latter was a closed system of crystal spheres. According to Newton, the world space is an empty space and bodies move around in it in a "savage" way, and do not head for their "proper place." The whole picture is no longer one of bodies moving in a purposeful way according to eternal traffic rules.

The modern conception thus encountered very great obstacles when it was considered from the point of view of the world system of the Middle Ages. It abandoned the concept of a "prime mover," and destroyed the foundations upon which the traditional proofs for the existence of God were based. The world picture needed to be completely rebuilt, not only in physics, but in the fields of theology and ethics as well. For example, the new conception involved the idea that the bodies in the sky were not different from those seen on earth. Christ was sent to save the earth, but if this earth were no different from other bodies seen in the sky, why should not God have sent a Savior to the other planets as well? For advocating a belief in the plurality of worlds, Giordano Bruno[16] was burned at the stake on February 17, 1600. The interpretation of things like mechanics has always been connected with the attempts of political powers to force certain principles upon science. We have already observed this in Plato and in the Middle Ages.

5. Newton's Laws of Motion

We shall now speak about mechanics as we have spoken about geometry—from the "purely scientific" point of view. We shall find interesting points in the comparison of mechanics with geometry. Do the principles of mechanics concern the external world, or do they, like the geometrical axioms, set up a formal scheme which does not, by itself, concern the external world? Let us first recall Newton's laws:[17] The first law (the law of inertia) states that if no unbalanced force acts on it, a body will remain at rest, or if in motion, it will continue to move with undiminished speed in a straight line. The second law (the law of force) states that an unbalanced force acting on a body will equal the product of the mass of the body times the acceleration which that force produces. We may recall at this point that acceleration here means any change in velocity whether of magnitude or direction.

What is the meaning of these two laws of Newton? We shall find that these laws as formulated by Newton do not contain their operational meaning. What is the operational meaning of the law of inertia? The cause of much misunderstanding among students of physics is the fact that most textbooks do not mention that this law is meaningless unless an operational definition of its terms is given. The same problem exists also with respect to the second law—"acceleration" with respect to what? It must be "acceleration with respect to a certain body," but this body is not mentioned, so that the law by itself is meaningless. In most elementary textbooks and courses on mechanics, the terms "rectilinear motion" and "acceleration" are bluntly used without even a hint that a physical system of reference to which these terms refer must be given. In other words, the physical body must be specified with respect to which the Newtonian laws are supposed to be valid. Then, and only then, are these laws statements about physical facts.

Nevertheless, you can develop the whole theory as to what motions bodies describe under certain forces with respect to these coordinate systems for which you have assumed that the Newtonian laws are valid. If you have coordinate systems with respect to which the Newtonian laws are valid, all logical consequences of these laws are also valid. We cannot tell, however, from such a formal theory to which body of our experience such a coordinate system (S) is to be attached. There are many large volumes in which all the consequences of the Newtonian laws are thus drawn without specifying the physical system of reference (S).

Why do men do this? For the simple reason that it is practical to separate the drawing of logical conclusions from the physical interpretation of the system. The situation is exactly the same as in geometry. The Newtonian laws play the role of axioms. In "Analytical Dynamics" conditional statements are derived of this type: "If Newton's laws are valid with respect to a system of reference (S), then the conclusions are also valid with respect to (S)." But we cannot find out what is valid in the realm of observable facts in this way. The first step toward a physical interpretation is the specifying of the physical body with respect to which terms like "acceleration" are to be defined.

The experiences of our daily life would lead us to believe that our earth was such a body (S). This is obviously not generally true. While in most experiments we make we observe that if we push a body, it will move in a straight line with respect to the earth, we see in the famous Foucault pendulum that its motion does not try to keep its direction with respect to the earth. There is a deviation from the original direction due to the rotation of the earth. A good approximation to the system (S) with respect to which the Newtonian laws are valid is the system of the fixed stars. This is the whole meaning of the statement that the earth is rotating—it is rotating with respect to the system (S) in which the Newtonian laws are valid, and the system of fixed stars does not rotate with respect to this system. We can ask what would happen if there were disorder

in the system of fixed stars?   By using the fixed stars as our system
(S), we assert that the fixed stars form a rigid system; but the fact
is that constellations are destroyed in time.   If constellations should
dissolve, it seems that a stone would be at a loss to know what the
straight line is that it should traverse under the law of inertia.
Practically, we can say that the system of fixed stars is the system
(S) to which Newton's laws refer, but if we believe, as Newton did,
that the law of inertia is one of the most universal laws of nature, it
cannot refer to an individual group of masses such as the fixed stars.
The only safe thing to say is that there are such systems (S) in the
same way as in geometry we say that there are rigid bodies for which
the laws of geometry are valid.

A system (S) with respect to which the Newtonian laws are valid
is called an "inertial system."   The first approximation to such a
system is the earth, and the second is the system of fixed stars.   If
we say that the Newtonian laws are exactly true, we are only saying
that we believe that there is an inertial system.   Eventually, all
that acceptance of the Newtonian theory implies is a belief that
there is a physical interpretation of the Newtonian laws, but the
system itself is a purely logical structure.   This point of view was
introduced by Henri Poincaré[18] and is called the doctrine of conventionalism.   It asserts that the laws of mechanics are linguistic
conventions.   Newton's first law says: A body on which no unbalanced force is acting performs a rectilinear motion, but how can we
know that no unbalanced force is acting on a body?   By observing
that it moves in a straight line we can be sure of this.   Newton's
first law thus becomes a convention about how to use the expressions
"unbalanced force" and "rectilinear motion."   This and similar
conventions are not forced on us, either by observation or by logical
conclusions.   In this sense, they are "arbitrary" conventions.

However, these arbitrary conventions must also be useful conventions; they are introduced in order to allow a good description of
the phenomena of motion to be formulated.   The Newtonian system
is a purely formal system, it is true, but a system which can be
applied.   Modern science describes observed facts not directly, but
by making use of formal systems; ancient science also proceeded in
this way, although this is not so obvious.   If we tried to describe
observational facts directly, we would have to do it by using the
words and expressions of our everyday language.   Such a descrip-

tion would not have a precise meaning, and the precision could not be improved by drawing exact logical conclusions. If we want to draw strict logical conclusions, and to obtain precise results, we must use words without physical meaning, symbols, as they are used in formal structures. Then we can obtain precise results as in pure mathematical geometry (Chapter 3); but these results say nothing about the physical world until we add "operational definitions." Then these statements obtain meaning in the physical world, but their meaning is again vague. The procedure of modern science combines the methods of strict logical conclusions with the method of sense observation by confining the logical deductions within a formal system (axioms and theorems) and producing the object of sense observations by applying operational definitions to this formal system.

One of the first logicians who described correctly the relation of formal systems to physical laws was Ludwig Wittgenstein.[19] He wrote about the "world" and Newton's laws: "The fact that it can be described by Newtonian mechanics asserts nothing about the world." In this pointed fashion, Wittgenstein meant to say that Newton's laws are a formal system and, without operational definitions, cannot be regarded as a description of the physical world. He continued:

> But *this* asserts something, namely that it can be described in that particular way in which it is described, as is indeed the case. The fact, too, that it can be described more simply by one system of mechanics than by another says something about the world.

Wittgenstein is making the point here that Newton's laws are not a description of the world, but a machinery which can produce such a description if it is used in the right way; hence, the laws of mechanics are of no use if no advice is given on how to use this machinery. Part of the advice consists in the operational definitions of the terms used in the laws. If we believe that all occurrences in the world can be derived from mechanics, this machinery must, if properly used, yield all true statements about the world. Wittgenstein wrote: "Mechanics is an attempt to construct according to a simple plan all *true* propositions which we need for the description of the world."

6. The Operational Definition of "Force"

If we regard the law of inertia, fundamental to modern mechanics, as an axiom of the formal system, this law is very simple. It says: A body upon which no force is acting performs a rectilinear motion with constant speed. However, this statement helps us not at all to deduce where an actual body will be in the actual world. It does not tell us how we can recognize whether or not a certain motion has these properties. We encounter difficulty when we try to give operational definitions of "rectilinear motion," "constant speed," and "no force is acting." Such a definition can be given by singling out a concrete rigid body (S) and stipulating: By "rectilinear uniform motion" we mean "rectilinear and uniform with respect to (S)," where (S) is an "inertial system" in the sense defined in Section 5. The law of inertia does not tell us which body is an inertial system; it simply says that there is such a system. It is the task of physics and astronomy to find the body of which theoretical mechanics claims the existence.

Newton's second law describes what departures from the rectilinear motion take place if a force is being exerted. This departure can be measured by the "acceleration" vector, which we denote by *a*, and which is determined by the change in speed and direction. If we denote by *f* the force vector, Newton's second law asserts that the acceleration is proportional to the force that is exerted upon the moving body by other bodies. The experiences of everyday life make it clear that by equal pushes from outside "big and heavy" bodies are less accelerated than "small and light" bodies. The proportionality between force and acceleration must contain a factor that was called by Newton the "quantity of matter" contained in the body. This quantity was given the name of "mass" of the body. According to Newton's conception, this quantity could only be changed by adding or subtracting some quantity; practically, this means by breaking up a body into fragments or by merging small bodies to make a large one. Newton formulated his second law by the simplest possible assumption: Mass times acceleration is equal to force. This can be expressed by the mathematical formula: $m \times a = f$, where m denotes the mass of a body and a the acceleration caused by f, the net or unbalanced force on the body. If we do not restrict ourselves to rectilinear motions, the acceleration a and the force f are "vectors" which have a certain magnitude and a certain direction. If we know the net force on a body and its mass, we can use this law to calculate the acceleration of the body, and we find $a = f/m$. And if we know the acceleration, we can calculate the motion of the body in the course of time (since the acceleration is the second derivative with respect to time of the displacement of the body).

Some say, as Poincaré did, that the law $m \times a = f$ is a pure convention, that "force" is just a name for $m \times a$. If we do not give an independent definition of "force," this law is certainly a pure convention, a definition of "force." In order to find out what this law says about our physical world, we must examine how the law is applied in practice. For example, a body may move in a small region of space. From the simple assumption that f is constant, it follows that the acceleration of the body will be a constant, g. Then it follows by a simple mathematical argument that the space traversed by the body in the time t is given by the expression $s = \frac{1}{2}gt^2$. Thus, we can predict or calculate what space will be traversed by the body in a certain time. The formula $s = \frac{1}{2}gt^2$ can be checked and verified by experiment and observation. By this observation it is confirmed that the simple assumption f = constant accounts for the motion of a body under the influence of gravity in a small region of space.

If we examine motion in a large region of space under the influence of gravity, e.g., the motion of the planets around the sun, we must assume that f is not constant but is given by the Newtonian law of gravitation. This states that the gravitational force f exerted on one body by another is proportional to $1/r^2$ where r is the distance between them. The vector f has the direction of the straight line connecting the two bodies. From this it can be derived mathematically that the one body follows an elliptic orbit around the other. This has been confirmed by observation. Therefore, if we have a simple formula for the force f in terms of the distance r of a body from other bodies (in the simplest case f is a constant), we can use the law $m \times a = f$ to compute the acceleration and thus the motion of the body. If we do not have such a simple formula, then $m \times a = f$ becomes completely empty, and f is just a name for $m \times a$. The factual content of Newtonian mechanics is not the formula $m \times a = f$, but the existence of specific expressions for f which are to be substituted into the formula and yield the acceleration of bodies as a simple expression in terms of the distance of these bodies from other bodies. This acceleration, of course, is acceleration with respect to a system in which the law of inertia is valid, to what we have called an "inertial system." The acceleration of a body with respect to an inertial system is given by a simple formula in terms of the distances of this body from other bodies—this is the real discovery of Newton.

In ancient physics it was believed that the motion of a terrestrial body could be described by knowing the place for which it was heading. For celestial bodies the motion was described as in perfect circles. No one had conceived the idea that the motion of a body should be described by its acceleration. Kepler found that the planetary orbits are not circles but ellipses. The latter were considered to be not so perfect as circles, but they were still thought to be fairly near perfect. Then it was discovered that the planetary orbits are not even ellipses—this would be the case only if there were only the sun and one planet—due to the interactions of other planets. The orbits are disturbed ellipses, very complicated curves for which we cannot give a simple equation. It was Newton's idea to apply a completely different description; what matters, according to him, is not the curve but the acceleration. It is given by a simple formula; it is proportional to $1/r^2$. If we should describe the curve itself, it would be very complicated. It we wish to bring order into the description of motion, we describe it in terms of acceleration, and not just any acceleration, but acceleration with respect to the inertial system.

We can see that the definition of "force" has an element of vagueness if we attempt to formulate it for all cases of motion. In the case of motion in the planetary system, we could simply say that the acceleration of a body with respect to the inertial system is proportional to a sum of terms each of which is inversely proportional to the square of the distance between the two bodies. The function of the distances defined in this way is the "force" acting upon this body, B. If, however, we consider not only the gravitational forces in the planetary system, but all possible forces, we can only formulate the following generalization: The acceleration of a body B with respect to the inertial system (S) can in all cases be expressed as a simple function of the distances and velocities of B with respect to other bodies. This "simple function" is called the "force" acting upon B, and has to be substituted for f in the formula $m \times a = f$.

There is certainly a "human element" in this definition of force. The criterion of what is called a "simple formula" depends on the psychological situation of a certain social group at a certain period of history; but given a certain situation there will often be, in practice, widespread agreement as to whether or not a certain specific formula is "simple." If one could not find such a simple formula, "force" could not be defined. The Newtonian law of force asserts that there is, in every specific case, a formula that would be recognized as "simple" by the scientists of our period, or that there is a hope that intellectual abilities will develop in the future to such a degree that scientists will find a formula that they will recognize as "simple." It is obvious from these considerations that, as we remarked above, the "factual" meaning of the Newtonian laws is highly involved and is connected with the psychological and social evolution of mankind. Our belief in the simplicity of these laws comes from the fact that we consider them only as parts of a formal system where they are only definitions, and disregard their "factual" meaning.

7. The Operational Definition of "Mass"

The definition of "force" given in the preceding section assumed that only one single moving body B is taken into consideration. Therefore, only one single value of "mass" came into play, and the influence of mass on motion did not manifest itself. If we look for an "operational definition" of "mass," we have to base it on experiments in which different bodies with different masses are moving under the influence of one and the same force.

We remember that Newton defined mass as the "quantity of matter" that is contained in a certain body. If we use only the language and experience of "common sense," we believe that we understand fairly well the meaning of the statement that in a certain volume of a body there is a certain "quantity of matter." This concept seems to be very clear if we assume that "matter" consists of a great number of equal small particles (formerly called "atoms") and therefore we mean by "quantity of matter" in a certain volume just the number of these equal particles. This concept of "mass" as the number of "atoms" was familiar to the Greek atomists and

Epicureans.   The Roman Epicurean, Lucretius,[20] wrote in his poem *On the Nature of Things* as follows:

> Why do we see one thing surpass another in weight though not larger in size?   For if there is just as much body in a bale of wood as there is in a lump of lead, it is natural that it should weigh the same, since the property of body is to weigh all things downward, while, on the contrary, the nature of void is ever without weight.   Therefore, when a thing is just as large as another, but is found to be lighter, it surely proves that it has more void in it; while, on the other hand, that which is heavier shows that there is more of body in it, and that it contains within it much less of void.

Obviously, the definition of "mass" as "quantity of matter" does not describe the operations by which we can measure how much "quantity of motion" is contained in a certain moving body.   Newton's definition is not an "operational definition" but refers to common-sense conceptions.   On the other hand, Newton's laws have been proved to be very useful in applied mechanics.   If a physical law is to be checked by experience, all its terms must be replaced by operational definitions.   Therefore, Newton and all those who have applied Newton's laws have actually used an operational definition of "mass" which we can discover by examining how Newton's laws have actually been applied.

According to the definition of "force" discussed in the preceding section, there is a "simple formula" that ascribes to the force acting on a body B a value f that depends upon the external circumstances of this body B.   We may write Newton's second law as $a = f/m$.   Therefore, if we have two different bodies with the masses m_1 and m_2, they develop different accelerations under the same external circumstances.   Under the same circumstances, the "simple formula" assigns to the forces the same value f.   In other words, two bodies with the masses m_1 and m_2 have different accelerations, a_1 and a_2, when given the same "kick."   Whatever the intensity of this "kick" or whatever the value of the force f, we can easily see that $f = m_1 a_1 = m_2 a_2$.   If we form the ratio of the accelerations, we will find that it is independent of f.   For two specific bodies, the ratio of their accelerations will always be the same; it is not dependent upon the external circumstances f, but depends only on the bodies themselves, $a_1/a_2 = m_2/m_1$.   This equation becomes a unique definition of "mass" if we choose an arbitrary unit of mass (say, ascribe to a cubic centimeter of water the value $m = 1$). This means that the product $m \times a$ depends only on external circumstances, not on the body itself. According to Section 6, it can, moreover, be expressed as a "simple function" of these external circumstances. This is the basic assertion of Newtonian mechanics.

On the purely mathematical side, $m \times a = f$ is a mathematical formula, a definition of f. We can derive many mathematical consequences if we add the formulae that express a by the increase of velocity; but to apply it to observable phenomena, we must give operational definitions to the terms. The operational meaning of "mass" is now found in the ratio of accelerations. If m is defined in this way, an operational definition of "force" can be given by the equation $f = m \times a$. It is a unique definition because $m \times a$ depends only upon the external circumstances of a body, and is independent of its mass m. In this sense, the formula $m \times a = f$ is certainly only a "definition" of "force" and not a physical law that could be checked by experience. However, if we substitute for f the simple law discussed in Section 6, the relation of $m \times a = f$, where a means the acceleration with respect to an inertial system, is a law about physical facts and no longer a mere definition of force.

We can distinguish two aspects in mechanics which correspond to mathematical and physical geometry (Chapter 1). We can mean by a in Newton's law, $m \times a = f$, the acceleration with respect to an arbitrary system (S). Then, if positions and velocities are given relative to (S), we can compute from Newton's laws the motion with respect to the same system (S). If, e.g., there is no force $(f = 0)$, we can conclude that our mass m moves relative to (S) in a straight line with constant speed. This statement is conditional in the same way as the theorems of mathematical geometry are conditional. It says: If Newton's law, $m \times a = f$, is valid relative to (S) and $f = 0$, it follows that the motion relative to (S) is rectilinear. This statement is true whatever may happen in the world; it is purely logical. Even if no system (S) existed with respect to which the Newtonian laws were valid and no situation in which the force f disappears, our statement would be true.

The second aspect considers specific physical systems (S) which are "inertial systems" and specific circumstances under which forces disappear. The description of these circumstances in terms of physical facts gives "operational meaning" to the Newtonian laws. Then we can check by actual observations to see if the conclusions drawn from Newton's laws and their operational meaning are in agreement with experience.   We can check by physical measurements whether or not a motion relative to a concrete physical system (S) is rectilinear and uniform.   If there is in the world a physical system (S) which is an inertial system, Newton's laws can be applied to the physical world.   To say that "Newton's laws are true" means that they are applicable in the same way as to say "the laws of geometry are true" means that they are applicable to the physical world.

The whole system of Newtonian mechanics hinges on the experiential fact that the ratio of the accelerations of two bodies produced by the same force does not depend on the external circumstances of these bodies; it is, in particular, not dependent upon the velocity of these bodies.   Then we are certain that if we define the "mass" of a body by the ratio a_2/a_1, it will be constant.   From Newton's definition it seemed self-evident that the "mass" of a body is constant and cannot depend on its velocity; it seems obvious from the common-sense image that is invoked by the expression "quantity of matter."   If we buy a "quantity" of meat or canvas, it seems obvious that this "quantity" is something intrinsic in meat or canvas and cannot depend on its velocity.   Actually, such a "constant mass" can only be introduced into mechanics if the ratio a_2/a_1 turns out to be, according to our experience, independent of the velocities of the bodies concerned.   Our actual experience has as its objects bodies of small velocity which means velocities which are small compared with the speed of light.   This ordinary experience shows that the ratio a_2/a_1 actually is independent of the velocities of bodies.

Before the twentieth century, physicists assumed that a_2/a_1 actually is constant, whatever the velocity of the bodies may be.   This follows only if we assume that "what is true for small velocities is also true for high velocities."   In the nineteenth century, the belief in Newtonian mechanics was so strong that practically no one doubted that the constancy of "mass" was a universal law of nature.   Man was not aware that the basis of this belief was either the identification of "mass" with "quantity of matter" or the experience that the ratio of the accelerations a_2/a_1 produced by the same force was, for small velocities, independent of the velocities of the bodies. In the year 1883, the Austrian physicist and philosopher, Ernst Mach, published a book[21] *Mechanics and Its Evolution*, which turned out to be in many respects a landmark in our understanding of the laws of motion. Mach presented a critical analysis of Newtonian mechanics and directed the attention of scientists to the fact that the "constancy of mass," if the operational definition $m_2/m_1 = a_1/a_2$ is used, is an experiential fact and by no means a "philosophical truth" which can be derived from "intelligible principles." There was a possibility that experiments would show a change of mass caused by external circumstances. As a matter of fact, toward the end of the nineteenth century, J. J. Thomson[22] derived from Maxwell's electromagnetic field theory the result that mass points behave like particles with electric charges.

In the twentieth century, the motion of fast electrically charged particles was systematically investigated, for example, in the cyclotron. If electrostatic forces are acting in the direction of the actual velocity, high-speed particles (*i.e.*, with speeds comparable to the speed of light) obtain accelerations which are noticeably smaller than the accelerations of slow particles in the same electrostatic field. This means that we cannot define by f/a a constant mass m which could be assigned to a body whatever its speed might be. The equation $ma = f$, where m is a constant, has no physical interpretation that would describe the actual motion of high-speed particles. The alterations in the "axioms" or the "formal system" which is applied to the motion of such particles will be discussed more precisely and elaborately in Chapter 5: Relativity.

In the present chapter, these alterations serve as a very helpful example by which we can illustrate the logical structure of science in general. If the new observations of high-speed particles had not been made, we would be tempted to say that Newton's laws are of "universal validity." This would mean that Newton's equations as a "formal system" ($ma = f$ where m is a constant) could be applied to all motions that have occurred and will occur in the universe. This belief in their "universal validity" was responsible for such statements as the one that these laws are the "real laws of motion in the whole Universe." We now know that only for low-speed particles is this "formal system" a convenient description of motion. Even today, however, there is no objection to building up

Newton's "formal system" into an elaborate theory of motion.   In this way we obtain a system of "Analytical Dynamics" which contains such statements as the following: If we know the initial state of motion of masses with respect to an inertial system (S) and if Newton's laws are valid for all speeds relative to (S), we can compute the state of motion relative to (S) at any time t, by using the "formal system" based on Newton's laws.   This if-statement or conditional statement remains valid even after we have learned that Newton's Laws are not applicable to high-speed particles.   If, however, we examine the problem of physical interpretation, we see that our if-statement can only have a valid physical interpretation if we assume that the speed of the particles is small compared with that of light.   If this is the case in all the motions, we can apply Newton's laws; but even for the highest speed the "if-statement remains valid" because it is a purely mathematical or logical statement.

8. Remnants of Organismic Physics in Newtonian Mechanics

In Aristotelian physics the world was a large apartment house erected according to a blueprint that was adjusted to the life and death of human beings.   The laws of motion were rules according to which bodies and souls moved from one room into another of this vast structure.   A living man was regarded as heavy and the "natural motion" of his body, if it was separated from the soul, was to fall toward the center of the earth; but the soul moved, as described in Dante's *Paradise*,[23] upward into the heavenly spheres.   By "place of body," Aristotle always meant a vessel in which this body was contained.   In his *Physics* he wrote: "As a vessel is transportable place, so place is a nontransportable vessel."   In the theories of Copernicus, and still more of Galileo, this apartment house lost its magnificent simplicity; a certain dissatisfaction entered the minds of all who were deeply concerned with astronomy.   We may quote as an example a man who was not at all an orthodox Aristotelian, but rather a staunch anti-Aristotelian, Francis Bacon.   He wrote in 1605:[24] "Astronomy, as it now stands, loses its dignity by being reckoned among the mathematical arts, for it ought in justice to make the most noble part of physics."

Organismic physics, which was felt to be satisfying to the human intellect, suffered a much more severe blow, however, when Newton advanced as the fundamental law of motion a law that is concerned with the motion of a body in empty space: the law of inertia. This law describes how a body would move in the void, but Aristotle[25] had written:

> Not a single thing can be moved if there *is* a void, . . . in the void things must be at rest, for there is no place to which things can move more or less than to another, since the void insofar as it is void admits no difference. . . . All movement is either compulsory or according to nature . . . but how can there be natural movement if there is no difference throughout the void or the infinite? Either nothing has a natural locomotion, or else there is no void.

Aristotle argues in a similar way that a compulsory motion cannot take place in a vacuum either, and then he continues:

> Further, no one could say why a thing once set in motion should stop anywhere; for why should it stop *here* rather than *here*? So that a thing must either be at rest or must be moved *ad infinitum*, unless something more powerful gets in its way.

A nineteenth- or twentieth-century scientist might conclude from these lines that Aristotle was very near to advancing the law of inertia, but that would be reading Aristotle with a mind that had been formed by studying modern physics. The ancient Greek found the idea that a body could move by its own speed uniformly into the infinite so absurd that he concluded: "It is clear from these considerations, that there is no separate void."

In contrast to this Aristotelian conception, in Newton's *Principles of Natural Philosophy* the "void" is the basic concept. A body in motion preserves its velocity relative to the "void." Newton stressed the point that this law does not refer to any physical body of reference, but to "absolute space," which is the same thing as what Aristotle would have called a "separate void." We have learned in Sections 6 and 7 that Newton's laws have an operational meaning only if the system of reference, the inertial system, is specified. We learned that in practice, to a good approximation, the inertial system can be identified with the system of the fixed stars. This would mean, in Newton's language: We must add to his laws the statement that the fixed stars are at rest relative to absolute space. This fact would have been just incidental for

Newton.   Whatever the fixed stars might do, any material body would preserve its velocity relative to absolute space.   We shall soon see that it is impossible to determine by any physical experiment the speed of the earth (or of any material body) relative to absolute space; therefore, this speed has no operational meaning.

Newton understood this difficulty very well.   In order to give an operational meaning to "absolute space," he preserved some elements of organismic physics.   As in Aristotelian physics it was assumed that there was a divine being in every moving sphere of the heavens, Newton assumed that absolute space was identical with the "sensorium of God."   This statement has been interpreted in different ways, but we get a very clear understanding of it from the diary of David Gregory.[26]   This pupil and intimate friend of Newton recorded in 1705 a talk about the question of what fills space that is empty of bodies.   He reported:

> The plain truth is, that he believes God to be omnipresent in the literal sense; and that as we are sensible of objects when their images are brought home within the brain, so God must be sensible of everything, being intimately present with everything: for he supposes that as God is present in space where there is no body, he is present in space, where a body is also present.

We could not understand the logical structure of Newton's physics, if we ignored the fact that he used in his law of inertia an "organismic" or, we may say in this case, a "theological" element.   In the late eighteenth and early nineteenth centuries, when the effort was made to purge physics of all theological elements, Newton's physics became illogical.   "Absolute space" became a mere word without the slightest trace of operational meaning.

There is, moreover, another theological element in Newton's physics which has been much more familiar to scientists and philosophers than the omnipresence of God as the foundation of "inertia."   When Newton applied his theory to the motions of the planetary system, he was able to derive from his laws the conclusion that the planets move around the sun in ellipses with the sun at one focus.   However, there are other regularities in the motion of the planets which cannot be derived from Newton's laws: The orbits of the planets and comets are all situated in approximately the same plane, and all are circulating in the same sense.   Newton's laws allow for arbitrary initial positions and velocities of all masses and would allow, therefore, for many irregularities which actually do not appear. Newton explained these uniformities by using typical arguments from organismic physics. He wrote:[27]

> But it is not to be conceived that mere mechanical causes could give birth to so many regular motions, since the comets range over all parts of the heavens in very eccentric orbits. . . . This most beautiful system of the sun, planets and comets, could only proceed from the counsel and dominion of an intelligent and powerful Being . . . .

It was very clear to Newton that his "explanation" of the regularities actually consisted in pointing out the close analogy between the regularity in the planetary system and the regularity produced by deliberate planning in human relations. He continued: "All our notions of God are taken from the ways of mankind by a certain similitude, which though not perfect, has some likeness, however." From this quotation it is obvious that Newton explained the planetary motions by an analogy with the behavior of human beings, just as Aristotelian physics did.

At the end of the eighteenth and the beginning of the nineteenth centuries, when the general climate of opinion became opposed to all arguments based upon "organismic science," attempts were begun to eliminate all remnants of organismic thinking from Newton's physics. It is well known that Newton's laws of motion tell us how to predict the future motions of a mechanical system at a certain instant of time if the "initial state," the positions and velocities of all masses in the system, is given. These data must be added to the Newtonian laws in order to compute future motions. Newton himself assumed, as we have learned, that these initial conditions were determined by a superhuman intelligence, somehow similar to human intelligence; if this intelligence is eliminated, there is a hole in the fabric of Newtonian physics. We must then say that the "initial conditions" are undetermined, and we must invent or discuss some principle from which they can be derived. At the end of the eighteenth century, as is well known, Kant and Laplace advanced the hypothesis that the planetary system had its origin in a rotating ball of gas. The regularities that Newton reduced to a human-like planning are thus ascribed to a rotation which all parts of the planetary system once had in common. This means that long before the formation of our planetary system the initial conditions of its masses were such that the angular momentum had a certain value denoted by a certain vector. We can, of course, also ask how this initial rotation began. We could again introduce an organismic element, or we could assume other initial conditions from which, according to Newton's laws, the common rotation could develop. There have been many attempts of this kind, but in any case, we have to make an arbitrary assumption about the initial conditions at some arbitrary instant of time. We could make the "minimum assumption" that the initial conditions describe a random motion of all atoms, as the Epicurean Lucretius[28] did in his book *On the Nature of Things*. But the assumption of random motion is also arbitrary, and the development of angular momentum from random motion requires delicate and somewhat debatable computations.

We shall now leave the problem of eliminating theological remnants in the initial conditions. These hypotheses about "initial conditions" or, in everyday language, about the "origin of the universe" have attracted interest mostly because of their "philosophical" interest. Little has been derived from them that could be tested by experience. On the other hand, the attempts to eliminate organismic and theological elements from the law of inertia have proved to be of great consequence in the field of "science proper." They have stimulated research for new physical laws.

We learned in Sections 6 and 7 that Newton's laws have an operational meaning only if the inertial system is described by physical operations. We learned that Newton described it as the "sensorium of God" and that the actual applications of mechanics were based on the "accidental fact" that the system of the fixed stars was, according to Newton, at rest in "absolute space." However, if we examine the two premises: The law of inertia is valid with respect to absolute space; and the fixed stars are at rest with respect to absolute space, we can draw the simple conclusion: The law of inertia is valid with respect to the fixed stars. In this way, the term "absolute space" is eliminated from the laws of motion, and along with it all organismic remnants. "Inertia" is now, to speak in the language of everyday life, the tendency of a launched stone to preserve its speed and direction with respect to the fixed stars. "Inertia" is then an interaction between material bodies just as much as gravitation is.   This point was stressed toward the end of the nineteenth century by Ernst Mach, first in a short paper in 1872, and then more elaborately in his book which we mentioned above.   He wrote in *Mechanics and its Evolution*[29] in 1883:

> The behavior of terrestrial bodies relative to the earth can be reduced to their behavior relative to the very remote celestial bodies (the fixed stars).   If we claimed that we know more about moving bodies than their hypothetical behavior with respect to celestial bodies, which is suggested by experience, we would be guilty of *dishonesty*.   If we say, therefore, that a body preserves its direction and speed *in space*, this is only a short advice to consider the whole world. . . .   The inventor of the principle [Newton] can permit himself the abbreviated expression [space], because he knows that no difficulties will turn up in the application of the law.   But he can do nothing if real difficulties occur, *e.g.*, there are no bodies which are at rest with respect to each other.

In this case, the elimination of the organismic elements has led to the discovery of new laws of interaction between material bodies. These new laws form the core of Einstein's theory of gravitation. They will be more elaborately discussed in the Chapters on Relativity (Chapters 5 and 6).

5

Motion, Light, and Relativity

1. Aristotle, St. Augustine, and Einstein

By regarding absolute, infinite space as a basic concept of physics, Newton introduced a difficulty which had not existed in ancient and medieval science. Aristotle's physics possessed a natural system of reference—the earth was the center with the spheres around it. It possessed also a natural clock, the rotation of the spheres. When Aristotle spoke about the place of a body, he always meant what we would call today the "relative place" within an environment of other bodies. Aristotle wrote:

> The existence of place is held to be obvious from the fact of natural replacement. . . . What now contains air formerly contained water, so that clearly the place or space into which and out of which they passed was something different from both. . . . Further, the typical locomotions of the elementary natural bodies—namely, fire, earth, and the like—show not only that place is something, but also that it exerts a certain influence. Each is carried to its own place, if it is not hindered, the one up, the other down. . . . Place is rather what is motionless: it is rather the whole river that is place (of a boat), because as a whole it is motionless . . . thus we conclude that the innermost motionless boundary of what contains is place.[1]

This whole well-ordered system of "place" was disrupted when Newton advanced the conception of a space in which there were no places, no boundaries, no "motionless" containers. It is interesting to consider that the Hebraic-Christian world view introduced on one occasion an absolute and empty space: the state of the world before creation. We note that St. Augustine, who was brought up in the tradition of Greek philosophy but later converted to Christianity, had some trouble in assimilating the new doctrine to his philosophical background. If the universe was entirely empty before the creation, the question arises of why God had idled such a long time before he decided to create the world. St. Augustine wrote in his *Confessions*:[2]

> If any excursive brain . . . wonder that Thou the God Almighty and All-Creating and All-Supporting, Maker of Heaven and Earth, didst for innumerable ages forbear from so great a work before Thou wouldst work it: let him awake and consider that he wonders at false conceits. For whence could innumerable ages pass by, which Thou madest not, Thou the Author and Creator of all Ages? Or what times should there be, which were not made by Thee? Or how should they pass by, if they never were? Seeing then Thou art the Creator of all times, if any time was before Thou madest Heaven and earth, why say that Thou didst forego working? . . . But if before Heaven and earth there was no time, why is it demanded "what Thou then didst?" For there was no "then" when there was no time.

Bluntly, this means that in the empty absolute space that existed before the creation, there was no time, and it does not make any sense to ask what God did "then" before he started the creation.

It is perhaps instructive to compare these statements of St. Augustine with some statements which Einstein made in order to explain in a popular and slightly facetious way the central contention of his theory of relativity. When Einstein first arrived in the United States in 1921, he was met in New York harbor by a group of journalists who wanted him to explain to them in one sentence the basic tenet of his famous theory. Einstein answered them:[3]

> If you don't take my words too seriously, I would say this: If we assume that all matter would disappear from the world, then, before relativity, one believed that space and time would continue existing in an empty world. But, according to the theory of relativity, if matter and its motion disappeared there would no longer be any space or time.

2. "Relativity" in Newtonian Mechanics

We learned in Chapter 3 that Newton's laws of motion, which verbally determine acceleration with respect to "absolute space," can only be used for the prediction of observable facts if they are interpreted as referring to an "inertial system" which is a system of physical bodies. As a first approximation, we can identify this system with the constellations of fixed stars as far as we can regard them as forming a rigid system. Newton had already raised the question of whether we can recognize by mechanical experiment performed in a certain room, whether this room is an inertial system or not, and if so, whether or not it is the only inertial system. If we denote by a_{in} the acceleration with respect to an inertial system (S), Newton's laws of motion can be written $ma_{in} = f$, if m is the mass of a body and f the "simple formula"[4] which gives us the Newtonian force. If we disregard for the moment nuclear forces which account for the motion of subatomic particles, the motion of masses of average size is determined, as a matter of fact, by only two types of force, electromagnetic and gravitational. In the first case, if the mass is increased (by adding more pieces of matter) the acceleration a_{in} will decrease according to the formula $a_{in} = f/m$. With sufficiently increased mass, the acceleration will become infinitely small. This is certainly the case if the force is of the electromagnetic type. If e is the electric charge, and E the electric field strength, we have $f = eE$ and $a_{in} = e/m \times E$. In a given field, the acceleration of an increasing mass relative to the inertial system will decrease toward zero. To the electromagnetic type of force belong all forces of cohesion, among them the forces which are produced if one gives a body a direct push or pull. There is, however, a different type of force. We have known since Galileo that, in the case of freely falling bodies, the acceleration with respect to the earth is independent of the mass. This means that f/m is independent of m or f is proportional to m. If $f = mg$, it is evident that $f/m = g$ is independent of m. This is the gravitational type of force. We shall disregard it at this point and speak only of forces which impress upon very large masses only very small accelerations.

While Newton's laws yield the acceleration a_{in} with respect to an inertial system, we shall now consider the acceleration relative to an arbitrary system (the "vehicle") which is produced by the force f which may, for example, be eE. We may assume, for the sake of simplicity, that all accelerations and forces have the same direction. We shall denote the acceleration of the "vehicle" relative to the inertial system (S) by a_{ve}, the acceleration of the mass m relative to the inertial system (S) by a_{in}, and the acceleration of the mass m relative to the "vehicle" simply by a, as the vehicle is an arbitrary system or, as Aristotle put it, a "movable place." Then we have, obviously, $a_{ve} + a = a_{in}$, and the equation of motion becomes $ma_{in} = ma_{ve} + ma = f$, or $ma = f - ma_{ve}$; for example, $ma = eE - ma_{ve}$. If we refer the motion to an arbitrary vehicle system, the acceleration a of a mass m is not determined by the Newtonian force (e.g., eE) by itself; we must add a term, $-ma_{ve}$, that we call the "inertial force," f_{in}. Then the laws of motion with respect to an arbitrary vehicle system are $ma = f + f_{in}$ where $f_{in} = -ma_{ve}$. We see immediately that we can formulate the laws of motion with respect to a system of reference that is not an inertial system by adding to the Newtonian force an "inertial force" $f_{in} = -ma_{ve}$. This force is not of the electromagnetic but of the gravitational type. If we divide the equation by m, we obtain for the acceleration a relative to an arbitrary system: $a = f/m + f_{in}/m = f/m - a_{ve}$. If the mass becomes very large, the acceleration becomes $a = -a_{ve}$. Since a is obviously measurable by ordinary operations of measurement, we can also measure $a_{ve} = -a$, the acceleration of any "vehicle" relative to an inertial system.

As we remember, Newton identified the inertial system with his "absolute space" that was not a physical body but was the "sensorium of God."[5] Newton had already directed attention to the fact that the acceleration of a very large body relative to an arbitrary system of reference provides us with the acceleration of this vehicle with respect to an "absolute space." If we give a push to a very large ball in a railway train compartment, we can find the acceleration a_{ve} of the compartment (relative to an inertial system) from the acceleration a of the ball with respect to the compartment. The most familiar case is the rotation of the vehicle. In this case we do not have to do with motions in the same direction; every acceleration or force is a vector $\boldsymbol{a}$ or f which has not only magnitude but also direction. On a rotating vehicle, the acceleration $\boldsymbol{a}_{ve}$ is centripetal. Therefore, the acceleration $\boldsymbol{a}$ of a mass relative to the vehicle is an acceleration away from the axis of rotation in a centrifugal direction. Newton, in describing his famous bucket experiment, considered the "centrifugal motion" of masses as a criterion for the rotation of the system of reference relative to absolute space. Newton claimed, therefore, that the "rotation of a room relative to absolute space" had consequences which could be observed by physical experiments (centrifugal phenomena).

The situation is different if the "vehicle" moves with constant speed q along a straight line relative to the inertial system. In this case $a_{ve} = 0$, and the acceleration relative to the vehicle is determined by $ma = f$, just as if the vehicle were an inertial system. In this case the speed q of the vehicle with respect to the inertial system can have an arbitrary constant value. This value does not manifest itself in the law of motion determining a. Therefore, if the initial positions and velocities of all masses relative to the vehicle are given, the law $ma = f$ allows us to compute all future positions relative to the same vehicle. The knowledge of q is unnecessary, and, of course, by observing the laws of motion relative to the vehicle, one can learn nothing about the speed q relative to the inertial system. This theorem that follows from Newton's law is called Newton's "Theorem of Relativity." We can formulate it in a positive or in a negative way. The positive formulation is: By knowing the relative initial condition of masses in a vehicle, we can predict their future relative motion without knowing the speed q of the vehicle itself. The negative formulation says: By observing the motions relative to a vehicle, we cannot find out the constant speed q of this vehicle, provided it is moving in a straight line relative to an inertial system. We can also say: a vehicle moving with uniform motion relative to the inertial system (S) is itself an inertial system that may be called (S'). From these considerations it clearly follows that the speed q of (S') relative to the inertial system, or, according to Newton, to absolute space cannot be derived from any physical experiment. This speed q has no operational meaning in physics and was given a meaning by Newton within the system of theology.

3. Newton's Relativity and Optical Phenomena

We have learned by mechanical experiments within a vehicle that its rectilinear motion with a constant speed q does not manifest itself. We can now ask what is the case if we observe *optical* phenomena with certain definite initial conditions relative to the same "vehicle" (S'). Will the rectilinear speed q of the vehicle have some influence on the outcome of the optical experiment? After the great success of Newtonian mechanics, the idea was prevalent that the phenomena of light had to be accounted for by mechanical theories of light. There were two theories of this type, both of which assumed the universal validity of Newtonian mechanics for all phenomena of motion. The first theory assumed that light consisted of small corpuscles emitted by the light source and falling into our eyes. These corpuscles followed exactly the laws of Newtonian mechanics; we speak of a "Corpuscular Theory of Light." The second type of theory assumed that the whole world space was filled with a subtle elastic medium, the ether, that also followed the laws of Newtonian mechanics. Light consisted of a propagation of waves through this medium.

According to the corpuscular theory, light is propagated with higher speed in water than in air (because of the greater mutual attraction in the denser medium). The wave theory leads to the result that light is propagated with lower speed in water than in air (according to Newtonian mechanics for solid and fluid bodies, waves are propagated with lower speed in denser bodies). In 1850, the French physicist Foucault[6] demonstrated by a clear-cut experiment that light moves with less speed in water than in air: thus the corpuscular theory seemed to be refuted by a "crucial experiment." The wave theory was generally accepted. This decision had a great bearing on the question of whether, in a room moving with a uniform speed q, this speed q had any effect on the optical phenomena relative to this room or, in other words, whether or not Newton's theorem of relativity was also valid for the phenomena of light. If we accept the corpuscular theory, the propagation of light is not different from the motion of launched balls. They behave relative to the vehicle in a way that is independent of the value of q, the speed of the vehicle. The theorem of relativity would also be valid for optical phenomena. But the situation looks different, if the wave theory is accepted. Then, light consists of vibration in the ether. If we want to predict the propagation of light in a vehicle moving with the speed q, we must make some assumption as to how the velocity of the ether particles is influenced by the motion of the vehicle.

The annual aberration[7] of the light coming from the fixed stars suggests the hypothesis that the ether filling the world space remains at rest. The light coming in perpendicular to the direction of the earth's velocity in its orbit is apparently deflected by an angle q/c (the angle of aberration) where c is the speed of light in a vacuum. The fact that this value remains the same under all circumstances suggests the assumption that no component of the earth's velocity q is added to the vibration of the ether particles. If we now compare the propagation of light emitted from a source at rest in the ether, system (S), and in the vehicle earth, system (S'), the initial conditions in the mechanical sense are not the same in both cases. In the first case the initial velocities of the ether particles are the velocities of small vibrations. But relative to the (S') all particles have, in addition to these velocities of vibration the common velocity $-q$ since (S') is moving with respect to the ether with the speed q. Therefore, we cannot expect that Newton's theorem of relativity is applicable to the propagation of light.

We can easily see what kind of departure from the relativity theorem we should expect on the basis of the ether theory of light. From a light source at rest in the ether, system (S), light will be propagated with the speed c relative to (S). If the source is at rest in the "vehicle" (S'), the light will also be propagated with the speed c relative to (S) since this speed depends only on the elastic properties of the medium and not on the method of excitation. Therefore, relative to (S'), the light from a source at rest in (S) will be propagated with the speed $c + q$ or $c - q$, depending on whether the direction of q is the same as or opposite to the direction of the light ray. This looks as if the speed q of a vehicle in uniform rectilinear motion q relative to the ether (S) would have an influence upon the propagation of light relative to the vehicle and, on the other hand, that from the observation of optical phenomena in the vehicle we could compute q. The attempts to test this possibility by experiment have comprised a great chapter in the history of theoretical and experimental physics. It was found that the phenomena of reflection and refraction could not provide results that were practically observable, let alone measurable. The climax was a proposal of the great British scientist James Clerk Maxwell. He originated an experiment which would yield a measurable effect of the speed q of the earth upon optical phenomena on the earth.

It had been found that no experiment would give an effect of the order of magnitude q/c. The experiment proposed by Maxwell would only yield an effect of the order of magnitude q^2/c^2, but it would nevertheless be measurable because measurement by interference of light waves could be applied.

Maxwell proposed[8] to investigate the reflection of the light emitted from a source at rest in the vehicle and reflected from mirrors at a distance L from the source. He compared the times spent by light rays which were propagated parallel to the motion of the vehicle (the earth) and perpendicular to this direction, respectively. If the vehicle were at rest $(q = 0)$, the time of a reflection would be $T_0 = 2L/c$, independent of the direction of the light ray. If, however, the vehicle is moving with the speed q, the time of reflection parallel to the direction of q has the value $T_p = T_0/(1 - q^2/c^2)$ and perpendicular to this direction $T_n = T_0/\sqrt{1 - q^2/c^2}$. This follows simply from the assumption that the speed of light relative to S (the ether) is always c and that the ether is not influenced by the motion of the system (S') through the ether. Obviously T_n is smaller than T_p. The difference $(T_p - T_n)$ is approximately $\frac{1}{2}T_0 q^2/c^2 = Lq^2/c^3$. Maxwell argued: Since the two light rays reflected from the two different mirrors could be brought to interference, the difference in the times $(T_p - T_n)$ could be measured by comparing it with the period of light vibrations. The only question that remained was whether in an actual experiment the time difference $(T_p - T_n)$ would exceed the margin of error. If it should, it would serve to compute the speed q. The realization of Maxwell's proposed experiment would provide the final confirmation of the current theory that the ether was not moved by the speed q of material bodies but transmitted light vibrations with a speed c.

Maxwell's suggestion was carried out, not long after its announcement (1881) by the American physicist Albert A. Michelson;[9] the result was negative. The expected time difference $(T_p - T_n)$ was greater than the margin of error; if the errors were disregarded, the observed time difference was nil. This would mean that the speed q of a vehicle could have no effect on the optical phenomena in the vehicle. It would mean, in other words, that Newton's theorem of relativity would also hold for optical phenomena, although it followed (as Maxwell proved) from the current ether theory and mechanics that it should not hold. Most contemporaries of Michelson's experiment, among them Michelson himself, looked for an interpretation within the dominating theory that light was a mechanical phenomenon and the ether a medium that obeyed Newton's laws of motion. Within this frame one could, of course, modify the assumption that the vibration of the particles was not altered by the velocity q of the vehicle moving through the ether. Michelson and most of his contemporaries again took up the hypothesis of Stokes that the ether particles on the surface of the earth add its speed q to their vibrations. In this case, there is, of course, no reason for a difference between T_p and T_n and the theorem of relativity would hold; but we have mentioned already that the theory of an "ether carried along by moving matter" is hard to reconcile with the aberration of starlight. We could, of course, invent special laws for the motion of ether which would allow us to explain aberration as well as Michelson's experiment, but the theory becomes very complicated. We can say that toward the end of the 19th century the mechanical theory of light entered into a state of great complication and confusion.

4. The Electromagnetic World Picture

The Michelson experiment was one instance in which the attempts to derive all physical phenomena from Newton's laws of motion led to trouble. It was certainly not "proved" that it was impossible to regard the propagation of light as a mechanical phenomenon, but it was certainly clear that a derivation from Newton's laws had yet to be made in a "simple" way. There was, in addition, the wide domain of "electromagnetic phenomena." They have been derived since the last decades of the nineteenth century from Maxwell's differential equations of the "electromagnetic field." Originally these "field equations" were regarded as describing a special mechanism that obeyed Newton's laws of motion; Maxwell himself invented a mechanism of this kind. The derivation of the electromagnetic equations from such mechanisms was never entirely satisfactory, however; it became more and more involved, although it has never been proved that a satisfactory derivation is impossible. Eventually, in 1889, Heinrich Hertz[10] said bluntly that the theory of electromagnetic phenomena was identical with Maxwell's field equations, just as Newton's theory of motion is identical with Newton's laws of motion. The reduction of the field equations to the equations of motion is irrelevant. During a certain period, physicists gave a "dualistic" presentation of their science. One part was regarded as "physics of matter"; mechanics, acoustics, heat: the other part, the "physics of the ether," contained electricity, magnetism, and optics. Very soon it became apparent that such a clear-cut division did not yield a satisfactory derivation of all experiences about the interaction between the motion of material bodies and electromagnetic wave propagation. This was clearly a result of the failure to account for Michelson's experiment on the basis of Newton's laws.

In 1890, the British physicist Sir Joseph John Thomson[11] showed that a particle of very small mechanical mass could possess a tremendous inertia, provided that its electric charge or speed were sufficiently great. This fact that could be derived from the laws of the electromagnetic field was originally verbalized as follows: Every electric charge possesses an "apparent mass" which behaves under the influence of a force like a "real mass." Later on, one ventured the hypothesis that there might not be any real mass at all and that inertia was a phenomenon of the electromagnetic field. From this hypothesis, the great Dutch physicist Hendrik A. Lorentz[12] derived that the apparent mass of a particle increases with its speed and increases beyond all limits if the speed approaches the speed of light. If we start from the "electromagnetic theory of mass," every mass has this property, and we can conclude that the speed of light enters as a constant into the equation of motion. Newton's laws must be altered in such a way that they contain the speed of light. The effect of a force upon a mass depends on the ratio v/c, where v is the speed of the mass and c is the speed of light.

If we return now to the Michelson experiment (discussed in Section 3) and apply to it the electromagnetic theory of matter, we see that we can avoid the contradiction between the theoretical derivation and the experimental result. From the application of Newtonian mechanics we obtained the result that $T_n < T_p$ while the experiment shows that $T_n = T_p$. The inequality resulted from the following argument: $T_p = T_0/1 - q^2/c^2$, and $T_n = T_0/\sqrt{1 - q^2/c^2}$, where $T_0 = 2L/c$. In this formula, L denotes the length of either equal arm of Michelson's apparatus when it is at rest. From Newton's mechanics it follows that they remain equal when the apparatus is moving at a high speed. If we assume that the mass of the par-

ticles of which the arms consist is "electromagnetic mass" and has
its origin in their electric charges, the motion in a certain direction
produces electric currents in this direction.   The forces exerted by
these currents upon each other account for a tension within the arms
which, in turn, is responsible for a deformation.   Lorentz made it
plausible that the result of these tensions could be a contraction of
the arms in the direction of motion.   Then the length of both arms
in motion would not be equal.   If we denote the length of the
moving arms no longer by L, but by L_p and L_n, respectively, the
times of reflection would become $T_p = 2L_p/c(1 - q^2/c^2)$ and $T_n =
2L_n/c\sqrt{1 - q^2/c^2}$.   Lorentz made some assumptions about the dis-
tribution of electric charges in the particles from which he could
derive the result that the formulae $L_p = L_0\sqrt{1 - q^2/c^2}$ and $L_n = L_0$
would be quite compatible with the laws of the electromagnetic
field.   But then we have $T_p = T_0/\sqrt{1 - q^2/c^2} = T_n$, which would
be in agreement with the negative result of Michelson's experiment.

We see now that from the "electromagnetic theory of matter"
new laws of motion could be derived that contain the speed of light c
as a constant and imply great deviations from Newton's laws if v/c
approaches the value one, but are almost identical with Newton's
laws if v/c is very small.   If we accept these new laws of motion, we
can account for the interaction between the motion of bodies and
light propagation as it is revealed by optical phenomena in moving
bodies and by the Michelson experiment in particular.

The acceptance of the "electromagnetic theory of matter" has
been a very important factor in the evolution of scientific and philosophical thought.   Since the rise of modern science (around 1600)
the prevailing opinion among scientists was the belief in a "mechanistic science," which means the belief that physical phenomena are
"understood" or "explained" only if these phenomena can be
reduced to Newton's laws of motion.   Obviously, the "electromagnetic theory of matter" has dropped this requirement.   It was
stated, starting with Heinrich Hertz, that man must cease trying
to reduce all physical phenomena to the laws of mechanics.   It was
required, instead, that all physical facts must be derived from Maxwell's laws of the electromagnetic field.   This meant a radical
change in the meaning of the conception of "understanding" or
"explaining."   The requirement of a reduction to Newton's laws
was raised because it was believed that these laws were "self-

evident"; the reduction to Newton's mechanics proved a reduction to "intelligible" principles in the sense of Aristotle's. Hardly anybody, however, would think that Maxwell's equations of the electromagnetic field were self-evident or intelligible. Therefore, the abandonment of a mechanistic explanation meant also dropping the request for a deduction from intelligible principles. Maxwell's equations of the electromagnetic field and Lorentz's hypothesis about the distribution of electric charges in "material" particles were accepted only because the observed facts about the motion of bodies and propagation of light could be derived. Thomas Aquinas' criterion for the "inferior" type of truth, the "scientific," not the "philosophical" truth, became the decisive criterion. Principles of physics were accepted if they could stand the test of logical consistency and empirical confirmation. The era of mechanistic physics was reaching its end, and the era of logico-empirical physics was beginning. Roughly speaking, we may say that the mechanistic era had extended from 1600 to 1900, and that the twentieth century opened with the logico-empirical conception of science in the making.

The philosophical interpretation of the electromagnetic world picture was the starting point of Lenin's book, *Materialism and Empirocriticism*, written in 1903.[13] The views that Lenin advanced in his fight against one kind of interpretation have become the cornerstone of the official philosophy of the Soviet Union, and particularly the philosophy of science that has prevailed in the teaching of science in Russian universities.

5. The Principles of Einstein's Theory

The Newtonian principles of motion were certainly not a result of man's experience with the motions which he observed in his daily life. They contained elements of lofty imagination like the law of inertia. Newton's laws, however, were certainly much nearer to man's daily experience than the laws of motion which were derived from the electromagnetic world picture. The introduction of the speed of light c into the laws of motion is very far from the observations of everyday life because it has no influence on observed motions unless they have speeds near the speed of light, which does not happen in any motion that is observed in technical mechanics or even astronomy. With the acceptance of the electromagnetic picture, the idea was dropped that the general principles of mechanics must reflect our daily experience with motions. This was closely connected with dropping the requirement that the general principles should be "intelligible."

The way was now clear for the launching of the new principles which have been characteristic of twentieth-century physics: the theory of relativity and the quantum theory. The goal of science in the twentieth century has been to build up a simple system of principles from which the facts observed by twentieth-century physicists could be mathematically derived. It was no longer required that these principles or some of their immediate consequences should be in agreement with our daily experience, or, in other words, with "common sense." What was required was a high-grade logical simplicity and agreement with the refined experiments of twentieth-century physicists. In the year 1905, Albert Einstein[14] advanced his theory of relativity, which was to be the first building stone in the structure of twentieth-century physics. His goal was to set up simple principles from which the interaction between motion of material bodies and propagation of light could be derived without introducing the hypothesis of the ether or the Lorentz hypothesis about the distribution of electric charges in material particles. The new principles led with logical consistency to the new laws of motion that contained the speed of light.

In order to find these new laws, Einstein started from the most prominent case in which the old laws of motion and light propagation had failed to yield the observed facts: the Michelson experiment. As we learned in Section 3, this experiment showed that Newton's theorem of relativity holds also for the phenomena of light propagation in moving vehicles, although according to Newtonian mechanics and optics this should not be the case. Einstein ventured, therefore, the hypothesis that the principle of relativity might be a principle of higher generality than Newton's laws of motion and the ether theory of light. From the latter, he took only one general result which seemed to be a plausible generalization of observed facts: There is a system of reference (F) in the world, with respect to which light is propagated through vacuum with a constant speed c, whatever the speed of the source of light with respect to (F) may be. This principle is called the *constancy principle* (Principle I). In addition to it, Einstein assumed the *principle of relativity* (Principle II), which can be formulated as follows: a vehicle system (F') may move with a constant speed q along a straight line with respect to (F). We start any optical or mechanical experiment with given initial conditions relative to (F'). Then our principle says that the outcome of the experiment is not dependent upon q, or, in other words, if the initial conditions relative to F' are given, the further motion and light propagation with respect to F' are determined; they do not depend upon q.

From these two principles, results can be derived that seem to be paradoxical or even self-contradictory. We consider a light ray emitted from a source at rest in (F). This light is obviously propagated with a speed c with respect to (F); the same source has a speed q relative to (F'). According to Principle I, the speed of propagation of this light ray is the same as if the source were at rest in (F'). But in this case, according to Principle II (relativity), the speed of propagation relative to (F') would be c. Therefore, one and the same light ray is propagated with a speed c relative to (F), and with the same speed c relative to (F'), although (F') has the speed q relative to (F). Light is propagated with the speed c relative to (F) or to any system (F'), whatever its speed q relative to (F) may be. This means that we can derive from Principles I and II a self-contradictory result, or, in other words, that the principles of constancy and relativity contradict each other. As a matter of fact, this "contradiction" is not a logical one; it occurs only if we add to Principles I and II a physical interpretation (or operational definitions) and assume that physical bodies follow the laws of Newtonian mechanics and that light waves are propagated through the ether. Briefly speaking, the system consisting of Einstein's principles (I and II) and the traditional laws of physics (mechanics and optics) is self-contradictory.

Einstein conceived a fundamentally new system of physical laws that would embrace mechanics and optics; it would provide a new derivation of the electromagnetic world picture in a new and simplified form. He did not start from Maxwell's differential equations of the electromagnetic field and Lorentz's theory of electrons (electric charges), but from the principles of constancy and relativity (I and II). If we assume, as Einstein did, that both principles are valid, they obviously have to be compatible. If so, the traditional laws of mechanics and optics cannot be correct. Therefore, Einstein's two principles imply a modification of the traditional laws of physics (mechanics and optics). Einstein's principles are equivalent to a new theory of the interaction between observable motion and propagation of light. The consequences will be presented more elaborately in the next section.

6. The "Theory of Relativity" Is a Physical Hypothesis

In order to understand Einstein's theory of relativity well, perhaps the most important point is to understand precisely how one can derive from a physical hypothesis the "Relativity of Space and Time." If we understand this argument thoroughly, we will not be fooled by current misinterpretations of the term "relativity." The apparent contradiction between the Principles I and II can be briefly formulated as follows: We consider (as in Section 5), a system of reference (F') (a vehicle) that is moving relative to the fundamental system (F) with a speed $q(<c)$. We consider now a source of light at rest in the vehicle (F') that emits a light ray in the direction of the motion of (F'). Then according to Principle I, the speed of the emitted light relative to (F) is the same as if the source were at rest in (F); this means that this speed is c. However, according to Principle II (relativity), the speed c' of light emitted by a source in (F') relative to (F') is the same as if source and vehicle were at rest in (F). This means that $c' = c$. One and the same light ray has one and the same speed relative to (F) and (F'). On the other hand, it follows from the most elementary laws of traditional mechanics that $c' = c - q$. This is obviously contradictory to $c' = c$ if $q \neq 0$. However, this does not prove that Principles I and II form a self-contradictory system, but only that Principles I, II, and the laws of traditional mechanics together are a self-contradictory system.

The conclusions which we draw from this "self-contradiction" depend entirely upon whether we regard traditional mechanics as a formal axiomatic system or as a physical empirical science. We discussed this distinction elaborately in Chapter 3 (Geometry) and Chapter 4 (Laws of Motion). If we regard Newton's laws of motion as a formal axiomatic system, we can derive from this system that $c' = c - q$. Then we have proved that Einstein's two principles and the axioms of Newtonian mechanics are together a self-contradictory system of axioms. This would be analogous to the geometry we obtain by replacing the Euclidean Axiom of Parallels by Lobatchevski's axiom, but keeping the theorem that the sum of the angles in a rectilinear triangle is independent of the size of the triangle and equal to two right angles. Then, of course, the statement of plane geometry about straight lines and angles would form a self-contradictory formal system. We can eliminate this self-contradiction in two ways, a purely formal mathematical way and a physical empirical way. To speak first of the formal way, we can find out how we must modify the statements about straight lines and angles in order to obtain a logically coherent system that includes Lobatchevski's axiom instead of Euclid's. To achieve this, we must replace the traditional theorems about the sum of the angles in a triangle by the more complicated one that the sum of the angles depends upon the area of the triangle and only equals two right angles for very small triangles. By doing this, we would not change anything in our statements about the physical world, but would only change the definitions of straight lines.

We could, of course, proceed in the same way in the case of Einstein's new Principles I and II. In this case, we could regard $v' = v - q$ as a formal axiom or a definition that connects the velocity v with respect to F with the velocity v' relative to (F'). If we replace this axiom by a new one $v' = f(v,q,c)$, we can achieve that for $v = c$ we obtain $v' = c$. The well-known relativistic theorem of addition is actually such a formula: $v' = (v - q)/(1 - vq/c^2)$. Obviously, if $v = c$ it follows that $v' = c$, independent of q, but this is certainly not what Einstein intended in his theory of relativity, although a great many presentations of this theory give this impression.

Actually, Einstein did not intend his Principles I and II to be "definitions" of terms. It was an essential point of his theory that he added operational definitions to the verbal formulation of his principles—in particular, of the basic term "velocity relative to a system of reference." In this way, Einstein converted his Principles I and II into physical hypotheses. This corresponds to the conception of geometry that we called "physical geometry" in Chapter 3. If we replace Euclid's "axiom of parallels" by Lobatchevski's axiom, we change physical hypotheses. Then the physical hypothesis that the sum of the angles in a triangle of light rays is independent of the size of the triangle would be contradicted by the new (Lobatchevski's) hypothesis. To restore compatibility among the principles of optics, we must replace the theorem about the sum of the angles by a more complicated theorem, according to which the sum of the angles differs the more from two right angles, the greater the area of the triangle. This means that if we start from Lobatchevski's axiom with its physical interpretation, we advance a physical hypothesis about the behavior of light rays. To say that this hypothesis is "valid" means that light rays behave very differently from the way in which they are supposed to behave according to traditional physics. The sum of the angles in a triangle actually depends upon the size of the triangle. This is a statement about the interaction between light rays and protractors or, more generally speaking, between light rays and mechanisms.

Viewing Einstein's conclusions, we are faced by exactly the same situation if we add to Einstein's Principles I and II the operational definitions of the "velocity of a material body and of light propagation relative to systems of reference." The operational definition of velocity v is based upon the operational definition of a spatial distance s and a time distance t, since $v = s/t$. The operational definitions of s and t are the standards prescribed in technical manuals for makers of precision instruments like clocks and yardsticks. We must ask ourselves whether these operational definitions include some special system of reference, like (F) or (F'). If we take for granted that the velocity v with which a clock or a yardstick is moving relative to (F) has no influence upon the readings of the instruments, the velocity v is irrelevant for the result of the measurement. But in Section 4, we learned that, according to the theory of electrons (elementary electric charges) advanced by H. A. Lorentz, a moving rigid body is contracted in the direction of motion. This would obviously happen to moving yardsticks. Since the spatial distance between two points is defined by the operation of laying yardsticks end to end, the result of the measurement depends upon whether the yardstick is at rest relative to (F), or has a velocity v. "Spatial distance relative to (F)" means the result of the measurement that uses a yardstick at rest relative to (F), while "spatial distance relative to (F')" refers to a yardstick at rest in (F') which has, therefore, a speed q relative to (F). Exactly the same definition is applied in the case of "time distance." Larmor derived from the theory of electrons that a standard clock moving with a speed q relative to the ether lags behind the clocks that are at rest relative to the ether. In the operational definition of "time distance relative to a system (F')," we must introduce a clock that is at rest in (F') and has, therefore, the speed q relative to (F).

If we add these operational definitions to Principles I and II (constancy and relativity), the contradiction between them disappears. The contradiction seemed to arise when we derived from Principle I (Section 4) that $T_p > T_n$, while from Principle II (relativity) it followed that $T_p = T_n$. If we add the operational definitions mentioned above to Principles I and II, the relation derived from Principle I refers to the time distance relative to (F), while the principle of relativity (II) refers to the time distance relative to the vehicle system (F'). The relation $T_p = T_n$ refers to the time relative to (F'), while the relation $T_p > T_n$ referred to the time relative to the fundamental system (F). If we denote the time distances relative to (F) and (F') by T and T', respectively, we have the relations $T_p > T_n$ and $T_p' = T_n'$, which are not contradictory to each other. Einstein's basic hypothesis was that Principles I and II are both valid. From this assumption it follows that T and T' must be different from each other, or, in other words, that the time distance between two events depends upon the speed of the clock by which this distance is measured. If a process takes place at a point P of (F) and takes one minute measured by a clock in (F), it will take less than a minute if we use a clock that is at rest in (F') and moves with a speed q relative to (F). Exactly speaking, the durations of the same process, if measured relative to (F) and to (F'), are in the relation $1/\sqrt{1 - q^2/c^2}$ (Sections 3 and 4). A very similar argument can be advanced about moving yardsticks. If we consider two points A and B at rest relative to (F'), then the distance AB will depend on whether the yardstick by which we measure AB is at rest in (F) or (F'), or has any speed q relative to (F). We denote the results of the measurement of this distance AB by L if the yardstick is at rest in (F), and by L' if the yardstick is at rest in (F'). Then we find from the simultaneous validity of Principles I and II that L' is smaller than L. To speak precisely, $L' = L \sqrt{1 - q^2/c^2}$.

From the principle of relativity, we can also conclude that the lagging of a moving clock is not restricted to a spring watch or a pendulum clock. The same thing also occurs if we use any type of clock mechanism, for example, the oscillation of an electron in a sodium atom, or the beating of the human heart. From this presen-

tation of Einstein's theory that has been called the theory of relativity, we learn that this theory is a system of hypotheses about the
behavior of light rays, rigid bodies, and mechanisms, from which
new results about this behavior can be logically derived.   It is very
misleading to say, as has frequently been said, that the electromagnetic theory of matter was a "physical theory" that "explained"
the negative result of the Michelson experiment, while Einstein's
theory of relativity did not explain it but merely "described" it by
a "new definition" of "space" and "time."   It is plausible to everybody that it would have been impossible to derive from new definitions new facts about the behavior of rigid bodies and light rays.
Actually, Einstein's two principles (constancy and relativity) are
hypotheses about such behavior, and it is obvious that theorems
about rigid bodies and light rays can be derived from them.   To be
sure, Einstein's theory of relativity starts exactly like Maxwell's and
Lorentz's theories about the electromagnetic field from hypotheses
about physical facts, except that the facts assumed in Einstein's
Principles I and II are of a much more general kind than those
formulated in Maxwell's equations of the electromagnetic field.
However, we must be aware that by introducing operational definitions of the terms, both types of theory are converted into
hypotheses about observable facts.

7. Relativity of Space and Time

In the philosophical interpretation of modern physics much attention has been given to formulations like: "The time distance between
two events has no absolute value (*e.g.*, one second), but its amount
is in every case defined with respect to a certain system of reference."
"This table has a certain length with respect to the earth, and
another length relative to the moon depending on whether the yardstick employed is at rest on the earth or on the moon."   Philosophers have often interpreted such statements by saying that according to the theory of relativity the table has no "objective length,"
but only a "subjective length" according to the observer; but such
an interpretation would be misleading.   We can give a better interpretation by digging a little deeper into the operational meaning of
the term "time distance."

We start, for example, from the statement that a lecture "lasts

one hour." This means that while the lecture lasted the hour hand
of the clock on the wall has turned by a certain angle (30°). The
clock itself is defined by the technical standards which are used by
clock manufacturers. To be sure, the duration is not defined by an
individual clock; it would be very wasteful to introduce a definition
of one hour that is dependent upon an individual instrument of
measurement. The "duration" must be measurable even by different types of clocks, e.g., by pocket watches and pendulum clocks,
or even by the human pulse rate. An operational definition of
"duration" is of practical use only if all these types of clocks give
identical values. Even the psychological estimate of duration
should give an approximate agreement. If the duration were to
be measured differently by the individual students and teachers or
by the clocks on church towers or the expectations of the audience,
the definition would be without value for human cooperation. The
agreement between different types of measurement is based upon
the validity of specific physical laws. A pendulum performs a certain number of oscillations, a spring watch unwinds by a certain
angle, a certain volume of water leaks out of a container, the human
heart beats a certain number of times, and an audience becomes
tired to a certain degree.

In order to judge whether a certain operational definition is "practical" or not, we must know the physical laws for the operations
involved. Therefore, every discovery of new physical laws urges
us to modify our operational definitions because such definitions are
practical only if they allow us to formulate the laws in a simple way.
A very simple example would be the discovery of the law that the
length of a solid body increases with temperature. Before we knew
this law, we could define the unit of time by the period of a pendulum
that had the length of one meter. This means again that the rod
of the pendulum was found to be equal to the standard meter in
Paris by direct coincidence of the ends; but because of the influence
of temperature, this definition would disagree with the definition of
the time unit by spring watches or the human pulse rate. At high
temperatures, there could be many more heart beats per unit of time
because the period of the pendulum would increase by the lengthening of the rod. In order to restore the unambiguity of the time
definition, we would have to modify the definition of the time unit
in the following way: The unit of time is the period of a pendulum

which can be brought into coincidence with the Paris meter-standard at the freezing temperature of water.   By using such an instrument of measurement, we can determine unambiguously how many units of time a certain lecture will last.   If we replace "freezing tempera-ture" in the definition by "boiling temperature," the same lecture will last a different number of time units.   In the first case, the lec-ture may last 3000 time units, in the second case, 3100 units.   We see in this example that operational definitions must be adjusted to the known physical laws.   The more laws that are known, the more complicated the definitions must become.   Instead of measuring a duration of time in time units without specification, we now have to measure in time units with respect to a certain temperature.

A very similar situation occurred when we concluded from the theory of relativity that the length of a yardstick and the rate of a clock depend upon the speed of these measuring instruments.   Then a statement like "The duration of this lecture is one hour" becomes ambiguous.   It says only that during the lecture the hand of a clock turns by an angle of 30°; to make it an unambiguous statement, we must state whether the clock used was at rest relative to (F) or had a certain speed q relative to (F).   If the clock was at rest relative to the vehicle F' (moving with a speed q relative to $[F]$), we formulate the result of our measurement briefly by saying: "The duration of this lecture is one hour relative to the system (F')."   The addition "relative to (F')" means a specification of the method of measure-ment like the addition of the temperature at which a measuring rod was found to coincide with the standard meter at Paris.   In the same way, a statement like, "This table has a length of one foot," must be made specific by an addition, "relative to a vehicle (F')," in order to make it unambiguous.   This addition means merely that the richer our knowledge of physical facts or physical laws, the more complex operational definitions have to become in order to formulate the new laws in a simple and practical way.

This state of affairs has frequently been described by saying: "Absolute length" is now a meaningless expression, and only "rela-tive length" has a meaning since it is helpful for the formulation of physical laws.   There is no objection to using such formulations if we understand them in the sense just described, as an advance in semantics necessitated by the advance in our knowledge in the field of physical facts and laws.   This formulation cannot be construed, however, as meaning that it is "impossible for science" to find the "real length" of a physical object and that the search for real length may belong in the field of "metaphysics" or a "philosophy of nature."[15]   The length defined in the theory of relativity is as "real" as the length defined in Newtonian mechanics.   In both cases, "length" is defined by an unambiguous operational definition, but because relativistic mechanics is more complex than Newtonian mechanics, the operational definition of "length" or "duration" is more complex too.   A definition of "length" defines a "real length" if it is useful for the formulation of physical laws.

We could, of course, ask whether one could define an "absolute length" and use this concept to formulate the physical laws implied by the theory of relativity.   This refers, in particular, to the laws about the dependence of time and space measurements on the speed of the measuring instruments.   We could, for example, call the length relative to our galaxy the "real length" or "absolute length" of a physical object, and then call the length relative to any other system of reference an "apparent length" or a "relative length." But in this case, we could not formulate the principle of relativity in the simple way in which it was advanced by Einstein: "The laws of physics have the same form relative to all vehicles (F') with a uniform rectilinear velocity q relative to (F'').   If we meant by velocity a velocity relative to our galaxy, such a brief formulation of relativity could not be given.

The "relativization" of space and time consists actually in the introduction of new operational definitions that are better adapted to the actual needs of the scientist.   The "relativization" of space and time is an advance in semantics and not, as has frequently been said, an advance in metaphysics or ontology.   We cannot say that "there is no real length" unless we start from an operational definition of "length."   If we compare different definitions of "length," we cannot judge them according to whether or not they meet the "real" conception of length, but only according to whether or not they are helpful in the formulation of the known laws of nature and in search for new laws.   It would be confusing to say that, according to the theory of relativity, "there *is* no real length," because the statement has no operational meaning unless we construe it to mean that the concept of "real length" or "absolute length" is not helpful for the formulation of general physical laws in a simple and practical way.   The same assertions hold for expressions like "absolute time" or "absolute velocity."

Our judgement about the usefulness of such expressions may change considerably if we consider not only the realm of physical facts in the narrower sense (*e.g.*, the motion of planets) but ask also for a general picture of the world and include the phenomena of human behavior as facts to be represented.   After Copernicus had advanced his system, everybody agreed that it was mathematically a simpler picture of the planetary system than the Ptolemaic one. On the other hand, it was also obvious that it was more complicated to make the Copernican system compatible with the generally recognized (Aristotelian) philosophy than the geocentric system.   This was a serious difficulty because Aristotelian and Thomistic philosophy were regarded as necessary for the formulation of religious and moral laws among men.   The belief that this world picture was supported by science gave to the believers a greater feeling of security. Whether fostering a feeling of security is of greater or smaller importance than providing simpler and more practical formulations of physical laws is a question that cannot be decided within science in the narrower sense; it does not belong in the domain of mathematical or physical problems.   The decision can only be made by investigations into the interaction between different branches of human action.   For example, we must investigate the relationship between man as a maker of science and man as a believer in political and religious creeds.   This means, in our case, that the usefulness of expressions like "absolute motion" cannot be judged from physics alone, but must depend on conclusions drawn from the "science of man," *e.g.*, from psychology or sociology.

8. The "Disappearance" and the "Creation" of Matter

Besides the doctrines about the "relativity of space and time," nothing in the theory of relativity has had so many repercussions in "philosophy proper" as its assertions that "matter" can disappear or can be produced.   We must not forget that the main target in the fight that "materialistic" science waged against traditional religion was the doctrine that "in the beginning God created matter from nothing."   Without going into the discussion of this philosophical conflict, we shall try to understand in what sense we can say that, according to the physical theory of relativity, matter can disappear or can arise from "nothing."

We learned in Chapter 4 that in Newton's mechanics the "quantity of matter" was measured by the mass of a body. We learned that to every material body a constant value, its "mass," could be assigned by a specific procedure of measurement. We learned also that this assignment is unambiguous only if Newton's laws of motion are valid; otherwise, there is no constant number that has all the properties that the "mass" of a body should have under the traditional definitions. We can easily see that the theory of relativity is not compatible with the assumption that Newton's laws of motion are universally valid—specifically, that they are valid for all velocities. We shall try to show, in a rather perfunctory way, that, in contrast to Newton's laws of motion, a material body cannot be accelerated to a speed that equals the speed of light or surpasses it. This follows simply from what we showed in Section 4.

We learned that from the theory of relativity, *i.e.*, from the simultaneous validity of the Principles I and II, it follows that these principles are compatible only if the rate of a moving clock lags behind a clock at rest in the ratio $1/\sqrt{1 - q^2/c^2}$, if q is the speed of the moving clock and c the speed of light; but if q surpasses the speed c of light, we would have to say that $q/c > 1$ and $\sqrt{1 - q^2/c^2}$ would become imaginary. It would be impossible to find any change in the rate of a clock that would make the Principles I and II valid simultaneously. Therefore, if we assume the validity of these principles, we exclude the possibility that a material body could be accelerated to the speed of light relative to (F). This would be, however, in flagrant contradiction to Newton's laws. If a constant force f is acting upon a mass m which has a velocity v, the "quantity of motion" (mv) increases per unit of time by a quantity $\Delta(mv)$ that is equal to the force $\Delta(mv)/t = f$. If m is constant, this means $m\Delta v = ft$. If t is sufficiently great, we can reach any increase of velocity $\Delta v = ft/m$. Therefore, we can accelerate a material body to any speed and, therefore, certainly to a higher speed than the speed of light. This means that from the validity of the principles of constancy and relativity we can conclude that Newton's laws of motion cannot be universally valid. They cannot be valid for a body that has a speed that is comparable to the speed of light or, in other words, they can only be valid for small velocities, where "small" means "small relative to the speed of light."

The definition of mass (quantity of matter) in Newton's mechanics is based upon the validity of Newton's laws. Only under this condition is the ratio of the accelerations of two bodies under the same circumstances constant. If we choose one of these bodies as having unit mass, this ratio is the "operational definition" of "mass"; but if this ratio depends upon the actual speed of the body, there is no constant m that has the property of being equal to the ratio of accelerations. This means that the Newtonian definition of mass has no counterpart in nature. It loses its usefulness as a term in the description of physical phenomena. It has often been said that the theory of relativity has "proved" that the mass of a body is a function of its speed. This method of speaking is slightly misleading, and has often been the basis of criticism directed against the theory of relativity. The correct way of describing the situation is approximately as follows: The operational definition of mass which has been used in Newtonian mechanics loses its usefulness and must be dropped. In order to keep up the continuity of physical science, we introduce again a term "mass," which is defined by an operational definition which cannot be identical with the definition of mass used in Newton's mechanics.

Since for "small" velocities of a body, Newton's mechanics is assumed to be valid, we could retain the traditional definition of mass, but restrict it to velocities that are "small" (compared with the speed of light). Then mass is again constant and connected with force by $m = f/a$, or $ma = f$. We can say that mass in the new sense is the resistance of a body against a change of velocity under the assumption that the actual velocity is vanishing or very small; it is the resistance against acceleration from rest and is called "rest-mass" (m_0). However, this rest-mass no longer has the property $ma = f$ when the velocity increases and becomes comparable with the speed of light. We know that for high speeds the ratio f/m must decrease, which is incompatible with the constancy of m_0. If we want "mass" to retain the property of being a ratio of accelerations as it was in Newton's mechanics (Chapter 4, Section 7), we must drop the property of constancy and assume that "mass" increases with velocity.

However, if we go a little further into the conclusions drawn in the theory of relativity, we soon note that the ratio f/a is not the same if the force has the direction of the actual velocity as if the force is perpendicular to the actual velocity. Which ratio should be called "mass?" When Newton introduced the conception of a "quantity of motion" or "momentum" defined by mv (mass times velocity), he could derive from his laws that the sum of all momenta (Σmv) in a system remains constant if no external forces are acting upon the system. Because of the great role of this law (conservation of momentum), it was decided to give the name "mass" to the ratio f/a, in the case where f is normal to the actual velocity, because this definition of mass "m" is identical with the definition that the sum of all $mv(\Sigma mv)$ shall be preserved through all interactions within a system, provided that no external forces are acting. The mass m defined in this way is called "relativistic mass." It increases with the increase of speed and is computed from the rest-mass m_0 and the speed of light c by the formula $m = m_0/\sqrt{1 - v^2/c^2}$. If v is not very great, this is approximately $m = m_0 + K/c^2$ where K is the kinetic energy $m_0 v^2/2$ of the particle.

These conclusions were drawn even before Einstein advanced his theory, from the electromagnetic theory of matter by J. J. Thomson,[16] H. A. Lorentz,[17] and M. Abraham[18] about the year 1900. But at that period, Newtonian mechanics was so taken for granted that to speak of a "mass" that was not a constant seemed to be abstruse or to be a violation of a parlance deeply imbedded in our way of speaking about the physical world. To the mass that was dependent upon the speed v was given the name of "apparent mass" in contrast to the "real mass" that was identified with the rest-mass. Actually, the question of which is the "legitimate successor" of the Newtonian "quantity of matter" cannot be decided. The "rest-mass" has inherited the property of "constancy" while the "relativistic mass" that is defined by f/a has inherited the property of being the ratio of force to acceleration. Hence, the question of which of them should be declared the "legitimate heir" of the Newtonian mass can only be decided upon the grounds of convenience, simplicity, and similar types of consideration.

The reasons for giving the "relativistic mass" $m = m_0 + K/c^2$ the traditional name of mass have become even stronger since so much interest has centered upon the conversion of rest-mass into energy occurring in nuclear processes. If we consider phenomena like the fission of a uranium atom, we can ask whether, in a split of a uranium nucleus into fragments, the sum of the rest-masses or the sum of the "relativistic masses" is being conserved.   We know very well today, not only as a conclusion from the theory of relativity, but also from direct measurement, that the fragments of the uranium nucleus have together a smaller sum of rest-masses than the original uranium nucleus.   We can show that it is not Σm_0 that has been conserved during the fission process, but $\Sigma m = m_0 + 1/c^2 \Sigma K_0$. This means that if ΣK_0 increases, Σm_0 must decrease.   This is what actually happens: By fission an increased kinetic energy of the fragments is produced and therefore a small loss of rest-mass occurs. The sum of rest-masses (Σm_0) is not conserved during an internal reaction.   Therefore, "rest-mass" has not the main property ascribed by traditional mechanics to "mass"; if "rest-mass" were identified with "mass," the fundamental law of the "conservation of mass" would not hold.   Again, we have the choice of whether to drop this law or to drop the identification of "mass" with "rest-mass."   The latter parlance leaves more theorems of Newtonian physics unchanged than the first one.   All these considerations show us that if we introduce "mass" as the object which has as many as possible of the properties of the old Newtonian mass, this is the only possible justification for the introduction of statements like "mass is not constant," or "mass can disappear."

6

Four-Dimensional and Non-Euclidean Geometry

1. The Limitations of Euclidean Geometry

We have seen that from Einstein's two principles[1] (constancy and relativity) we can draw conclusions concerning the phenomena which occur in a rigid body that is moving at high speed. If we assume that these principles are compatible, we must also assume that a rigid body moving with a speed v in a certain direction with respect to F will become shorter in the direction of motion but will not alter its cross section perpendicular to the direction of motion.

Very soon after Einstein had advanced his theory, it was shown that these results were in disagreement with the properties of a rigid body that had previously been taken for granted. This can be shown by considering a rigid circular disc that is rotating with constant angular velocity ω around an axis that is perpendicular to the disc and pierces the center of this circle. If the radius of the disc is r, every point of the perimeter will move with a speed $v = r\omega$. Let us consider now the part of the disc between the edge and a circle of radius r' which is a little smaller than r. This part has the form of a circular ring. Let us consider now a segment of this ring that is so short that it can be regarded as approximately rectilinear. The motion during the rotation is, during a small time interval, approximately a rectilinear motion of a rod. According to Sections 6 and 7, of Chapter 5, this rod is contracted in the ratio $\sqrt{1 - v^2/c^2}/1$, where

149

$v = \omega r$. Hence the whole perimeter of the circle is contracted in this ratio.

Next we shall consider a part of the disc which consists of a narrow strip around a radius of the disc (the spoke of a wheel). During the rotation of the disc this strip behaves as a rod that moves in a direction normal to its length; hence the length is not affected by the rotation. Here we have the case that the perimeter P of a circle of the radius r is shortened by the motion while the radius itself is not affected. The greater r is, the smaller the ratio P/r becomes. For very small radii, $v = r\omega$ is small compared with c and the contraction can be disregarded. This means that for small r, the ratio P/r has the value 2π(where $\pi = 3.14159$). In Euclidean geometry P/r has this value for all possible values of r. Therefore, a rotating disc of a material which we describe in traditional geometry as "rigid" does not obey the Euclidean laws of geometry if we define length, as it is defined in Euclidean geometry, by the juxtaposition of rigid measuring rods.

This consideration was advanced very soon after Einstein had proposed, in 1905, his theory of relativity. The conclusion was drawn that the theory of relativity was absurd since it was incompatible with plane geometry, but Einstein argued that it must be concluded that in a rotating rigid body Euclidean geometry was not valid. In other words, there are no "rigid bodies" that are rotating relative to the inertial system if we define a rigid body by the property that it obeys Euclidean geometry. In the conception of a rigid body, we have a similar situation as in the conception of "mass." There is no body that would have all the properties that traditional physics and geometry ascribe to a "rigid body." If we add to the axioms of Euclidean geometry an operational definition of "straight line," then the axioms become physical statements. We may choose as the operational definition of a straight line the "edge of a rigid cube," where the latter is defined by the technological rules by which such a cube is produced. Then the axioms and theorems of geometry are statements about the behavior of rigid bodies. But if we do not introduce the technological rules of manufacturing, the axioms of geometry can be regarded as definitions of a "rigid body." Then, according to traditional physics and geometry, there are empirical bodies that by these definitions are "rigid bodies." According to our argument at the beginning of this section, however, a rotating disc does not fulfill these laws, and cannot be regarded as a rigid body. This means, further, that there is no rigid body that would fit the old definition, according to which a rigid body should be rigid under all circumstances, at rest or in rotation. This means that it should obey the Euclidean axioms under all circumstances.

If we assume, now, the validity of the theory of relativity,[2] a rigid body in the old sense exists only under very special circumstances; specifically, a body can be rigid if it is at rest relative to the fundamental system. Now, however, we can give a new definition of a rigid body, which would be identical with the old definition if the body were at rest, while for a rotating body the Euclidean axioms would be replaced by the axioms of non-Euclidean geometry. The departure from Euclidean geometry would be different in different parts of the rotating body. Near the axis of rotation the departure would be small, whereas far from the axis the departure could be considerable. Again we are in the same situation as in the case of "length" and "mass." The Euclidean axioms and the laws of motion in the theory of relativity are not compatible with each other. We have, therefore, the choice: either we keep the old definition of rigid bodies, which cannot then cover rotating bodies; or we start from the behavior of rotating bodies, in which case we must introduce new rules for the behavior of rigid bodies, *i.e.*, non-Euclidean geometry.

Again we may ask which definition is the definition of a body that is "really rigid?" We might say: A body is "rigid" if it has all the properties that the traditional definition ascribed to a "rigid body." Then we would have to say that it is "not rigid" when in motion. We could also call a body "rigid" if it fulfilled at rest the axioms of Euclidean geometry, but if when rotating it fulfilled the theorems of non-Euclidean geometry required by the theory of relativity. In this case, such a body would be "rigid" under all circumstances.

2. Relativity of Acceleration and Rotation

According to the theory of relativity, the uniform motion of a room relative to the inertial or fundamental system (F) does not produce phenomena relative to this room from which the speed v of the room relative to (F) could be computed. We found, on the other hand, that an accelerated or rotating motion of a vehicle (F') could be detected by observing phenomena relative to (F'). We could, for example, observe the effect of centrifugal forces, Foucault's pendulum (Coriolis force) and, as we learned in Section 1, by checking the validity of Euclidean geometry on a rotating disc, we could find the angular velocity of the disc. We must, obviously, measure the ratio, perimeter/radius (P/r), for radii of different lengths. The changes of P/r would increase in proportion to the angular velocity.

Since the hypothesis of a quasi-material medium, the ether, had been dropped, rotation relative to (F) meant nothing but rotation relative to empty space. According to the fundamental idea of relativity, a uniform motion of a vehicle relative to empty space could have no effect upon the laws of physics relative to this space, but it seemed that a vehicle in accelerated or rotating motion relative to empty space would show an influence of this motion upon the laws of physics in this vehicle. This seemed not impossible to Newton. He regarded "empty space" as a cosmic object, as identical with the sensorium of God, and thought it very plausible that an acceleration relative to this important object should have observable consequences; but scientists who preferred to ban metaphysical and theological argument from physics would not accept this method of accounting for the effects of rotation and acceleration. In 1872 Ernst Mach[3] published a critical analysis of Newtonian mechanics in which he stressed the point that our experiences with centrifugal and Coriolis forces produced by the "rotation of the earth" actually prove that these effects can be computed from the angular velocity ω of our earth relative to the system of fixed stars (our Milky Way) and do not compel us to introduce the concept of a rotation relative to an empty space or "absolute space."

Mach suggested, to put it briefly, the reformulation of Newton's laws of motion by replacing the "absolute space" or "sensorium of God"[4] by the system of the constellations of fixed stars, regarding it as a rigid system of reference. Then the law of inertia would state that a body upon which no force is acting will move in a straight line at constant speed relative to the fixed stars. If, on the other hand, a vehicle rotated, as our earth does, the rotation relative to the fixed stars would produce centrifugal and Coriolis forces relative to the vehicle. The rotation of the plane of the Foucault pendulum would, for example, be ascribed to the rotation of the fixed stars around the earth. Since, according to Newton's theory of planetary motion, the forces exerted by the fixed stars upon our earth are negligible and, moreover, do not have the direction of centrifugal force, Mach's suggestion amounts to the following hypothesis: The fixed stars exert upon the material bodies on our earth an influence that cannot be accounted for by Newton's law of gravitation. This suggestion seemed at that period very daring and even absurd. The great German physicist, Max Planck,[5] the originator of the most revolutionary hypothesis of the twentieth century, rejected Mach's reformulation of Newtonian mechanics emphatically. He had definitely approved Einstein's theory of relativity, but he regarded Mach's theory, according to which the rotation of the Foucault pendulum is due to an action emanating from the fixed stars, as a fantastic assertion which has its source in Mach's theory of knowledge. Planck regarded Mach's theory of the Foucault pendulum as a definite proof of the absurdity of his philosophy.[6]

However, Einstein started[7] a new analysis of Newtonian mechanics which eventually vindicated Mach's reformulation. While the traditional presentations have regarded the gravitational field as just one among the fields of force for which Newton's laws are valid, Einstein pointed out that motion in a field of gravitation is in many respects different from motion in electromagnetic and other fields and that, in particular, motion in a gravitational field is analogous to motions relative to accelerated or rotating vehicles. Einstein made his start from the "homogeneous" field of gravitation, where the forces have the same direction and intensity everywhere. This is approximately the case in every living room or laboratory. Since Galileo, it has been known that a mass m moves downward everywhere with the same acceleration g, whatever the value of the mass m may be. From the general Newtonian formula $a = f/m$, it would follow that the acceleration is in inverse ratio to the size of the mass. If acceleration is to be independent of the mass, we must assume that the force f is proportional to the mass; this means that $f = mg$, and hence $a = f/m = mg/m = g$. In any other field, the force f is determined by the field and is independent of the mass upon which it acts, but in the gravitational field, the force is proportional to the mass m and produces, therefore, an acceleration a that is independent of the mass.

Therefore, in the homogeneous field of gravity, the law of motion is purely geometrical. We can predict the geometrical form of the trajectory of a launched projectile from the initial conditions without knowing the mass m, but in the electromagnetic field, for example, the law of motion is a "dynamic" one; we cannot predict the geometrical form of the trajectory without knowing the mass of the body upon which the field is acting. This characteristic of motion in the field of gravity is also the characteristic of motion relative to an accelerated or rotating vehicle (F'). If the vehicle is an inertial system, and no force is acting upon a mass m, we can predict the geometrical form of the trajectory; it will be a straight line, whatever the mass may be. If the vehicle has a constant acceleration relative to the inertial system, the trajectory will be a parabola whatever the mass. If the vehicle is rotating, the mass will traverse trajectories determined by "centrifugal forces." In any case, if no force is acting, the geometrical form of the trajectory is determined by the acceleration of the vehicle, while the mass of the moving body is irrelevant. We can see now that the behavior of a mass relative to an accelerated vehicle, if no force is acting, is completely analogous to the behavior of a mass relative to an inertial system under the impact of a gravitational field. These special cases were, for Einstein, the basis for advancing his general *Principle of Equivalence*. It states that every motion relative to an accelerated vehicle (in the widest sense of the word) can also be interpreted as a motion relative to an inertial system under the influence of a gravitational field.

This principle now connects two problems: the general theory of motion in a gravitational field, and the general theory of motion relative to a vehicle (F') that is not an inertial system. The theory which offers the solution of these problems is called the *General Theory of Relativity*,[8] which is also the general theory of gravitation. In the "restricted" theory of relativity, the speed of a vehicle could not be computed from the observation of physical phenomena relative to this vehicle. In the "general" theory of relativity, the acceleration or rotational velocity cannot be computed either. For, according to the principle of equivalence, every phenomenon that can be ascribed to the acceleration of the vehicle can also be ascribed to a gravitational field. If we observe a phenomenon of centrifugal force, for example, the flattening of a rotating fluid, we can say equally well that it is rotating relative to the inertial system, or that it is at rest in the inertial system and is flattened by the gravitational force of the rotating sphere of fixed stars.

According to Newton's mechanics and Einstein's "restricted" theory of relativity, the speed of a vehicle (F') relative to (F) has no influence upon the mechanical and optical phenomena relative to the vehicle (F'). This state of affairs is often formulated by saying: "There *is* no absolute speed," or "Every speed *is* relative." According to Einstein's general theory of relativity, the rotation or acceleration of a vehicle (F') cannot be computed from the phenomena relative to the vehicle (F'). This state of affairs, in turn, has been described by saying: "There *is* no absolute acceleration or no absolute rotation." This way of speaking is analogous to some statements of everyday life, such as: "There *is* no three-headed calf." By speaking in this way, the theory of relativity seems to deny the existence of some "entities," to impoverish our world, and to make the statements of physical science more vague and less direct. Actually, a statement such as "There *is* no absolute rotation" means exactly that there is no opportunity for using the term "absolute rotation" in a discourse which intends to give a simple formulation of the laws of physics. This point is particularly important for the understanding of "philosophical" discourse. One often reads statements like "There *is* a soul" or "There *is* no matter," or "There *is* no free will," all of which have the syntactical form of "There *is* no three-headed calf." If we learn the actual physical meaning of "There *is* no absolute speed" or "no absolute rotation" by understanding the theory of relativity, we shall also learn how to understand the meaning of the philosophical statements about matter and soul. The meaning of statements about the mind or soul was recently investigated and lucidly presented by Gilbert Ryle,[9] an Oxford philosopher.

3. Curvature of Space

In Section 1 we learned that a rigid disc which is rotating does not obey Euclidean geometry; the greater the angular velocity, the greater the departure from Euclidean geometry. This means[10] that the greater the angular velocity, the smaller the area of the unit triangle. Moreover, if the angular velocity is given, the departures grow larger in proportion to the increase in the linear velocity $v = r\omega$. This means that the departures from Euclidean geometry in a domain of the disc (rotating with an angular velocity ω) are greater the more removed this domain is from the axis of rotation. This means again that the area of the "unit triangle" depends upon its distance from the axis. While in the non-Euclidean geometry discussed in Chapter 3, Section 6, the unit triangle has the same size in the whole plane, in our case it is different in different domains of the plane. The departure from Euclidean geometry is measured by the departure of the angle sum of a triangle from two right angles. If the angles are α, β, and γ, the "defect" Δ is defined by $180° - (\alpha + \beta + \gamma)$. Since the defect depends upon the area of the triangle, we define a quantity Δ/A, where A is the area of the triangle with the defect Δ. $K = \Delta/A$ is called the "curvature of space." This measure for the departure from Euclidean geometry depends only on the distance of the domain from the axis and not on the area.

We learned previously that by rotation of a rigid body the geometry within this body becomes non-Euclidean. But we learned in Section 2 that every effect that can be produced by the rotation of a body can also be produced by a field of gravitation within a body that is not rotating. As an example we discussed the centrifugal phenomena on the earth that are usually described as the effect of the earth's rotation with respect to the fixed stars. According to Einstein's principle of equivalence, they can also be regarded as the effect of rotating masses around a resting earth. In the same way, the departure from non-Euclidean geometry in a rotating disc can also be accounted for by assuming that the disc is at rest but large masses are rotating around the disc and producing a field of gravitation. Then the departures from Euclidean geometry are the effect of this gravitational field. In the vicinity of large masses where the gravitational field is strong there will be a great curvature of space; the unit triangle will be small.

We must avoid some misunderstandings which have arisen concerning the effect of gravitation on the geometry of space, and which have their origin in a philosophy that is separated from science. The first difficulty has its source in the expression "curvature of space." It is admitted that a "surface" could be "curved in space," *i.e.*, it could depart from a plane surface; but how could the three-dimensional space itself be curved? This difficulty springs from the ambiguous meaning of the word "curvature." If we examine the surface of a sphere, we can measure its curvature in two ways. We can measure it by measuring the departure of the surface from a tangential plane, but we can also consider a triangle drawn upon the sphere, and measure the departure of the sum of the angles $(\alpha + \beta + \gamma)$ from two right angles. On the sphere, $(\alpha + \beta + \gamma)$ is greater than two right angles. The "excess" of the triangle $(\alpha + \beta + \gamma) - 180°$ divided by its area is equal to the curvature of the spherical surface. We can, therefore, measure this curvature in two ways. If we consider a physical triangle (made of light rays or rigid rods) in space, we can measure the "defect" or "excess" in different places in the space and obtain in this way the "curvature" of the space. But a direct measurement of this curvature, like the curvature of a curved surface, is not possible. I can compare a curved surface with a plane since both are situated in the same three-dimensional space. I can observe in this way a departure of the curved surface from the plane and call this departure "curvature," but I cannot observe besides our "three-dimensional curved space" a "three-dimensional plane space," both situated in the same "four-dimensional space" and find a departure of the curved from the plane space. I can, however, in the "curved three-dimensional space" construct surfaces and triangles upon these surfaces. Then I can measure the angle sum of these triangles, and find out whether their angle-sum is 180° independent of their size. If there is a departure of $(\alpha + \beta + \gamma)$ from 180°, a "defect" or "excess," even if the surfaces are as "plane" as possible, we would say that our "space" is "curved."

The curvature of a three-dimensional space means, therefore, the "defect" or "excess" of triangles and, in other words, the departure from Euclidean geometry. The "curvature" is observable and measurable by a method which is also used to measure the curvature of a spherical surface in our ordinary space. But the second kind of measurement, the departure from a plane, cannot be applied to a "curved space." We have again the situation that after the "discovery" of the fact that in a gravitational field Euclidean geometry is not valid, we introduce the term "curvature of space" in order to describe the gravitational field in a convenient way. This "curvature" has, as we learned, a precise operational meaning. The "curvature of space" can be measured by different kinds of operations which all yield the same result, but only a fraction of these definitions are analogous to the operational definition of curved surfaces.

It would be very misleading to say that "curvature of space" introduces an element into physics that cannot be described by measurements performed upon ordinary bodies made of steel, wood, or stone. There have been attempts to interpret the introduction of curved spaces as an introduction of spiritual elements into physics.

4. Is the World "Really Four-Dimensional?"

If we keep to the "classical" system of physics based on Newtonian mechanics, we can describe everything that happens in the universe by "point-events." Starting from a Cartesian system of reference (F), each event takes place at a certain point x, y, z, at a certain time t shown by a clock located at x, y, z. We say that this "point-event" has the event-coordinates x, y, z, t relative to (F). The theory describes the motion by expressing x, y, z as functions of t; this means a motion along a curve in the ordinary three-dimensional space. A certain value of t is assigned to every point x, y, z. We can also interpret the equation $x = x(t)$, $y = y(t)$, $z = z(t)$ as a curve in the four-dimensional x, y, z, t space. Motion along a curve in the three-dimensional space is mathematically equivalent to a static curve in the four-dimensional space. In this sense, the great French mathematician Lagrange[11] had already called mechanics a "geometry in four dimensions." Every mass-point traverses a trajectory in our three-dimensional x, y, z space. At a certain instant of time $t = t_0$, the masses fill a certain part of our three-dimensional space, at another instant they occupy other parts of this space.

Describing the same facts in a different way has been suggested. We picture every "event" by a point in the four-dimensional x, y, z, t space. This is possible because every event is described by a quadruplet of values ascribed to x, y, z, t. Then every position of a mass point at a point x, y, z, at a time moment t is an "event." We could argue that the four-dimensional continuum with all its events has existed since eternity. Our life is only a change of place (an actual change of the t = const. plane) in the four-dimensional space where we shall meet the events which have awaited us. We can compare the situation with the case of a motion picture film that does not move. Instead of having the film move relative to the observer, the observer would move relative to the film, and have the same experiences as if the film moved as usually.

It has been suggested that we might speak about our universe by asserting that the "real" universe is a four-dimensional one that exists now in this very moment. This means that the future "exists" now, and all that we do during our lifetimes is to move through the four-dimensional continuum and become aware gradually of its three-dimensional cross sections. Obviously, one cross section of the four-dimensional corresponds to every instant of time t. This way of speaking has one obvious disadvantage: If "now" means $t = t_0$, then a future instant of time $t = t_1$ defines a different cross section of the four-dimensional continuum. To say that the four-dimensional continuum "exists now" implies that all cross sections "exist now" or, in other words, that the cross section $t = t_0$ is identical with the cross-section $t = t_1$. Otherwise, it could not exist "now." If we allow for this confusing way of speaking, the assertion that the "four-dimensional space-time continuum" has always existed and we are merely traveling through it asserts no more than the statement that the three-dimensional space continuum changes in time.

When the theory of relativity was introduced by Einstein in 1905, it was soon found that the principles and theorems of this theory could be formulated very conveniently by using the four-dimensional space-time continuum, the world of events. Hermann Minkowski[12] noticed and presented this emphatically in 1908. Let us consider two events x_1, y_1, z_1, t_1, and x_2, y_2, z_2, t_2. We mean by x_1, y_1, z_1; x_2, y_2, z_2 coordinates relative to a system of reference (F) that is measured by yardsticks at rest relative to (F). In the same way, t_1 and t_2 are time distances measured by clocks at rest in the system (F), but this does not say that the events themselves are tied somehow to the system (F). They may be lightning strokes in space that coincide, for example, at the time t_1 with a point x_1, y_1, z_1 of (F). The two events are separated by a spatial distance

$$S = \sqrt{(x_2 - x_1)^2 + (y_2 - y_1)^2 + (z_2 - z_1)^2}$$

and a temporal distance $T = (t_2 - t_1)$. What is the "operational definition" of $(t_2 - t_1)$? The symbols $(t_2 - t_1)$ denote the difference of the readings of two identical clock mechanisms that are at rest in (F). They may be checked by being put first at the origin O of (F) and found to be moving at the same rate, but how are they brought to the points x_1, y_1, z_1 and x_2, y_2, z_2 respectively? We know (from Chapter 5, Section 7) that clocks change their rate according to the speed with which they move. We try, therefore, to bring the clocks to their points x_1, y_1, z_1 and x_2, y_2, z_2 by moving them very slowly with a speed that is almost zero. If we use such clocks, the laws of physics become simple and practical. A body upon which no force is acting moves from P_1 to P_2 with constant speed, a light ray emitted at P_1 is propagated to P_2 with the constant speed c, etc., etc. We say then that these clocks are synchronized. In the description of the method of how to synchronize clocks, however, we spoke about clocks that were moved infinitely slowly relative to the system of reference (F). Therefore, the definition of "synchronized" becomes unambiguous only if we specify it by calling it "synchronized with respect to the system (F)."

If we consider now a vehicle system (F') which has the speed q relative to (F), we can describe the same event relative to (F'). This means that the yardsticks and clocks are at rest in (F'). We denote the space-time coordinates of the same two events relative to (F') by x_1', y_1', z_1', t_1' and x_2', y_2', z_2', t_2'. Again we denote the distance $S' = \sqrt{(x_2' - x_1')^2 + (y_2' - y_1')^2 + (z_2' - z_1')^2}$ as the "spatial distance" and $(t_2' - t_1') = T'$ as the "temporal distance" between the two events. According to Einstein's principles of the theory of relativity, each light ray has the speed c relative to (F) and also relative to (F'). Therefore, we have $S/T = c$, and $S'/T' = c$. This means that if $(S^2 - c^2T^2) = 0$, we know that also $(S'^2 - c^2T'^2) = 0$. It is easy to show that this can only be so if $(S^2 - c^2T^2) = (S'^2 - c^2T'^2)$. We can single out pairs of events which, relative to (F), have no spatial distance $(S = 0)$ and which have no temporal distance $(T = 0)$. It is obvious that because of the influence of motion upon lengths of yardsticks and rates of clocks, the spatial distance S with respect to (F) is not equal to the spatial distance S' relative to (F'). In the same way, T and T' are different from each other. If we assume that S and T are different from each other, the two events have a spatial and temporal distance relative to (F), but we can choose the speed q of a vehicle system (F') such that S' or T' may be zero. In the first case we have $(S^2 - c^2T^2) = -c^2T'^2$. In this case, the two events happen at one and the same point of (F') at different instants of time $(t_2' - t_1') = T'$. Such a vehicle can obviously only be found if $(S^2 - c^2T^2) < 0$ or $S < cT$. But if $(S^2 - c^2T^2) > 0$, we can find a vehicle (F') such that $T' = 0$ and $(S^2 - c^2T^2) = S'^2$. In this case, both events happen relative to (F') at one and the same instant of time $(t_2' = t_1')$ at a distance S' from each other. This means that if two events happen simultaneously (at $t_1' = t_2'$) relative to (F'), there is a temporal distance between them relative to (F) $(t_2 > t_1)$.

We can adequately describe these facts by using the four-dimensional space-time continuum. If we have two events, and we call them two points in this four-dimensional continuum, we can speak about these events in a way similar to that in which we speak about two points in the ordinary three-dimensional space. We speak, for example, about two points in a plane, the x, y plane. The two points have the rectangular coordinates x_1, y_1 and x_2, y_2. If we say that both have the same x-coordinate $(x_1 = x_2)$, this says nothing about the two points. We can choose another coordinate system (by a rotation of the original one) and find that for the coordinates x_1', y_1' $x_2'y_2'$ with respect to the new system, x_2' is different from x_1'. Saying that $x_1' = x_2'$ says something about the relation of the points to a specific coordinate system, but nothing about the points themselves. In the same way, if two events (points in the four-dimensional space) have no time-distance $[T' = (t_2' - t_1') = 0]$ relative to (F'), this is merely a statement about the relation of the two events to (F'), but says nothing about the events themselves. Just as for points in a plane, we could introduce a system of reference (F) for which $t_2 > t_1$.

Hence, by using the four-dimensional way of speaking we can describe the facts of relativistic physics in a simpler and more elegant way than by separating time and introducing a three-dimensional space. We find that the spatial distance S, as well as the temporal distance T, of two events, depends on the system of reference. Either of them can even disappear if we choose a certain system of reference. The combination $(S^2 - c^2T^2)$ of the two distances, however, has the same magnitude relative to all vehicle systems (F'), whatever the value of q may be. Since the description of events in the language of the "four-dimensional space" is more convenient than in the language of a three-dimensional space, we are inclined to say that the four-dimensional space is more "real" than the three-dimensional one. While S and T do not "exist" independent of the system of reference, the combination $(S^2 - c^2T^2)$ does "exist" on its own. H. Minkowski wrote in 1908, that a combination of space and time, the four-dimensional continuum, is what "really exists," while time and space, if separated, are only "appearances." But even if we call the four-dimensional space-time continuum more real than time or space separately, this means not more and not less than to say that this formulation is a practical and convenient presentation of the theory of relativity.

There is no doubt that if we call the four-dimensional space-time continuum a "reality," we are encouraged to adopt Lagrange's assertion that mechanics is a four-dimensional geometry, and to say that the four-dimensional continuum "exists now," and that therefore all future events exist now, and the "future" consists in our moving through the four-dimensional space-time continuum. But exactly as before Minkowski's formulation of the relativity theory, we must also admit that the use of the word "now" in the formulation is rather misleading. By "now" we mean the cross section of the four-dimensional space-time continuum that is defined by $t = t_0$. Therefore, it is self-contradictory that any future instant of time $t > t_0$ can exist "now."

Use has often been made of this four-dimensional space-time continuum in order to "prove" that the future is "predetermined." If an event E happens with respect to (F) at $t = t_0$, the same event can happen with respect to (F') at an earlier instant of time $t' > t_0$. This state of affairs has been expressed thus: A small interval of time before t_0 nobody knew whether or not E would happen; as a matter of fact, however, it had happened in (F'); therefore it was predetermined that it would happen later in (F). Everything that is to happen in the future has, in fact, already happened, and can therefore not be prevented from happening. This would be a doctrine of "predetermination" or "predestination" in its most radical form. As a matter of fact, however, the event E happens only once, and the real state of affairs is simply as follows: This event coincides with a certain point of (F) when the clock which is at rest at this point shows $t = t_0$, while the clock at rest in (F') that coincides with the same event shows during the coincidence a time $t' < t_0$.

The four-dimensional formulation of relativity is a useful instrument for the presentation of physical events, but it cannot be interpreted in our everyday language by simply speaking about the four-dimensional space-time continuum as we have been accustomed to speak about our ordinary three-dimensional space.

7

Metaphysical Interpretations of Relativistic Physics

1. Metaphysical Interpretations of "Inertia"

There is no doubt that Newton's theory of motion has been very useful for deriving the observable motions of material bodies and for operating mechanical devices. There has always been the question, however, of whether or not these laws actually "explain" the observed motions, whether or not they are "intelligible" in the Aristotelian sense. Unless these requirements are met, Newton's laws of motion are liable to the general objection that science tells us nothing about real causes and that it gives us only formulae which are of practical value but are by themselves senseless and, as Ralph Waldo Emerson put it, "inhuman." It is no wonder that again and again there have been attempts to show that Newton's laws of motion or the law of the indestructibility of matter can be derived by "seeing with the intellect" or "metaphysical insight." It is particularly instructive to investigate the results of this "metaphysical intuition" in order to discover whether or not they are perhaps, as in other cases, actually the results of attempts to understand Newton's principles of mechanics by analogies with experiences that are familiar to us from our daily life.

Aristotle advanced a proof that the statement, "A body launched in any direction will, if no external force acts upon it, move along a straight line into the infinite with constant speed," is absurd and contradictory to intelligible propositions.[1]   According to Aristotle's physics, the speed of a body is inversely proportional to the density of the medium in which it moves.   If a body should move into empty space, the density of the medium would become zero and, therefore, the speed infinite; but then the body would reach great distances in no time, which is absurd.   However, when Galileo and Newton advanced the law of inertia as a fundamental principle in mechanics, it became possible to derive a great many facts that could be checked by experiment.   Newton's mechanics became the cornerstone of all astronomy and all engineering mechanics.   It still remained to be shown, however, that these principles were not of the "inferior" type (in the sense that St. Thomas would have considered them so[2]) that could only be confirmed by their consequences and not by reasoning.   There have been repeated attempts to prove that the principles of Newton's mechanics are "intrinsically clear" or, in other words, that they can be "seen by the intellect."

In order to acquire a good understanding of how the principles of mechanics have been "proved" by "seeing with the intellect," we may discuss two characteristic examples: the law of inertia and the law of the indestructibility of matter.   We shall study the arguments of two philosophers of very different types: Immanuel Kant[3] and Herbert Spencer—the first a so-called critical idealist, and the second a strict empiricist who would even be called by some a materialist.   Kant[4] attempts to prove the intelligible character of the law of inertia, which he formulates as follows: "Every change of matter has an external cause," and this he regards as equivalent to Newton's formulation.   Before advancing the proof itself, Kant says: "We take over from general metaphysics the proposition that every change has a cause; at this point we have only to prove that every change of matter must have in every case an external cause." This proof is given in the following way:

Matter, an object of the external senses only, is determined only by the external conditions in space and suffers no change except by motion.   Hence (according to the principle of Metaphysics), a change of one motion into another one, or from rest into motion, must have a cause.   But the cause cannot be an internal one because matter is not determined by internal reasons.   Hence, any change in it has an external cause, *i.e.*, it remains at rest or continues moving with constant speed if not forced by an external cause.

If we compare this argument with the approach of modern science to the law of causality,[5] we see that Kant's syllogism is not very convincing. Everything depends upon what we understand by a "state of motion." If we call a "change of place" a change of motion, then uniform motion without permanent external cause would be impossible. But if we understand by "state of motion" just "speed," we can prove that only a change of speed needs an external cause. But to identify, as Newton did, "state of motion" with "velocity" or "speed" is a physical hypothesis which can be confirmed by its consequences, but not by any metaphysical intuition.

Since Kant obviously felt that his argument, although it has the form of a logical conclusion, did not sound very convincing (probably because terms like "change of motion" were used without adding operational definitions), he added to his proof a paragraph of "remarks" by which the proof would be made more convincing. The interesting point about these "remarks" is the fact that they contain analogies between moving bodies and some very familiar common sense statements taken from daily life. Kant wrote:

> The inertia of matter is nothing but, and does not mean anything but that which is lifeless. . . . *Life* means the quality of a substance to determine itself by an internal principle to act or the quality of a material substance to determine itself to motion or rest, as a change of its state. We don't know any other internal principle of a substance, to change its state, but desire and generally no other internal activity but thinking with everything connected with it, the emotion of pleasure and displeasure, of concupiscence or will. These motives and actions do not belong to what is given by the external senses and therefore not to the qualities of matter as matter. Therefore, all matter, as such, is lifeless. This is what the law of inertia says, and nothing else. . . . Upon the law of inertia (besides the permanence of substance) the possibility of science proper is based. The opposite of it, and therefore the death of all Natural Philosophy, would be Hylozoism (the assumption matter has life). From the same conception of inertia, as the absence of life, it follows that inertia of matter must be a positive tendency to keep its state. Only living beings are inert in this sense because they have an idea of another possible state, detest it and struggle against the change.[6]

The situation of everyday life which Kant used as an analogy to the inertia of matter is the contrast between matter itself and the artisan who operates on the material in order to produce a certain result that he envisages in his mind.   Matter is passive and inert, but man is active and uses his mind.   The characteristic feature of matter is its complete passivity, and this quality is, according to Kant, responsible for its inertia.   This analogy certainly makes the law of inertia more "human," but we would be badly misled if we believed that it gives the "explanation" of inertia.   Although this introduction of the concept of "life" into physical science does make it more "human," it certainly has little to do with the actual law of inertia in mechanics.   It even gives the misleading impression that for living organisms the law of inertia would not be valid.

Kant's argument can be reformulated in such a way that no analogy with living organisms is involved.   We may regard the "living organism itself" as a mechanical system of mass-points, while by "matter" we may mean a single, isolated mass-point. Then the "internal forces" in the system are "external forces" for the individual mass-point, and we can understand how such a system can change its state from rest to motion; external forces act upon the individual mass-points.   A motion of the system may arise provided it does not contradict the law of conservation of momentum.   If, however, we consider a mass-point isolated in space, there is no external force and the state of motion cannot change because any change would contradict the "metaphysical truth" that there cannot be any change without an external force.

In this form, the "proof" of the validity of the law of inertia is very similar to the most "refined" proof that has been given.   This is the proof given by the great British physicist James Clerk Maxwell.   This proof looks very convincing, but at second glance we find that it is actually no proof, but that it only stresses pure analogy with common-sense experience.   Maxwell writes[7] about the law of inertia after having presented the experimental evidence:

> But our conviction about the truth of this law may be greatly strengthened by considering what is involved in a denial of it.   Given a body in motion, let it be left to itself and not acted upon by any force. What will happen?   According to Newton's laws it will persevere in moving uniformly in a straight line.

Now Maxwell examines the assumption that the speed might vary.

If the velocity does not remain constant let us suppose it to vary. The change of velocity must have a definite direction and magnitude . . . determined either by the direction of the motion itself or by some direction fixed in the body. Let us suppose, *e.g.*, the law to be that the velocity diminishes at a certain rate. . . . The velocity referred to in this hypothetical law can only be the velocity referred to a point absolutely at rest. For if it is a relative velocity its direction as well as its magnitude depends on the point of reference. . . . Hence the hypothetical law is without meaning unless we admit the possibility of defining absolute rest and absolute velocity.

This, however, is impossible.

The denial of the law of inertia would thus imply, according to Maxwell, the assumption that it makes sense to say of a certain system of reference that it is at absolute rest or has a certain absolute speed. He stresses the point that the human mind cannot conceive what absolute position in space is; therefore, the denial of the law of inertia "is in contradiction to the only system of consistent doctrine about space and time which the human mind has been able to form." This only system is, of course, based on the conception that "position" and "velocity" only have meaning relative to a certain system of reference.

If we examine Maxwell's argument and keep in mind everything we have learned about the "scientific aspect" of inertia, we can easily see that every "proof" that has been advanced to make "inertia" intelligible is actually not a "proof" but a metaphysical interpretation of inertia. It is clear that not only does the denial of the principle of inertia imply a system of reference that is at absolute rest, but that the affirmation of this principle is meaningless also if we do not refer it to a system of reference that is at rest. The principle of inertia claims that a mass upon which no external force is acting remains at rest or moves along a straight line. But "rest" and "moving along a straight line" are meaningless unless we have a system of reference with respect to which the mass is to be at rest or moving along a straight line. Hence, from Maxwell's proof we can only infer that "if a mass-point has an initial velocity with respect to a system of reference (*S*), it will continue moving with the same velocity with respect to (*S*)," But this statement is certainly wrong if we take it as a statement of physics. From what we learned by studying the science of motion, a mass would *not* preserve its velocity with respect to a rotating coordinate system.[8]   Hence, there is no proof, no "seeing by the intellect," no "metaphysical intuition" from which we can learn with respect to what system of reference a body preserves its velocity.   From Maxwell's argument, there follows a purely mathematical statement about a fictitious coordinate system without any operational meaning.

Later, Ernst Mach[9] stressed the point that Newton's laws should not be referred to a "system at absolute rest, Newton's absolute space," but to a physical inertial system which is, as a first approximation, coincident with the system of our galaxy.   Then Maxwell's argument had to be rephrased in the following way: "If a mass has a velocity relative to the fixed stars which would diminish, there must be a law according to which it diminishes."   But in this assumption there is no contradiction, whatever this law may be. As a matter of fact, in antiquity, after the time of Aristotle, the generally accepted theory was that a mass which had a speed relative to the galaxy will come to rest by itself, since it is the natural state of terrestrial bodies to be at rest.   However, this "proof" of the law of inertia is not correct for another reason.   Maxwell argues that the velocity cannot change because no law of change can be conceived that would be consistent with our general conception of time and space.   In this argument, it is taken for granted that the laws of motion should be formulated by describing the changes in velocity.   However, one could also assume, following the historic path of physics, that the laws of motion are to be formulated as changes of position.   Then one would conclude, following Maxwell's line of reasoning, that the position of a particle cannot change unless external forces are acting because no direction is determined in which it should move.   An argument like Maxwell's tacitly assumes that a direction is determined by the present velocity; this statement presumes, however, that the velocity is relevant, and not only the position.   In other words, the assumption is made that the "state" of our mass is not determined by position only, but by position *and* velocity.   This assumption is, however, almost identical with the assumption that the law of inertia is valid.

From these considerations, we can learn that all these "proofs" of the law of inertia are not proofs at all.   But what are they? Should we simply say, as scientists would be inclined to do, that these proofs are false? From the purely scientific aspect they are certainly wrong. But, on the other hand, they are "metaphysical interpretations" of the law of inertia. They try to interpret this law by analogies taken from fields of daily life experience, common-sense analogies. They speak of "velocity" as our daily life language speaks, without specifying a system of reference. They assume that if a body is known to us by sense-experience, we know the "state" of the body. They neglect the essential point that the term "state of a body" does not belong to the common-sense description of the body, but is a part of the scientific language that has been built up in order to formulate the laws of physics in a convenient form. The example of "inertia" is for this reason very instructive. We learn that the common-sense analogies that we invent, in order to "humanize" the physical laws and which are afterwards called "metaphysical interpretations," have two characteristics: They neglect or minimize operational meaning and disregard the fact that the "state" of a body is an artificial concept deliberately produced by the scientists in order to formulate the physical laws in a simple and convenient way.

2. The "Indestructibility of Matter" as a Metaphysical Interpretation

It has not always been regarded as a matter of common-sense experience that matter cannot be produced or annihilated in the sense that the mass (or weight) of a body cannot be altered without adding or removing a piece of the body. Herbert Spencer[10] gives several examples of people who did not believe in the indestructibility of matter: "I knew a lady who contended that a dress folded up tightly weighed more than when loosely folded up; and who, under this belief, had her trunks made large that she might diminish the charge for freight!" Spencer gives some other examples, all referring to ladies. He seems to have believed that ladies have longer preserved the common-sense opinions that belonged to an earlier stage of science. He makes the point that along with the advance of science the "indestructibility of matter" has become a belief accepted by common sense.

Although Herbert Spencer was a strong advocate of empirical thinking and an avowed opponent of "metaphysical intuition," he actually had an intense longing to trace science back to "intelligible principles." It is very instructive to learn by what argument he attempted to make the "indestructibility of matter" an "intelligible principle." As a starting point, he argued that without this assumption no positive science would be possible.

> For if, instead of having to deal with fixed quantities and weights, we had to deal with quantities and weights which were apt, wholly or in part, to be annihilated, there would be introduced an incalculable element, fatal to all positive conclusions. Clearly, therefore, the proposition that matter is indestructible must be deliberately considered.[11]

In earlier periods, common-sense experience certainly did not lead to the statement that matter is conserved. Such familiar experiences as the burning of wood or coal induced the belief that these bodies disappear and flames are produced of which one did not know whether they were matter or not. There were two interpretations that contradicted one another: The burning of coal was interpreted as the emission of phlogiston, and it was later interpreted as the addition of oxygen. "The current theology," wrote Spencer, "in its teaching respecting the beginning and end of the world, is clearly pervaded by it," meaning the belief in creation and the annihilation of matter. Hence, this belief has not been at all times repulsive to common sense. "The gradual accumulation of experience, however, and still more the organization of experiences, has tended slowly to reverse this conviction, until now [1860], the doctrine that matter is indestructible has become a commonplace." As it has happened in a great many cases, so also in the case of the indestructibility of matter, after this assumption had been demonstrated to be practically helpful in interpreting the data of experience, it became obvious that to assume the opposite (destructibility) would be contrary to common sense.

Spencer asked the question, "whether we have any higher warrant than the warrant of conscious induction [from experience]." Spencer was convinced that we have a "higher warrant." He thought that by self-introspection, by describing our stream of consciousness, we can prove by experience that it is psychologically impossible to imagine the annihilation of matter. He wrote:

> Careful self-analysis shows this to be a datum of consciousness. Conceive the space before you to be cleared of all bodies save one. Now imagine the remaining one not to be removed from its place, but to

lapse into nothing while standing in that place. You fail. The space which was solid you cannot conceive becoming empty, save by transfer of that which made it solid.

Spencer pointed out that it is impossible to imagine that matter is compressed into nothing; what one imagines is always the decrease in the distances between the parts.

> While we can represent to ourselves the parts of the matter as approximated, we cannot represent to ourselves the quantity of matter made less. To do this would be to imagine some of the constituent parts compressed into nothing; which is no more possible than to imagine the compression of the whole into nothing. . . . The annihilation of matter is unthinkable for the same reason that the creation of matter is unthinkable.

Spencer's argument is particularly instructive because he elaborates the "impossibility of imagining the annihilation of matter" in a detailed psychological way, while most authors just claim that the annihilation of matter is contradictory to the testimony of our "inner eye" or to the "intellectual intuition."

From Spencer's argument it is also clear that by the impossibility of imagining or thinking of the annihilation of matter we actually mean the impossibility of finding in our common-sense experience a fact that one could call annihilation of matter. Therefore, what Spencer proves is the fact that we do not find in our daily life experience an analogy to the annihilation or creation of matter; hence, the doctrine of the indestructibility of matter is a metaphysical interpretation of Newtonian physics. As we learned,[12] there are in modern physics phenomena that are interpreted as "annihilation of matter," e.g., the conversion of an electron-positron couple into an elementary portion of radiant energy (photon) or the loss of mass accompanying the formation of helium from hydrogen. We have actually learned that by "packing" two hydrogen nuclei and two neutrons tightly a helium nucleus originates, the mass of which is smaller than the sum of the masses of the constituents. In nuclear physics we speak of the "packing effect."

It is instructive to note that Spencer ridicules, as quoted at the beginning of this section, a lady who believed in a "packing effect" a hundred years ago. He points out that the lady was not intelligent enough to understand that a "packing effect" is unthinkable and unimaginable. A person not trained in science, like the lady in Spencer's anecdote, would possess a range of common-sense experience which allows for the annihilation of matter by tighter packing. When a person has become better trained in scientific thinking, he will understand that a "packing effect" is unthinkable. From all this, we can learn that the lady ridiculed by Spencer was right, because a "packing effect" is possible, and the philosopher was wrong who believed that he could prove that destructibility of matter was unthinkable. To believe that a lady or, for that matter, anybody else can by strenuous training of his mind improve his common sense seems to be erroneous. If the lady quoted above had followed Spencer's advice, she would eventually have succeeded in grasping the idea that annihilation of matter and a "packing effect" are unthinkable.

The advances in atomic physics that led to the discovery of the "packing effect" were not achieved by great efforts in imagining the annihilation of matter, but by attempts to build up a system of symbols, a conceptual framework from which the observable phenomena could be derived. Among the concepts that formed the system was the concept of a decrease of mass by tightening the structure of an atomic nucleus. The problem is not to imagine directly how matter can disappear, but to derive observable phenomena from the statement that matter disappears. There is no doubt that every observable phenomenon is also thinkable and imaginable. Certainly the packing effect dreamed up by the economy-conscious lady seemed to be very small, and she was probably ridiculed for giving any importance to this small loss of weight. But today we know that the future of our world may hang upon a small fraction of mass since this loss of weight is the "secret of the hydrogen bomb."

3. Metaphysical "Implications" of the Theory of Relativity

In the presentation of Einstein's theory of relativity from the logical and empirical aspect[13] it was pointed out that its logical structure is not essentially different from any physical theory; it starts from a formal system to which operational definitions are added and derives from it observations logically statements which are checked by actual observations. These observations are of exactly the same kind as any observations in traditional mechanics or optics; they consist essentially in observing the coincidence of marks on different scales.   The theory can be presented as a system of physical hypotheses or as a system of definitions, as any physical theory can be presented.   The system of definitions is a device by which the hypotheses can be formulated in a simple and practical way.   However, a great many authors have said again and again that the theory of relativity is not a physical theory in the ordinary sense of this word, but a philosophical or metaphysical doctrine that explains new physical facts without advancing new physical hypotheses.   It proposes a new view of space and time and brings the observing scientist himself into the picture of the physical world.

Moreover, many notable authors, philosophers, religious leaders, educators, and even scientists have claimed that under the impact of the theory of relativity the general views of the position of man in the universe have changed radically.   The mechanistic world picture that had been prevalent since the seventeenth and eighteenth centuries served as a considerable support in the march toward a materialistic philosophy.   This trend in the eighteenth and nineteenth centuries looked to a great many observers almost irresistible. However, in the twentieth century, the impression gained ground that this mighty trend was stopped by twentieth-century physics, especially by the theory of relativity and quantum theory.   It was obvious to many authors that the trend toward materialism had been stopped, and a sharp turn toward idealism had been taken. Edmund Ware Sinnott, a prominent biologist at Yale University, recently published a book, *Two Roads to Truth*, in which he attempted a reconciliation between science and religion based upon the results of contemporary science.[14]   He writes:

> After the revolution introduced by relativity, quantum mechanics, and nuclear physics, science was forced to modify some of its earlier conclusions.   The plain truth is that the Universe is a much more complex system than it seemed to be in Newton's time. . . . Scientists accept now without surprise ideas that would have seemed preposterous not long ago.   This change has been reflected in a more open-minded attitude on their part towards idealistic philosophies. For three centuries a confidently advancing science seemed to undermine the very foundations of faith, and religion was forced to modify its position in many ways or lose the support of its more thoughtful

partisans.   The tide, however, has begun to turn, and an aggressive idealism is going over from the defense to the attack.

A similar philosophical interpretation of the theory of relativity is given by the prominent sociologist of Harvard University, Pitirim Sorokin.[15]   He stresses and resents the fact that our culture, since the rise of modern science (about 1600), has become more and more a "sensate culture," by which he means that the chief interest is centered upon sensory phenomena.   He compares it unfavorably with the "ideational culture" prevailing in the Middle Ages when spiritual and ideal values were the main goals of human striving. Sorokin sees a characteristic of a culture in the concept of time which prevails.   In a "sensate culture" there is a "sensate time" that can be reduced to quantitative measurements, while "ideational time" has a characteristic quality connected with the evolution of the universe.   Sorokin points out that in the twentieth century some signs of reaction against the exclusive emphasis on "sensate time" have begun to appear.   One is the restoration of qualitative time by the French philosopher Henri Bergson,[16] who distinguishes between "quantitative time" as the physicists conceive it and what he calls "duration," which is qualitative time and is used by Bergson to describe the evolution of organisms.   Another sign, mentioned by Sorokin:

> . . . is the Minkowski-Einstein "space-time continuum" which in a sense is also a revolt against the extreme "sensate time." . . . This means that signs of revolt against sensate time are not absent.   This revolt is in harmony with other "revolts" against sensate culture mentality at the end of the nineteenth and in the twentieth century, in all departments of culture.

Sorokin directs our attention to the fact that the basic concepts that are used in the science of a certain period are not independent of the concepts which are used in the formulation of the cultural values of the same period.   Speaking of twentieth-century science, he stresses "the dependence of the fundamental conception of science upon the transformation of the whole cultural mentality."   Moreover, it has been pointed out that by this "turn of the tide" the old gap between science and religion, which had seemed unbridgable, now seemed to become narrower.   The building of a bridge has become possible.   The great bearing upon our general world view which was ascribed to the new physical theory by a great many authors in the fields of education, religion, and even politics has put the physical scientists themselves "on the spot." Many have been glad that their cherished science has been recognized as a support of their cherished moral and religious beliefs; but a great many others have objected that relativity is an honest physical theory which attempts to give a description of observable phenomena and cannot provide a decision in the fight between materialism and idealism, and still less between religion and its enemies.

The question has arisen of how a new theory can be, on the one side, merely an improvement in our scheme for the prediction of observable phenomena and, on the other side, a weapon in the battle for or against idealistic philosophy or religion. The great majority of today's physicists have been successfully trained to keep their special fields as separate from philosophy as they can. On the other hand, the great majority of philosophy students have been trained to believe that to achieve a good understanding of philosophy only a very superficial knowledge of physics is necessary. However, when the question is raised of how a physical theory can be interpreted as a support or as a refutation of materialism or idealism, the physicists are actually put "on a spot." If a philosopher who feels that his knowledge of physics is insufficient consults an "expert" on physics, he will very rarely obtain a satisfactory answer.

There are, of course, physicists who are prepared to accept even the vaguest argument, provided that it does not pretend to be scientific, but claims to be "philosophical" and is in agreement with the philosophy which the physicist has absorbed as a child. But if he speaks "as a physicist," he will usually say that all these philosophical "implications" about idealism or materialism are just "nonsense" to which an honest scientist should not pay any attention. Unfortunately, this "nonsense" has a powerful effect upon human behavior, and a physicist who is not able to give his students a precise account of the philosophical repercussions of relativity does not fulfill the duties of a physics teacher in a democratic society. The physicist is inclined to apply the word, nonsense, to all attempts to derive philosophical world views from physical theories because he feels that these consequences are not, strictly speaking, logical consequences of the scientific statements that constitute these theories, *e.g.*, the theory of relativity. However, we can understand the meaning of these philosophical implications very well if we do not regard them as logical consequences or as inductive generalizations of physical relativity, but as metaphysical interpretations of Einstein's theory.

As a matter of fact, the theory of relativity is particularly well fitted to become an example for the philosophical or metaphysical interpretations of science. If we keep this in mind, we can understand that the philosophical implications of relativity can be drawn in various ways that very often even contradict one another. This is understandable if we recognize that these interpretations are not uniquely determined by the theory; they are rather analogies to the theory of relativity drawn from the world of our everyday experience. Bertrand Russell[17] writes:

> There has been a tendency, not uncommon in the case of a new scientific theory, for every philosopher to interpret the work of Einstein in accordance with his own metaphysical system, and to suggest that the outcome is a great accession of strength to the views which the philosopher in question previously held.

As a matter of fact, there have been philosophers who characterized the theory of relativity as a mere description of observations without penetrating to the true laws of nature, while other authors have claimed that the theory of relativity is not a physical but a metaphysical theory that tells us about the innermost laws of the universe. Some authors have hailed the theory of relativity as the final victory of idealism over materialism, while others have accused the theory of relativity of being a crude form of materialism. This metaphysical interpretation plays a very large role in the attempts which have been made to popularize the theory of relativity, to interpret its meaning to the layman. Lincoln Barnett[18] writes:

> Physicists have been forced to abandon the ordinary world of our experience, the world of sense perception. . . . Even space and time are forms of intuition which can no more be divorced from consciousness than our concepts of color, shape, or size. Space has no objective reality except as an order or arrangement of the objects we perceive in it, and time has no independent existence apart from the order of events by which we measure it.

This is certainly true, but it is true for every physical theory because in every theory the world of our direct sense observations is replaced by a formal system, connections between symbols, which are in turn connected with sense impressions by "operational definitions." We can derive metaphysical interpretations from these irrefutable sentences by stressing some special analogies. We can stress the point that all statements about length or duration are no longer statements about "objective time or space," but are statements about our impressions. This seems to minimize the role of matter, to enhance the role of mind, and to amount to a refutation of materialism in science. But we can just as well stress the point that "space" and "time" had been, before Einstein, spiritual objects, and that they have now been replaced by readings on material clocks and yardsticks. This would amount to a materialistic interpretation.

The antimaterialistic interpretation has won ground by the argument that in the theory of relativity[19] the conservation of matter no longer holds; matter can be converted into nonmaterial entities, into energy. This statement has been regarded as a support of the creed of some religious groups that "matter has ceased to be" according to modern science, and even as a support of the famous words of Mary Baker Eddy,[20] founder of Christian Science: "There is no life, truth, intelligence, nor substance in matter."

The British philosopher Herbert Wildon Carr[21] hails Einstein's theory of relativity for having secured definitely for the mind its place in the objective physical world. He points out that before Einstein the general belief was that "nature can only affect the mind in the shadowy dream-like form of an idea." In the theory of relativity, however, as we have seen,[22] the laws of mechanics and optics cannot be formulated without introducing explicitly the mind of the observing scientist. Carr states:

> Now, when the reality is taken in the concrete as the general principle of relativity requires us to take it, we do not separate the observer from what he observes, the mind from its object, and then dispute as to the primacy of the one over the other.

The antimaterialistic trend in twentieth-century science has been maintained by, one would say, the public opinion on science. The *Encyclopædia Britannica* says, for example,[23] "Contemporary science is tending away from materialism and mechanism towards recognition of other than mechanical factors in the phenomena, even the physical phenomena, of Nature."

In countries where the Hebrew-Christian tradition is the ruling ideology, materialism has been generally regarded as harmful for desirable human conduct, and, therefore, the refutation of materialism by the theory of relativity has been regarded as a great achievement. If we wish to judge the utterances on the theory of relativity that come from the countries that are under the control of Communist governments, we must keep in mind that, according to the doctrine of the ruling party, desirable conduct of man can only be derived from the philosophy of Dialectical Materialism. This official philosophy is, in a great many respects, different from what we are accustomed to call "materialism." However, there has been a strong tendency among the Soviet authors to agree with the opinion that the relativity theory contradicts materialism. This implies, of course, a denunciation of relativity theory as a "reactionary theory" that would lead to undesirable political conduct. There are two points which have been singled out as targets of attack by Soviet authors: the abandonment of the ether as a material medium for the propagation of light, and the abandonment of the proposition that the earth is "really" moving and that the Ptolemaic system is "really" wrong.

Both these views are branded as antimaterialistic because they imply that physics is not a doctrine about the objective motions of material bodies but a doctrine about bringing order into our sense observations. This doctrine has often been connected with the name of the Austrian physicist and philosopher Ernst Mach, who has become a steady target of attack in Soviet literature. The *Large Soviet Encyclopedia*[24] says, in its article on *Ether:*

> The special theory of relativity takes refuge in a purely mathematical description and refrains from the investigation of the medium in which the electromagnetic events occur. It refrains at the same time from putting the question for the objectivity of physical phenomena, *i.e.*, it accepts, for the question about ether, the viewpoint of Ernst Mach."

The Russian physicist and philosopher Arkady Klimentovitch Timiryasev[25] wrote:

> The orthodox modern scientist does not dare to doubt Einstein's theory. For he regards it as an absolute truth. He holds definitely

the view that the Copernican and the Ptolemaic systems are one and the same thing. This standpoint is inacceptable for everybody who does not succumb to fashion in science. The identification of the Ptolemaic and the Copernican system is not a conclusion that has been drawn by idealistic philosophers from the theory of relativity. This identification is the starting point of the whole Einsteinian theory. This theory has this starting point in common with Mach who chose it under the influence of his reactionary philosophy.

Timiryasev, as many Soviet authors do, states very firmly that the "idealistic" conclusions are not arbitrary metaphysical interpretations of Einstein's theory, but that Einstein himself built up his theory deliberately in such a way that these conclusions can be drawn in order to support the Hebrew-Christian tradition that has developed in fighting materialism. Timiryasev thinks that Einstein's theory could not be purged of this interpretation without being destroyed. He writes: "If one attempted to fight this reactionary interpretation of the Theory of Relativity one would have to transform this theory radically. It is debatable whether after this reconstruction much of this theory could remain." Pointing out the specific reasons why he calls the theory of relativity an "idealistic" one, he writes:

> Einstein does not like the statement that rotation with respect to the ether or with respect to absolute space is the cause of centrifugal force.[26] We do not *see* the ether and there are no milestones in the ether or in absolute space. Therefore, ether and absolute space are, according to Einstein, merely fictitious. For they are not complexes of sensations. If we say that the rotation relative to the ether or absolute space is the cause of centrifugal force, we commit a crime against the law of causality. For by this law, according to Mach and Einstein, only observable things are admitted as causes. [This is certainly not in agreement with the scientific aspect of causality presented before.[27]] They replace absolute space by the System of Fixed Stars. They are observable, therefore not merely fictitious. Mach's theory is served by Einstein and—the essential thing—there is an opportunity of getting rid of the Copernican system.

From this argument it is plain to see that calling Einstein's theory "idealistic" and a "refutation of materialism" is nothing more than the stressing of an analogy. One point, however, is obvious: Mach and Einstein reject rotation with respect to absolute space as a cause of centrifugal force because absolute space and the ether are not observable physical bodies. Many would be inclined to call this attitude advocated by Mach and Einstein a materialistic attitude. We must not wonder, therefore, that the theory of relativity has been accused under some conditions of being a support of materialism. This accusation had to be raised by those groups who were, for some political reasons, enemies of Einstein's theories and, at the same time, enemies of materialism. It is, therefore, to be expected that we shall find this attitude among the representatives of the philosophy that was advocated by the ruling party under the Nazi government in Germany. In an indoctrination talk[28] a lecturer said:

> The formulation that the phenomena of nature obey a general Principle of Relativity is nothing but the expression of a radically materialistic attitude of mind and spirit. . . . The Theory of Relativity could only be hailed by a generation that had grown up in cultivating materialistic ways of thought.

On the other hand, some physicists in the Soviet Union, who dislike the derogatory label of "idealistic" attached by official writers to the theory of relativity, have credited Einstein with having eliminated the "idealistic and metaphysical" elements from Newton's mechanics, and have stressed the materialistic points in Einstein's theory. The prominent Russian physicist Sergei Ivanovitch Vavilov writes:[29]

> For Newton space existed objectively as an empty stage where the world process was played. It was partly filled with matter, partly void. Absolute time existed for Newton also independently as a kind of "pure motion." This pattern is obviously incompatible with Dialectical Materialism and inacceptable. Actually, Newton's metaphysical doctrine on space and time, the metaphysics of which has been little noticed, has lived behind the stage until our present time. The historic merit of Einstein consists in his criticism of the ancient metaphysical ideas on space and time.

However, the popular opinion was expressed by an editorial in the *London Times* in 1919 after the astronomical confirmation of Einstein's theory of gravitation: "Observational Science has in fact led back to the purest subjective idealism."

4. In What Sense Does the Theory of Relativity Refute Materialism?

We have learned how strongly leaders in education, politics, and religion have been inclined to see Einstein's theory of relativity as a weapon for the refutation of materialism, and to make of it an effective instrument in the guidance of men.   We are now going to examine in a more precise way the reasons that have been advanced against materialism, and ascertain to what degree they are actually conclusions drawn from the scientific aspect of that theory.[30]   We may, perhaps, present and examine four main reasons:

(1) The real universe is not three-dimensional Euclidean, but four-dimensional non-Euclidean.

(2) Matter can be converted into a nonmaterial entity, radiant energy.

(3) Nonmaterial entities like the curvature of space can produce motions of heavy material bodies.

(4) The theory does not deal with the objective motion of material bodies, but with mental states, the impression made by physical objects on individual observers.

The argument for (1) runs in general as follows: Materialistic science, before 1900, had assumed that there was nothing real in the world except "matter" in the ordinary sense of the word, according to which matter is a three-dimensional, clumsy substance.   It could be treated according to the traditional rules of Euclidean geometry and was totally different from what one might call a "spiritual" substance like the human "soul" or "mind."   The theory of relativity, however, has shown that clumsy, three-dimensional matter is only an appearance, whereas the "reality" behind it is something infinitely more subtle, a four-dimensional space-time continuum which does not even obey the laws of Euclidean geometry, but possesses a "curvature."   We must, however, consider that in the sentence, "Our universe *is* four-dimensional," the word "is" has a completely different meaning from the one it has in the traditional proposition "Our universe *is* three-dimensional."   In the latter "is" is used in the present tense, but when we speak of a four-dimensional continuum, the word "is" has a very complex meaning.

If we analyze precisely what the theory of relativity says about the role of the four-dimensional continuum, we find the following assertion: The description of our physical experience can be presented in a simpler and more practical way by using the four-dimensional continuum than by using three-dimensional space and one-dimensional time separately.   If we say "in a more practical way," we mean that the four-dimensional presentation has been very helpful in finding new physical theories that have been observed to be in good agreement with twentieth-century experiments and observations.   It would hardly have been possible to find Einstein's theory of gravitation without the help of the four-dimensional symbolism.

Hermann Minkowski, the author of the four-dimensional presentation in relativity theory, wrote in 1908:[31] "From henceforth space in itself and time in itself sink to mere shadows, and only a kind of union of the two preserves an independent existence."   If we use this and similar propositions, we must not forget that they are only used in an analogical sense.   We speak of the four-dimensional space-time continuum as if it were an object of our everyday experience, and say that it is "real" in the sense in which a physical thing is real.   Hence, all statements which assert that the four-dimensional continuum is more real than the world of three-dimensional physical objects either uses the word "real" in the purely scientific sense as equivalent to "practical" or uses it as a mere analogy to the common-sense use of "real."

To say that the four-dimensional interpretation of relativity theory conflicts with materialism is misleading; the correct thing to say would be that it is impossible to formulate the general principles of relativity theory in terms of our common-sense language because in this language we cannot say that two events A and B happen simultaneously for one system of reference.   "The new situation in the thought of today," says Alfred North Whitehead,[32] "arises from the fact that scientific theory is outrunning common sense."   The introduction of the relativity of simultaneity is a heavy blow to common-sense language on the level of the general scientific principle.   "The earlier science," writes Whitehead, "had only refined the ordinary notions of ordinary people."   Relativity theory, however, has radically remodeled these notions.

In this sense, and in this sense only, we must understand the statement that relativity theory has refuted materialism.   The relativization of simultaneity, Whitehead says, "is a heavy blow at the classical scientific materialism, which presupposes a definite present instant, at which all matter is simultaneously real.   In the modern theory there is no such unique present instant."   Since the expression "all the matter in the Universe at the present instant" is the basic concept of traditional materialism, this doctrine would imply that all principles of science can be formulated in the language of everyday experience.   Since the theory of relativity has shown that this is not the case, it has "refuted" materialism; it has shown that the common-sense meaning of "matter" cannot be the conceptual basis of all science.

The theory of relativity is not presented in common-sense language because in this language one cannot, for example, say that a table has different lengths with respect to different systems of reference.   If we wish to present the theory of relativity in common-sense language, we can do it only in an analogical way.   We must, for example, as mentioned above, say that our Universe "is" in reality four-dimensional.   If we forget that this language is not a scientific, "universal" language, we can get into deep water.   One conspicuous example of this confusion is the interpretation of relativity theory as a support of "fatalism" or "predestination."   The argument runs like this: It can happen that an event, *e.g.*, the death of a person, that occurs in the present for us, occurred in the past with respect to another system of reference, and has been, therefore, predetermined.   This argument makes use of analogical language.   The terms "in the present" and "in the past" are used as they are used in common-sense language; but actually, the event in question happens only once, and, to speak in strictly scientific language, the clocks coinciding with the event show different positions of the hands if they have different speeds.

The argument (2) was discussed previously.[33]   It is certainly true that "matter" is not conserved if we use the term "matter" as it is used in common-sense language and in older physics.   It is certainly true that "materialism is refuted" if we mean by "materialism" the opinion that matter, in the common-sense meaning of the word, has filled the Universe since eternity and will remain for eternity.

The argument (3) according to which motion of material bodies can be produced by the "curvature of space," a nonmaterial property, can again be correctly regarded as a refutation of "materialism" if we mean by this word the opinion that all laws of nature can be expressed in common-sense language. The statement that the "curvature of space" is a "nonmaterial" property contains the term "nonmaterial" that is taken from common-sense language, according to which everything that is not palpable, like a stone or an elephant, is nonmaterial.

Summing up, we can say that all three arguments, (1), (2), and (3) refute the opinion that all laws of nature can be expressed in common-sense language; this refutation is certainly due to Einstein's theory of relativity. This was stressed by Einstein himself[34] when he pointed out that the general theory of relativity has shown as the characteristic feature of twentieth-century physics that the system of concepts by which the general laws of nature can be formulated are much more remote from the system of concepts by which our everyday experience is conveniently described than had been assumed in the eighteenth and nineteenth centuries. Einstein emphasized the fact that:

> The distance in thought between the fundamental concepts and laws on one side and, on the other, the conclusions which have to be brought into relation with our experience grows larger and larger, the simpler the logical structure becomes—that is to say, the smaller the number of logically independent conceptual elements which are found necessary to support the structure.

Concerning argument (4), we must investigate in what sense relativity theory is a doctrine about sense impressions, while Newton's mechanics deals with objective facts. There is no doubt that we can formulate every report about experiments as a statement about sense observations, whatever the theory by which we coordinate the experiment may be. We are, of course, accustomed from childhood to describe our experiences by introducing "physical bodies" instead of sense observations. We say that we see a "table" and not a complex of colored spots. If we say that a table has a "length of three feet with respect to a system of reference (S)," we mean by it that the table covers the distance between two marks (zero and three) on a rigid yardstick that has the same speed as the system (S). This is a statement about the behavior of rigid bodies exactly like the statement, "this table has a length of three feet" in traditional physics. The only difference is that in relativity mechanics the speed of the yardstick is specified because the result of the measurement depends upon it.

It should be noted that in these statements we speak only about tables and yardsticks, but not at all about "living observers." The observer may have any speed; he will always observe the same coincidence between marks on the yardstick and the edges of the table. There is no subjective element in the statements of relativistic mechanics. The appearance of subjectivity has only been introduced in attempts to formulate the propositions of relativity theory by some analogy to common-sense statements. Instead of saying "length with respect to a system of reference," which is not an expression from our common-sense language, we have used the expression "length *for* an observer in the system (S)." Then we could say that the table has a "different length *for* different observers." This expression "for an observer" is formed by analogy to the way in which we express in common-sense language the fact that an object may look different to different observers for reasons of different perspective, or optical illusion, or weakness of the eyes. By using this analogy, we introduce into the presentation of relativity theory analogies to common-sense language that are certainly useful because they bring in a certain intuitional element. However, these analogies become harmful if we forget that they are analogies and regard them as strictly scientific statements. This distinction can be easily understood when to every statement is added its operational meaning.[35] Then it will be noticeable that the common-sense analogies either become meaningless or obtain a meaning that is identical with the presentation in strictly scientific language. The "observer" will, for example, disappear completely or be replaced by a yardstick or a clock.

It is said that the theory of relativity has reduced physics to statements about mental phenomena and, therefore, refuted materialism. If this is correct for relativity theory, it is true for all physical theories. In this sense, every physical theory deals with sense impressions which are mental entities, and in this sense every physical theory refutes materialism. However, it is important to understand that an additional argument in favor of idealism or against materialism cannot be found in the theory of relativity.

5. Is the Theory of Relativity Dogmatic?

It has frequently been stated that the characteristic of modern science is that the general statements, theories, or principles are forced upon us by the observed facts, whereas in antiquity and the Middle Ages the general statements or principles were regarded as intelligible in themselves and were accepted by everybody who understood their meaning. The German physicist, Johannes Stark formulated the following distinction:[36]

> The pragmatic spirit advances continuously to new discoveries and new knowledge. The dogmatic spirit leads to muzzling of experimental research and to a literature which is as effusive as it is unfruitful and tedious, intrinsically akin to the theological dogmatism of the Middle Ages, which was opposed to the introduction of pragmatic natural science.

The author cites as a typical example of the dogmatic spirit in science Einstein's theory of relativity because it starts from the dogma that the "speed of light is the same in all systems of reference" and attempts to adjust all physics to this dogma. This procedure, according to Stark, is not pragmatic because all our familiar experience is incompatible with this dogma and can only be brought into agreement with it by extremely artificial hypotheses like the contraction of yardsticks and the lagging of clocks as a consequence of motion.

Seen from the "scientific aspect," the theory of relativity is no more dogmatic than any other physical theory. We have directed the attention of the reader several times to the fact that the general principles of physics have been since 1600, and particularly since 1900, very remote from statements that can be formulated in common-sense language. Among these statements are, in particular, the principles of Einstein's theory of relativity. It is certainly clear that the principles of constancy and relativity that form the backbone of Einstein's relativity theory[37] are not derivable from the facts of our experience, and not even from the experiments devised and carried out by physical scientists. However, if we call them dogmas, we should also have to call the principle of inertia "dogma." When it was advanced by Galileo and Newton, it was also very remote from common-sense experience, and could only be justified because the conclusions drawn from it were in agreement with actual observations.   In a textbook of "Natural Philosophy," (*i.e.*, physics) written a hundred years ago, the author says very correctly that the law of inertia is scientifically confirmed by the fact that one can derive from it the frequency of a pendulum as a function of its length, and that we know no other principle from which this frequency could be derived.   In the same way, the principles of constancy and relativity do not derive their strength from their own plausibility or intelligibility, but from the agreement of their consequences with actual observations.   Hence, if Einstein's principles are dogmas, the law of inertia and Newton's laws of motion altogether are dogmas, too.

As a matter of fact, every physical theory is "dogmatic" as a formal system and "pragmatic" as a system of statements about actual experiments.   There is, of course, the qualitative difference that a theory seems to be the more "dogmatic," the more remote the principles are from actual sense-experience.   This means, in turn, that the principles seem to be more dogmatic, the more they cover recent experience that is far from the experience of everyday life.   However, the statement that Einstein's principles are more "dogmatic" than the theories of traditional physics has been interpreted by analogies taken from our everyday experience.   "Dogmatic" has been interpreted as meaning that these dogmas, in this case the principles of constancy and relativity, are forced upon the scientists by authorities, as occasionally, political or religious dogmas have been forced upon people.   Following this common-sense analogy, it has frequently been said that the followers of relativity theory have ceased to look for the laws of nature by careful and pious research; they no longer look for the laws of nature by scrutinizing nature, but by forcing nature to accept the laws that physicists wish to impose upon her.

As has happened after every radical change in science, there developed, beginning about 1900, a "philosophy" that interpreted the increasing remoteness of science from common-sense concepts as a kind of devilish trend by which man is separated more from his "natural" background.   As we shall see in greater detail later on, every metaphysical interpretation serves some moral, religious, or political purpose.   Generally speaking, it supports some plan to guide human behavior toward desirable goals.   In this regard, the gradually increasing distance between the concepts of theory and the concepts of direct sense-experience has been interpreted as a "wicked" attempt to undermine man's connection with nature and in this way to lead man to moral and religious perversity.

We may quote as an example a German philosopher, Ludwig Klages,[38] who published a book which was at the time widely read and influential in the kind of anti-intellectualism that has always been cherished by some intellectuals. It belonged to the kind of philosophy that softened up German and other continental Europeans for the future acceptance of Nazi philosophy. According to Klages, the theory of relativity reveals, even to dim eyes, the hidden driving power of the twentieth-century intellect in its apparent search for knowledge. "For this intellect the so-called laws of nature are no longer searched for; they are dictated according to pleasure and whim. The intellect is as certainly antilogical as its attitude towards reality is detached arbitrariness."

At the time of the Nazi regime in Germany, the belief that Einstein's theory was "dogmatic" almost belonged to the "party line" of the ruling party. In an article in the *Journal for the Whole of Natural Science*, which advocated the party line in the sciences, the author describes the unhealthy features which have developed in twentieth-century physics. According to him:

> The basic relation between experiment and theory has been shifted in favor of the latter. Moreover, this theory has been worked out in a purely formalistic way, without regard to the human forms of thoughts and intuition, without strictly methodical thinking. . . .

As the most conspicuous instance of this dogmatic form of theory, the author quotes Einstein's theory of relativity.

> Its starting point is a dogma, the principle of the constancy of the speed of light. The speed of light in vacuum is to be constant, independent of the motion of the source of light and of the observer. It is maintained erroneously that this is a fact of observation.

The advancing and advocating of "dogmatic" theories like Einstein's theory of relativity was attributed frequently to a particular type of wicked mind, which attempted to divert man from Nature and direct him into an artificial trap prepared by his enemies. According to the party line, the distinction between the "pragmatic" and "dogmatic" type of science was accounted for by racial differences between scientists.

8

Motion of Atomic Objects

1. Newton Was No Newtonian

In the first half of the nineteenth century, it seemed to have been established with a high degree of certainty that all physical phenomena (in the widest sense of the term) were to be described by means of a scheme of moving particles obeying Newton's laws of motion. This means that the particles move spontaneously with constant speed along straight lines relative to an "inertial system"; however, when a "force" is acting, their deflection is inversely proportional to the constant "mass" of each particle. The special theory of relativity changed this pattern of description by introducing changes of mass dependent upon the exchange of energy between a system of particles and its environment. The general theory of relativity does not use an "inertial system" as its basis; it regards the motions that Newton's scheme ascribes to a "gravitational force" as spontaneous motions.

If we regard these "relativistic changes" of the Newtonian scheme from the point of view of today's physics, they seem to be only slight modifications. Hence, it is of the utmost importance to understand how the Newtonian scheme was radically changed when science came to grips with the motions of very small particles. We shall see that these changes are so fundamental that we can hardly say whether we can properly retain even the expression "motion of a particle."

Within the domain of everyday language, the meaning of the

statement "motion of a particle" seems very clear. Let us, however, imagine a homogeneous elastic medium of very low density
filling the whole world space. At only one point, P_0, the medium
has a high density. In time this density maximum may be located
at other points, P_1, P_2, of our medium. It "propagates" itself from
P_0 to P_1, P_2, etc. No particle of the medium is moving, only a
certain property (density) is propagated; but the process as a whole,
propagation of a density maximum through a medium, cannot be
distinguished by observation from the real motion of a particle.
Both processes are described by equating the three coordinates x, y,
z, of the particle, or the density maximum to functions of the time t:
$x = x(t)$, $y = y(t)$, $z = z(t)$.

If we describe the two phenomena in our everyday language, we
say that in one case a "particle is moving," while in the second case
"no particle is moving." Both descriptions are common-sense pictures of the observed phenomena, and if we have nothing but the
recording of the three functions, $x(t)$, $y(t)$, $z(t)$, we cannot decide
which of the two pictures is the "true" one. The experimental
arrangement by which we test the presence of a particle at a certain
point x, y, z is identical with the arrangement for testing the presence of a density maximum. In order to make such a decision, we
must know the laws of the "real motion" and the laws for the
"propagation of density." Then we can derive from these laws
observable consequences that may be different in each case. The
decision between the two hypotheses is a decision between two systems of laws. If we describe a certain physical phenomenon as
"moving particles," it means very little unless we add the laws
according to which these particles are moving.

The motion of medium-sized bodies has been regarded as governed
by Newton's laws of motion practically ever since Newton published
them. However, he was by no means certain whether these laws
would cover the motion of all possible particles, for example, of very
small particles. According to ancient hypothesis, known to the
Greeks, light is emitted from radiating bodies as a swarm of small
particles far below the size of the bodies in our daily environment.
Newton never committed himself to the assertion that these small
corpuscles move according to his three laws of motion. In his
Opticks[1] he expressed himself very cautiously:

> By the rays of light I understand its least parts. . . . For it is manifest
> that light consists of parts, both successive and contemporary; because

in the same place you may stop that which comes one moment, and let pass that which comes presently after. . . . The least light or part of light which may be stopped alone without the rest of light, or propagated alone, or do or suffer anything alone, which the rest of the light does not or suffers not, I call a ray of light.

Newton does not specify a light ray as the trajectory of a particle or of a density maximum. He does not derive the shape of this trajectory from any law of motion of a particle or otherwise. He sets up a series of axioms determining the shape of these trajectories. He defines "refrangibility of the light rays" as "their disposition to be turned out of their way in passing out of one transparent medium into another." He also formulates the axiom that "the sine of incidence is either accurately or very nearly in a given ratio to the sine of refraction." If we call this ratio n (index of refraction) and the angles of incidence and refraction α and α', respectively, the axiom says $\sin \alpha / \sin \alpha' = n$. If light passes from air into water, we learn from experiment that $\alpha > \alpha'$ or $n > 1$. This axiom implies that the shape of a light ray in a homogeneous medium (air or water) is a straight line.

However, in order to describe the laws of the bright and dark fringes that occur when light rays pass through a thin, transparent plate, Newton introduces an axiom about the refraction of incident light that passes a boundary between two media:

> Every ray of light in its passage through any refracting surface is put into a certain transient state, which in the progress of the ray returns at equal intervals and disposes the ray at every return to be easily transmitted through the next refractory surface. . . . The returns of the dispositions of any ray to be transmitted I will call its "fits of easy transmission" and the space it passes between every return and the next return, the "interval of its fits."[2]

This interval is, of course, what we call today the "wave length" of a light ray.

We certainly cannot say that, according to Newton's *Opticks*, the motion of particles of light is determined by Newton's three laws of motion. If we use the current terminology, we can scarcely determine whether this theory is a corpuscular or a wave theory of optical phenomena. However, Newton felt very clearly that it would be desirable to derive the phenomena of light rays from an hypothesis that they are trajectories either of particles or of density maxima in a medium. In his *Opticks*, he stated explicitly that his design was "not to explain the properties of light by hypotheses, but to propose and prove them by reason and experiments." It is interesting to note that Newton calls the method that we used in our presentation of geometry in Chapter 3 "proof by reason and experiment," while by an "explanation by hypotheses" he meant, in the case of *Opticks*, the derivation of observable facts from either a theory of the motion of particles or a theory of the propagation of density maxima.

In his "queries"[3] at the end of his *Opticks*, Newton discusses both possibilities. The more obvious hypothesis for Newton, certainly, was to assume that the light corpuscles move according to his three laws of motion. In this case, we have to assume a law for the forces that are exerted by the particles of a medium (air or water) upon the traversing corpuscles of light. Newton derived in this way the law of refraction. If the speed of the corpuscles in air is c and in water c', he found that because of the forces exerted by the water particles the light is accelerated in water and $c' > c$. Then Newton could prove that $\sin \alpha / \sin \alpha' = c'/c > 1$. The rectilinear form of a light ray in a homogeneous medium then follows simply from the law of inertia.

However, if we also wish to derive the bright and dark fringes occurring when a ray of light passes a thin transparent plate, we cannot derive this result from the three laws of motion since these laws do not account for what Newton called the "fits of easy transmission." Hence, Newton suggested an additional law:

> Nothing more is requisite for putting the rays of light into fits of easy reflexions and transmissions, than that they be small bodies (particles) which by their attractive powers, or by some other force, stir up vibrations in what they act upon, which vibrations being swifter than the rays, overtake them successively, and agitate them so as by turns to increase and decrease their velocities and thereby put them into those fits.[4]

This result can certainly not be derived from Newton's three laws of motion since the period (interval) of the fits does not occur in any law of force. Hence, there is no doubt that Newton did not think that his laws of motion were sufficient to derive the motion of light corpuscles. He suggested an addition which has definitely some similarity to de Broglie's wave mechanics.

2. The "Crucial Experiment" Versus the Corpuscular Theory of Light

In addition to the corpuscular hypothesis, there existed in Newton's time an alternative hypothesis that derived the law of refraction and other optical laws from the assumption that light is a propagation of density maxima or, more generally, of condensations in an elastic medium filling the whole world space. If in such a medium a maximum density at one point P is produced, this density will expand in spherical surfaces of equal density with P as a center. Therefore, it is not easy to understand how such a point of maximum density can move along a straight line as light rays do.

This hypothesis, that light consists in the propagation of condensations or compressions through a medium, was elaborated by Huyghens[5] into a theory from which could be derived the laws of reflection and refraction, including the special case of rectilinear rays. In order to achieve this result, Huyghens had to introduce an hypothesis which tells us under what conditions compressions can destroy or reinforce each other. This hypothesis has been known as Huyghens' principle and has since been taught in all courses of elementary physics. To Newton, however, an hypothesis seemed not worthy to be recommended, if by it, "the simplest phenomena," a rectilinear ray, cannot be explained in a simple way. He wrote:[6]

> Are not all hypotheses erroneous, in which light is supposed to consist in pressure or motion propagated through a fluid medium? . . . For pressure or motion cannot be propagated in a fluid in right lines beyond an obstacle which stops part of the motion but it will bend . . . beyond the obstacle.

From the mechanics of fluids, it follows that the speed of propagation (c) in air is greater than the speed (c') in water; $c' < c$. Huyghens derived the statement that light rays passing from air into water are refracted according to the law $\sin \alpha / \sin \alpha' = n = c/c' > 1$. The index of refraction n is greater than 1, as it is observed and as it is derived from the corpuscular hypothesis. However, from this hypothesis follows $n = c/c'$, while from the corpuscular hypothesis follows $n = c'/c$. It is obvious that we could decide between the two hypotheses if we could measure whether, in fact, $c > c'$ or $c' > c$, whether the speed of light is greater in water or in air.

The possibility of such a measurement seemed very remote in the time of Newton and Huyghens, however. When terrestrial methods for measuring the speed of light were developed, Arago suggested, in 1838, a "crucial experiment" that would decide, once and forever, whether "light is a material body," or whether it consists in the propagation of a disturbance through an elastic medium. The concept of propagation of pressure or density was at that time replaced by the propagation of transverse vibrations, but the problem of the speed of propagation remained essentially the same. Arago wrote about his project:[7]

> The system of experiments that I am going to describe will, as it seems to me, allow us to make a choice between the two rival theories. It will decide mathematically one of the greatest and most debated questions of natural philosophy.

By "mathematically" Arago means to say that after having observed the results of his planned experiments one could derive the decision "logically." If from a theory A it follows that a certain bright spot moves to the left, and from the rival theory it follows that the spot moves to the right, we have but to observe in which direction the spot does move. If it moves to the right, it follows logically that it does *not* move to the left. Then again logically it follows that the theory A leads by correct methods of conclusion to false results. Again, it follows logically that A is wrong. However, the corpuscular theory A and the undulation theory B do not exhaust all possible theories. Therefore, in our case, the validity of B does not necessarily follow from the falsity of A.

The experiment was actually carried out by Léon Foucault in 1850, and we give Arago's own description, as quoted by Foucault:[8]

> Two light-emitting points placed near each other on the same vertical line light up at the same instant of time opposite a rotating mirror. The rays emitted from the upper point arrive at the mirror after passing through a tube filled with water; the rays coming from the lower point reach the mirror without having passed through any medium except air. We suppose that, as seen by the observer, the mirror turns from the right to left. If the theory of emission (particle theory) is right, if light is matter, the upper point will seem to be at the left of the lower point (after reflection by the rotating mirror); it will seem to be at its right if, on the contrary, light is the result of vibrations propagated through an ethereal medium.

Arago attempts now to formulate the assertion that his experiment is an "experimentum crucis" (crucial experiment) as explicitly as possible:

Does the image of the upper point appear at the left of the other image?
Light is a body.

Is the contrary the case? Does the image of the upper point appear at the right?
Light is an undulation.

When Foucault performed the experiment, he found that the image of the upper point appeared at the right. From this result he concluded, in accordance with Arago's logical argument, that light does *not* consist of moving particles. We cannot, of course, conclude that light is a wave motion in a medium because we cannot prove that this theory is the only alternative to the particle theory. On the other hand, no other theory had been elaborated except the undulation theory in the form that was developed by Young[9] and Fresnel:[10] Light is the propagation of transverse waves in an elastic medium. Hence, practically speaking, Foucault's experiment of 1850 was regarded as a definite confirmation of the Young-Fresnel undulation theory.

If we argue strictly and logically, the experiment proves only that the undulation theory *can* be right. If we say that the "crucial experiment" shows that the undulation theory is *very probably true,* this statement is correct only if we enumerate under what assumptions it would be so. The assumption is, obviously, that a theory that has been confirmed by a great many experiments and refused by none is very probably valid. But this, again, is only correct under this assumption: It is very improbable that to a very well confirmed and never refuted theory there is an alternative theory with the same qualities. Arago's "crucial experiment" proves the validity of the undulation hypothesis only if we assume that there is no other alternative to Newton's corpuscular theory of light than the undulation theory of Young and Fresnel. There is certainly no reason to believe that this is so from the point of view of logic. From the empirical point of view, it is certainly true that the number of hypotheses and theories which have been thoroughly elaborated and confirmed by experience has always been small. Therefore, it would seem that by excluding one after the other by crucial experiments the "true" hypothesis can eventually be obtained.

This belief in the small number of theories probably has its source in an analogy between theories and organisms. If we look at animals, for example, we see that there is a finite number of species, with finite differences between them. If we compare, *e.g.*, an elephant and an ostrich, they are very different, and there is no continuous transition between them. It is easy to envisage a "crucial experiment" by which we can decide whether a certain animal is an elephant. We have only to make sure that it has a trunk, since no other animal has a trunk. Such a superficial experiment is sufficient to prove that an animal is an elephant. There are no animals that have trunks like elephants and look otherwise like ostriches. On the other hand, if we find an ostrich feather on an animal, we can be certain that it is an ostrich because there is no animal that has one ostrich feather and is otherwise an elephant. This is probably what Duhem[11] meant by claiming that there are "crucial experiments" in biology but not in physics.

From all these considerations it is clear that Arago's "crucial experiment" did not "prove" the validity of Fresnel's undulation theory of light except under very definite assumptions. If we say, however, that it eliminated once and forever Newton's particle theory of light, we must also understand that this "elimination" was only achieved under very definite and rather arbitrary assumptions. What was actually "eliminated" by the Arago experiment was every hypothesis that implied a greater speed of light in water than in air. However, this does not follow from the particle hypothesis by itself, but from the additional hypothesis that the particles move according to Newton's laws, and are accelerated when they pass through water. If this increased speed turns out to be in disagreement with observation, we cannot tell whether this increased speed would also result if the particles moved according to other laws, or were attracted by water according to other laws. As a matter of fact, the Arago experiment "eliminates" only the combination: particle hypothesis plus laws of motion plus laws of force in water. We could, therefore, without any logical contradiction assume that the particle hypothesis itself is not eliminated. It could still be upheld if other laws of motion or other laws of force in water were assumed.

To sum up these considerations about Arago's "crucial experiment"; it does not eliminate the possibility of a particle theory that would account for all the phenomena that have been derived from Fresnel's theory of undulation.   However, such a new theory was not advanced before a new crucial experiment was devised that put the choice between particle theory and undulation theory to a new test, by which, in turn, the undulation theory was eliminated.

3. A Second "Crucial Experiment"

Following Foucault's experiment in 1850, the undulation theory of light was generally accepted, with the modification that after the work of James Clerk Maxwell[12] and Heinrich Hertz[13] the elastic vibrations were replaced by electromagnetic vibrations.   In 1902, a "crucial experiment" was again carried out, one that might decide between the particle theory and the undulation theory of light; its originator, Phillip Lenard,[14] was not as aware of the importance of his experiment as Arago had been.   As a matter of fact, Lenard's experiment was not recognized as crucial until 1905, when Einstein[15] directed the attention of physicists to it.   We shall view the experiment, however, as if it had been devised as a crucial experiment designed to provide the basis of a decision between the two theories of light.   We know (from Section 2) that the result of Arago's experiment did not prevent the possibility of a second "crucial experiment" that could give the opposite result.

We assume that light is emitted from a radiating point P and absorbed by a plane screen, perpendicular to the rays, that has the distance r and the small area a.   If the energy emitted from P per unit of time is L (luminosity), the energy that hits a per unit of time is $La/4\pi r^2$ (since the distribution of light at the distance r is a sphere with the surface $4\pi r^2$).   We assume now that we are able to measure the radiant energy that hits the screen per unit of time. If we assume that the undulation theory is correct, L is a constant and $La/4\pi r^2$ becomes infinitely smaller if r increases.   If we move the screen farther and farther away, the light energy hitting it per unit of time tends toward zero.   If, however, we assume that P emits "particles," e.g., n particles per unit of time, each possessing energy l, the result is different; then the energy hitting the screen per unit of time is $nla/4\pi r^2$, but the energy absorbed by the screen can never

be smaller than the energy of a single particle $(n = 1)$.   It can be
zero if all particles pass to the side of the screen $(n = 0)$.

We can, therefore, make the following experiment: If we move
the screen of area a farther and farther from P, either the absorbed
light energy will become smaller and smaller below all limits, or it
will reach a minimum and then suddenly drop to zero.   In order to
carry out this experiment, we must be able to measure very small
amounts of energy, which can be done by means of the photoelectric
effect.   If light falls upon the surface of some metal, electrons are
emitted from the surface, and the energy of these "photo-electrons"
is a measure of the light energy absorbed by the surface.   As we
mentioned, Philipp Lenard found[16] that the energy absorbed by the
metal always remains above a certain level, the energy of an incoming "particle of light."   The undulation theory requires that this
absorbed energy tend toward zero as the distance increases from the
source of light.   Lenard's experiment showed conclusively that the
lower limit of the radiation absorbed by the screen as the distance
increases is independent of the distance and dependent only upon
the color (frequency) of the light.   This experiment "eliminated"
the undulation theory in its classical form given by Fresnel, and
proved that a "particle theory" is possible.

The "particle theory" that is assumed in this argument does not
contain, as the Newtonian theory did, any laws according to which
particles are attracted by matter, but only the assumption that they
move in straight lines in empty space.   If we sum up the results of
both "crucial experiments," we must note that both the particle
theory and the undulation theory, in their classical forms, are
eliminated by these experiments.   The correct theory must be
different from both.   Since a great many conclusions drawn from
either the one or the other of the "classical" theories turned out to
be in agreement with experience, the new theory would obviously
contain some features of the particle theory and some features of
the wave theory.   This means that the new theory would, under
some circumstances, lead to the same results as one of the older
theories, but there is no logical reason to say that the new theory is
a kind of "sum" of the older ones or, as some have said, that light
must be both "wave" and "particle" at once.

When Einstein,[17] in 1905, made the point that this "second crucial experiment" "eliminated" the undulation theory, he attempted to modify it as little as possible. However, he had to introduce such alterations as would put the "renewed" undulation theory in agreement with Lenard's experiment. From the undulation theory in its classical form, it has been concluded that the energy of vibration over a spherical wave surface has the same value for equal areas, but this value decreases as the distance from zero increases. This result is refuted by Lenard's experiment. Einstein assumed that the energy is not homogeneously distributed over the wave but is concentrated in parcels, called *light quanta* or *photons*, which are part of the electromagnetic radiation and move with the speed of light. This is, of course, a clear contradiction to the fundamental laws of the electromagnetic field. Hence, at a great distance from the source of light a screen of area a can never be struck by an energy smaller than that of a photon. The total energy absorbed by the screen is the sum of the energies of all absorbed photons; hence, at great distances per unit of time, either one photon is absorbed or none.

Lenard's experiment shows that light emitted by a source, hitting metal and producing emission of photoelectrons, actually follows this pattern. While eliminating the theory of undulation, it confirms the particle theory in the sense that light is emitted from a source in certain parcels that have been called "light quanta" or "photons." Of course, the experiment only shows that it is admissible that light is emitted as a swarm of particles, but the motion of these particles must not follow the Newtonian laws in order not to be in disagreement with Arago's crucial experiment. In order to alter the undulation theory as little as possible, Einstein assumed that light should remain a propagation of waves, now electromagnetic waves, but that the energy should not be distributed in the wave homogeneously. There should be condensation of energy that will propagate like parcels in such a way that the screen can never be hit by less than one parcel. Each parcel has the same energy, provided that the frequency of light remains the same. The Lenard experiment showed that the energy of each parcel is proportional to the frequency of light: $E = h\nu$, where E is the energy of one parcel, ν the frequency of the light supposed to be monochromatic, and h a universal constant, called Planck's constant.

4. The Laws of Motion for Light Quanta

Since photons are part of the electromagnetic wave, they move with the speed of light. This is very different from the particles occurring in Newton's mechanics; the latter can be at rest or have any speed, whereas photons always travel at the speed of light, and can never be at rest. The number of photons absorbed by a screen determines the total light energy which is absorbed. Hence, all optical phenomena are now to be described as motions of photons. We have already mentioned that to say that light consists of "particles" has no precise meaning unless we formulate the laws of motion of these particles. As we have learned by now and shall learn more elaborately, photons follow laws that are very different from the laws of motion that govern Newtonian particles. Hence, it is more or less a matter of taste whether we give to photons the name "particle" or not.

The most important optical phenomena are well covered by the undulation theory. It is clear that the introduction of photons must not alter this situation. Every optical phenomenon is ultimately describable as a distribution of bright and dark regions on a screen or by the physico-chemical effects of brightness or lack of it on a certain body. Under ordinary circumstances, we must require that the photons follow such laws of motion that they produce the same distribution of bright and dark areas as is derived from the undulation theory. As we learned in our second "crucial experiment," the undulation theory leads to incorrect results if the density of the light energy becomes very small. The photoelectric effect shows a deviation from the undulation theory if a small number of photons hits the observed area.

The most characteristic results that are derived from the undulation theory are interference and diffraction. They agree with the results of observation. These experimental observations have, as a matter of fact, been the most convincing proof of the undulation theory. We shall not go into broad generalities, but will describe a typical diffraction experiment: the passage of light through two slits in a diaphragm. The slits may be separated by the small distance a, and the wave length of light may be λ. In order to test the effect upon light of a passage through slits, a screen is erected at a distance D beyond the diaphragm. If the bundle of light rays is perpendicular to the diaphragm, we observe on the screen a pattern of bright and dark fringes. There is a bright fringe in the center and other bright fringes parallel to it, separated by dark fringes. If the wave length λ is small compared with the distance a between the slits, the distance between the bright fringes on the screen is approximately $D\lambda/a$.

The undulation theory derives this phenomenon (diffraction) in the following way: From each slit waves are propagated beyond the diaphragm. These two sets of waves interact by superposition of oscillations, by "interference." Where the crests of waves meet we have brightness, where a crest meets a valley we have darkness. Mathematically speaking, where the paths of light coming from both slits have a difference of one wave length or an integral multiple of one wave length (λ, 2λ, 3λ, $\cdots$) we have brightness, but where this difference is an odd multiple of a half wave length ($\lambda/2$, $3\lambda/2$, $\cdots$) we have darkness.

We must now describe this fundamental optical phenomenon by using the concept of "photons." The screen is not hit by a homogeneous wave front, but by parcels of energy called photons. What we called before "dark fringes" on the screen we now consider as regions where no photons hit, while bright fringes are regions where a great number of photons hit the screen. Hence, we say that the incoming bundle of light rays is divided by the diaphragm into two bundles passing through the two slits; then, from each slit, swarms of photons pass into the space between diaphragm and screen. At the slit, the photons are deviated in such a way that most of them hit the screen in the bright regions, while only a few arrive in the dark regions. To speak more precisely, most photons continue on their path, perpendicularly, onto the screen; a considerable number are deviated by a distance $D\lambda/a$ to the right and to the left, while a smaller number are deviated by the distances $2D\lambda/a$, $3D\lambda/a$ $\cdots$.

If we attempt to derive a law for the motion of a single photon from this result of the classical theory of undulation, we get into great trouble. We may consider the slits S_1 and S_2 at a mutual distance a. If a single photon passes through a slit S_1, we know only that among a swarm of photons passing partly through S_1 and partly through S_2, most will hit the screen at the bright areas, and few at the dark areas. But if we consider only the single photon passing through S_1, we know only that it is to hit the screen at such a point that the final pattern of hits by the swarm is the classical pattern of the undulation theory. This pattern is determined by the distance a of the slits, but the initial conditions of our experiment which deal with the single slit S_1 do not contain a. If we arranged the analogous experiment, using the slit S_2, we would wind up with the same difficulty. It seems to be impossible to regard the classical pattern of diffraction by two slits as the superposition of the motions of single photons that pass through the single slits according to a specific law of motion.

On the other hand, we can easily imagine and perform the following experiment: We have a source of photons that emits them very slowly, for example, one per second; we observe their passage through our two slits and their arrival upon the screen. Then the pattern of the fringes with the distance between the fringes $f = D\lambda/a$ is clearly determined by the initial conditions of our experiment: the source of light, the two slits, and the screen. If we wait until a great number of photons have passed, we can predict that as the result of these many hits we shall observe the fringes separated by the distance $D\lambda/a$. If we block one of the slits, this pattern disappears; we observe another pattern (that depends on the width of the slit) which we will pass over here. The pattern of fringes for two slits is not the superposition of the two patterns for single slits. Hence, there is no law of motion that would determine the trajectory of a single photon and allow us to derive the observed facts that occur when photons pass two slits.

This means a very fundamental change in the general pattern of physics, in particular, the laws of motion. While Newtonian physics, including the theory of relativity, builds up all physics on the basis of the trajectories of particles, we note that the new theory of light leads to the concept of particles (photons), but does not reduce the observed facts to the trajectories traversed by these particles. All we can do is to derive from the given experimental arrangement the facts that we are to observe, without being able to describe them in terms of trajectories of particles. If we consider the light source L and a bright spot on a screen produced by the photons emitted from L, the new modified undulation theory cannot answer the question whether the photon that produced this spot passed the diaphragm through slit S_1 or S_2. We shall discuss this whole argument more thoroughly and elaborately when we develop the new theories about the motions of material particles that are to replace Newton's laws of motion.

5. The Laws of Motion for Very Small Material Particles

Laplace, in his famous statement about the omniscient spirit who knows the initial positions and velocities of all particles in the universe, stated that this spirit could predict from the Newtonian laws of motion the future of the universe, provided that he was also a perfect mathematician who could integrate the differential equation of motion for arbitrary initial conditions. Laplace says explicitly that this sweeping statement takes for granted the condition that all particles, even small atoms, follow the same laws of motion which Newton set up to derive the motions of celestial bodies. The advance of atomic physics in the twentieth century seems to indicate that the motions of subatomic particles, such as electrons or nuclei, are actually not derivable from Newton's laws.

The most familiar example is the motion of electrons around the nucleus of the hydrogen atom according to the theory of Niels Bohr.[18] According to Newton's laws, the negatively charged electron would move around the positive nucleus in a circular motion with any radius, but Bohr showed that the spectral lines emitted by hydrogen can only be derived correctly if we assume that only circular motions of specific radii are possible. This means that there must be laws that exclude most of the orbits that are in agreement with Newton's laws of motion. To give a simple example: in the case of the hydrogen atom, the only orbits that are compatible with the observed spectral lines are those whose angular momentum is an integral multiple of a certain constant that is equal to $h/2\pi$, where h is Planck's constant, which occurred in the previous Section in the expression for the energy of a light quantum of the frequency ν. This energy $h\nu$ is observable in the photoelectric effect, and, therefore, h can be calculated. This restriction could not be derived from traditional mechanics. The question arose of how to modify Newtonian mechanics in such a way that among the possible trajectories some specific ones could be selected as the only possible ones.

The solution is due to a Frenchman, Prince Louis de Broglie,[19] who was at the time of discovery a learned historian and an amateur physicist. His attention was directed to the photons; they were

particles with motions that were not determined by our laws of
mechanics, but by the laws of diffraction. If the wave length is
small with respect to the obstacles or openings, the trajectories of
photons are rays in the sense of geometrical optics; they can be regarded as trajectories of particles.   In passing through small openings
or around small obstacles, however, the paths of photons are determined by the theory of undulation, and eventually the conception of
path has to be eliminated altogether.   De Broglie's idea was to treat
the motion of small particles in the same way.   His salient point was
the fact that in optics only a very special type of phenomenon can be
described by bundles of light rays and their orthogonal surfaces, the
wave surfaces.   The most general optical phenomena are described
by the "wave equation," a differential equation of the second order.
The solution of this equation can be interpreted as a bundle of
rays and an orthogonal family of wave surfaces when all obstacles
or slits are of a size that is very large compared with the wave
length of light.   De Broglie thought that perhaps the trajectories
of particles in Newtonian mechanics play the same role as the paths
of light rays in optics; perhaps these trajectories describe only a
very special type of motion.   There may be phenomena in mechanics
that are described by a generalized mechanics which is in the same
relation to Newton's mechanics of trajectories as the general
undulation theory of light is to the optics of light rays or of photons
traversing trajectories.

To make this generalization, de Broglie assumed that a type of
wave could be introduced (called by him "matter waves," later
Broglie waves) that accounts for the trajectories of material particles by a theory of diffraction in the same way that the general
theory of undulation accounts for the paths of light rays.   It was
obvious that a moving photon had a wave length λ, the wave length
of the light wave of which the photon was a part.   In order to
ascribe a wave length to a moving particle of mass m and speed v,
a new hypothesis was necessary, and was actually advanced by
de Broglie; it was simple and natural.   When Einstein introduced
the photons, he directed his attention to the mechanical momentum
of the photon.   From the general theory of the electromagnetic
field, it follows that every portion of electromagnetic energy E propagated with a speed c exerts a pressure (light pressure) upon a material body that is hit; hence, it imparts a linear momentum p to this

body. It follows mathematically from Maxwell's theory of the electromagnetic field that this momentum is $p = E/c$. Einstein could conclude that the momentum of a photon (energy $E = h\nu$) is $p = h\nu/c$ or h/λ since the wave length is linked to the frequency ν by $\lambda\nu = c$. If, on the other hand, a particle of mass m is moving with a speed v (that is small compared with the speed of light c) the momentum of this particle is, according to Newton's laws, $p = mv$. De Broglie's hypothesis was simply that a particle's motion was determined by a radiation, the photons of which had the same momentum as the particle. This means that the wave length λ of this radiation is determined by $p = mv = h\nu/c = h/\lambda$ or $\lambda = h/mv$, known as "de Broglie's equation." De Broglie's law

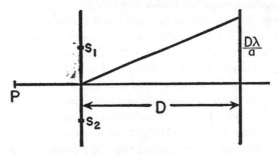

Figure 34

for the motion of particles then was: If small particles of mass m and the speed v are moving through slits in a diaphragm or around obstacles, they behave like photons of a wave length $\lambda = h/mv$, the "de Broglie wave length."

Let us describe a typical experimental situation: a source of electrons (charged particles) P, a diaphragm with two slits S_1 and S_2, and a screen parallel to the diaphragm and at the distance D. P may be situated symmetrically with respect to S_1 and S_2 (Figure 34). If the distance S_1S_2 is small, a particle passing from P through S_1 or S_2, and hitting the screen, will move nearly perpendicular to the diaphragm. In order to make the hits of particles observable, we assume that the screen is covered with zinc sulfide; then any contact will produce a brilliant spot. What happens when a beam of electrons is emitted from P, passes through the slits, and hits the screen? If the particles have the mass m and the speed v, a de Broglie wave of the wave length $\lambda = h/mv$ is emitted from P, passes the two slits, and produces a diffraction pattern on the screen consisting of a bright central fringe, a parallel bright fringe at the distance $D\lambda/a = Dh/mva$, and other weaker fringes. If the number of electrons striking per unit of time is very great, the fringes are produced almost immediately; but if the radiation is thin, if only at long intervals electrons strike the screen, scintillations will appear most frequently in the central region, very rarely at the dark regions between the fringes, and fairly frequently in the region of the first bright fringes at the right and the left at the distance Dh/mva. If the experimental arrangement is given, we can compute unambiguously the result of the experiment, the statistical distribution of scintillation points on the screen.

Does this mean that we know the law of motion of a single particle? There is certainly no law that will tell us where a single particle will strike the screen and, moreover, if we choose a hit on the screen, there is no way to find out whether this particle passed through the diaphragm through S_1 or S_2. Therefore, we may say that the theory of de Broglie determines the observable results of observable initial conditions, but does not determine the "trajectory of a particle." For a better understanding, it might perhaps be well to single out two special borderline cases. We shall assume first that one slit (S_2) may be blocked. If we neglect the diffraction produced by a single slit, the passing of electrons through S_1 does not produce any alternation of bright and dark fringes. The electrons hit the screen as frequently in the central region as in the "dark region." The scintillations will be distributed homogeneously over the screen. On the other hand, if the two slits are very far distant from each other, that is, if the distance a is large relative to λ, the distance from the first fringe to the central one tends towards zero, and the fringes merge with the central one. All particles hit the screen in the central zone, and outside it there is darkness. The outcome of the experiment is computed from the observable initial conditions by using as a mathematical device the interference of waves emitted from the two slits S_1 and S_2. The result tells us how many scintillations we shall observe in any area of the screen, but tells us nothing about the paths of the particles from the source through the diaphragm to the screen.

9

The New Language of the Atomic World

1. Heisenberg's Uncertainty Relation[1]

The laws of motion for small particles are formulated[2] in such a way that they connect observable initial conditions with observable results; the laws say nothing about "moving" particles. The scientist has always felt the need for retaining the traditional laws of motion as long as possible. They have been absorbed by our common-sense language, and it is certainly very helpful to use this same language as much as possible. In using common-sense language, the imagination of the scientist works with more ease and detachment than in using an abstract language where every result must be found by step-by-step formal reasoning.

If we again consider the swarm of particles passing through the diaphragm through two slits S_1 and S_2, we can describe the situation in the following way: A wave passes through both slits and produces interference on the other side. To say that one particle passes through both slits would be an odd way to speak about a small particle; therefore, one would do better to say that the particle passes through the diaphragm, but that the exact location of its passing is described by an uncertainty a because the distance between the slits through which it passes is a. In Newtonian mechanics, the initial conditions of motion are the position (coordinate) and velocity (or momentum). If these data are given, we can predict the future motion from Newton's laws of motion.   In the case described above, the application of de Broglie's theory to the passing of particles through the two slits, we learned that the uncertainty in the position (or in the x-coordinate) is a.

Is there also an "uncertainty" in the initial velocity—to speak precisely, in the x-component of the velocity?   We learned from the theory of diffraction that most particles continue to move either perpendicular to the screen beyond the diaphragm ($y = 0$), or are deviated by an angle $\varphi = \lambda/a = h/mva$.   Hence, the respective x-components of momentum, p_x, are $p_x = 0$, and $p_x = p\varphi = h/a$. From this it is evident that the "uncertainty" in the value of momentum is $p_x = h/a$.

Hence, if we denote the uncertainty in the x-coordinate by Δx, and the uncertainty in the x-component of momentum by Δp_x, we have $\Delta x = a$, $\Delta p_x = h/a$, and, therefore, $\Delta x \cdot \Delta p_x = h$.   This is the famous "uncertainty principle" first announced by the German physicist, Werner Heisenberg.   It states that the product of uncertainties of coordinate and momentum of a particle is equal to Planck's constant h($h = 6.55 \cdot 10^{-27}$ erg-sec.).   We see here that the formula can be stated very well without using the psychological term "uncertainty."   It is a formula about an empirical physical fact, diffraction of particles by two slits.   The formula states simply: If the slits S_1 and S_2 are separated by a distance a, the particles which are deviated from the direction perpendicular to $S_1 \rightarrow S_2$ have a momentum h/a in the x-direction.

If we eliminate the theory of diffraction which uses a mathematical scheme, the superposition of waves, to derive the result and try to formulate the laws of this phenomenon in terms of passing particles, we may say: If the location of the particles in passing through the diaphragm is determined by the setup of the experiment with an uncertainty a in the x-direction, the momentum (in the x-direction) of the particles within or behind the diaphragm is determined with an uncertainty h/a.   We see that the "uncertainties" are determined by the setup of the experiment, and not by any subjective events in the observer's mind.

It is instructive to examine the two extreme cases in the uncertainty of position.   The first one is that the uncertainty disappears ($a = 0$); then we have only one slit.   If we again neglect the diffraction due to the width of the slit, the particles are diffracted equally in all directions behind the diaphragm. The scintillations cover the screen with uniform density. Then the directions of the momenta are equally distributed; the direction of the momentum of a single passing particle is completely undetermined, or completely uncertain. The other extreme would be that the distance between the slits is so great that λ/a can be neglected. This means that $\varphi = 0$; the momenta of all particles are directed perpendicular to the diaphragm (Figure 35). In the first case, the x-component of momentum is completely undetermined; therefore, the direction of the velocity of a particle is undetermined. In the second case, the position of the particle at the x-axis is not, or is very little, determined because the distance a between the slits is large. However, the direction of momentum or velocity is strictly, or nearly strictly, determined; it is perpendicular to the diaphragm.

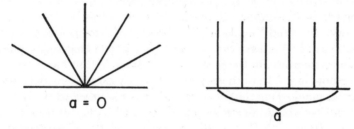

Figure 35

In each of these cases, we can make use, to a certain degree, of the Newtonian law of inertia in order to predict future motion from the present state. In the first case, we know the precise position of the particle in passing through the screen; then we can conclude that the particle may move in any direction without any discrimination between the directions. In the second case, we know the initial direction but nothing definite about the position when passing through the slits; then all particles will continue moving in the direction perpendicular to the diaphragm until they strike the screen. In the first case, we use only the position of the particles to predict the future movement by using Newtonian mechanics; in the second case, we use only the velocity. If we are concerned with neither of the extreme cases, our prediction looks like this: we are given the uncertainty of position a, then follows the uncertainty of momentum h/a, which, in turn, leads to the diffraction of electrons passing through two slits. If $a = 0$, the direction of the diffracted particle is completely undetermined; as uncertainty a increases, diffraction lines become more and more definite until, eventually (for large a), all particles move without any uncertainty of direction.

It is obvious here that the initial conditions for moving particles in Newtonian mechanics are completely different from the initial conditions for electrons passing through slits. In Newton's mechanics, the initial conditions are the position and velocity of each particle. In the mechanics of small particles based on de Broglie's wave mechanics, the initial conditions are the uncertainties of position determined by the holes in the diaphragm and the uncertainties of momenta connected with them. In the two extreme cases, the initial conditions are as follows: in the first one, the precise position of a particle with a complete uncertainty of momentum; and in the second one, the precise direction with a wide uncertainty of position.   If we wish to express this in a slightly different way, yet keep "weakened" principles of Newtonian mechanics, we can say: in the first case we have a particle that has a position but no momentum, in the second case a particle that has momentum but no position.   In both cases we can draw directly from Newton's laws the conclusions that correspond to these cases where one state variable (coordinate or velocity) is given, while the other may have any value.

So far, we have attempted to speak only about observable physical facts.   By the word "uncertainty" we have referred not to the state of some physicist's mind, but to the margin within which the coordinate of a particle would be contained if we described the observable facts in terms of particles. If we wish to keep strictly to the observable facts, however, we must eliminate one expression which we have used without definition in terms of observable facts, "the momentum of a particle in the x-direction."   We introduced the angle φ by which an incoming particle is deviated.   The angle φ, we assumed, can be measured by observing the location of the scintillations on the screen.   Then we assumed that the particle is moving from the slit to this point in a certain direction which is the direction of the linear momentum p of the particle.   Its projection upon the screen, p_x, is the momentum in the x-direction.   However, we know that there is no particle, in the Newtonian sense, moving from the slit to the screen.   Therefore, if we stick to observable quantities, the angle φ cannot be interpreted as the direction of the momentum of a particle.

Niels Bohr[3] suggested defining the x-component of momentum in the following way, which provides a real "operational" definition: If a particle passes through a slit with a velocity v under the angle φ, the momentum in passing through the slit is $mv\varphi$ in the x-direction. Therefore, the particle will impart to the diaphragm a momentum $mv\varphi$ in the x-direction. In describing the pattern of diffraction, the location of the scintillations of striking particles on the screen, we have tacitly assumed that our whole setup (source of particles, diaphragm, and screen) is rigidly connected with a framework that is, in turn, rigidly connected with an inertial system that we can, for our purpose, identify with our earth; then the momentum imparted by the passing particles will not move the diaphragm. In order to allow a measurement of this momentum, Bohr suggested linking the diaphragm to the framework by elastic springs; then the particles passing through the screen will impart to the diaphragm a velocity relative to the framework. This velocity of a medium-sized body is observable, and from it the momentum of the particle can be computed. This provides a new definition of the x-component of the momentum of the passing particle. This definition does not assume that the particles move from the slit to the screen according to Newton's laws of motion.

Now we can formulate the uncertainty relation in such a way that it does not refer at all to the scintillations on the screen, but only to the position and momentum of the particle passing through the diaphragm. If we give the diaphragm a rigid connection with the frame, and only one slit is open, we can find from the pattern of scintillations on the screen the exact location of the slit relative to the framework. If we have two slits, we can find from the diffraction pattern the position of the particle in passing through the slits, including the uncertainty a. If, however, we connect the diaphragm not rigidly but by an elastic spring with the inertial framework, the particles will in passing impart momentum to the diaphragm. Then we can measure the momentum of the diaphragm by the lengthening or shortening of the spring, and can compute from Newton's laws the momentum (in the x-direction) of the particle. If we know, in addition to the momentum, the position of the slits (relative to the framework) when the particle passes through, we would know for one instant of time the position and momentum of the particle, and could compute from Newton's laws the path of each particle until it produces a scintillation on the screen.   By observing a great many particles we would, eventually, derive the diffraction pattern by tracing the Newtonian trajectories of all particles.

We would then be able to obtain the diffraction pattern in two ways.   First, we could use, as we did originally, the interference of waves passing through the slits; or secondly, we could use the trajectories of the particles.   Then we could find out the path of the particle that produces a specific scintillation and the slit through which the particle passed.   Bohr showed, however, that this is an illusion, and that an experimental arrangement that allows us to measure the momenta of the particles on passing through the screen frustrates every attempt to measure the position of the slit relative to the inertial framework.   This is easy to see if we keep in mind that the two colliding bodies (diaphragm and particle) do not comprise a medium-sized body following Newton's laws.   They are "an atomic object" like the particle itself and follow the "uncertainty relation."   From this relation it follows that a precise position (relative to the framework) cannot be assigned to such an "atomic object" unless the momentum is completely undefined. If we define a momentum with a finite uncertainty Δp_x, the uncertainty Δx of position will be $\Delta x = h/\Delta p_x$; but if the position of the diaphragm and therefore of the slit relative to the frame has an uncertainty, the scintillations on the screen will no longer produce the simple diffraction pattern that we found in the case of a diaphragm rigidly connected with the frame.   The pattern of fringes will be blurred.   This means that if we are able to observe the diffraction pattern, we can measure the positions of the particles in passing through the screen, but we cannot measure the momenta. If, however, we can measure the momenta of the particles in passing through the diaphragm, we cannot observe the diffraction pattern.

2. Bohr's Principle of Complementarity

This state of affairs was formulated and generalized by Niels Bohr, in his famous "principle of complementarity."[4]   If "atomic objects" are emitted by a source at P, pass through the slits of a diaphragm, and produce scintillations on a screen, we cannot describe this phenomenon by stating what law governs each individual particle's traversing a path from P to the screen. Every description of such a path would require that at each instant of time the position and momentum of the particle could be numerically given. If we speak of "describing the phenomenon," we mean that we describe the source of emission, the diaphragm with its slits, and the scintillations on the screen. These descriptions do not contain any terms or expressions other than those used in the language of our everyday life.

We have learned from the remarks in Section 1 that we cannot describe these phenomena by introducing trajectories of particles each of which would pass through an individual slit. However, the development of our language from childhood, and a growing familiarity with elementary physics and mathematics urges us to introduce the expressions "position" and "velocity" (or "momentum") of a particle into the language that we use in the description of these phenomena. Niels Bohr has shown how the terms "position of a particle" and "momentum of a particle" can be introduced in a limited way, since their use in the familiar way seems not to be feasible. When the diaphragm is rigidly connected with the frame, the position of the particle passing through the diaphragm is defined by the position of the slit. If there are two slits, this position is defined with an uncertainty $\Delta x = a$. The momentum of an individual particle in passing through the slit is not defined at all. From the position of the slits, we can predict the position of the diffraction pattern on the screen. If the particles have a mass m and a speed v perpendicular to the diaphragm, they will produce bright and dark fringes with the distance $D\lambda/a$ between two bright fringes (where D is the distance between diaphragm and screen). As the momentum of an individual particle parallel to the diaphragm cannot be observed or computed, we cannot find out through which slit an individual particle will pass, or even has passed.

If, however, the diaphragm is not connected rigidly with the frame, but by an elastic spring, the momentum of an individual particle can be measured too[5] (as we saw in Section 1), but in this case the positions of the diaphragm and of the slit relative to the frame cannot be defined exactly. If we consider the collision between atomic object and diaphragm, we have to do with a two-body problem. One of these two bodies is an "atomic object"—

hence, the system of these two bodies is also an atomic object. As such, it obeys the uncertainty relation according to which a precise value of momentum does not allow any statement about position. If the momentum can be defined within a certain margin Δp_x, the position can be computed with an uncertainty $\Delta x = h/\Delta p_x$. This means that the measurement of the momentum of an individual particle, as described above, brings about an uncertainty in the position of the diaphragm. Hence, the diffraction pattern can be computed only with a certain uncertainty. If this uncertainty is of the order of magnitude of the distance between two bright lines in the diffraction pattern, this pattern is completely blurred and practically nonexistent. We shall prove later that the uncertainty in the position of our slits relative to the framework is actually so great that the diffraction pattern is destroyed if we attempt to discover the path of the "atomic object" between source and screen.

From these considerations, Bohr has drawn extremely important consequences concerning the conceptual pattern by which we can describe the motion of atomic objects. In Newton's mechanics a moving particle possesses at each instant of time a position and a velocity, whatever the state of motion of the surrounding bodies may be. But from our remarks about the principle of uncertainty in Section 1, we can conclude, so Bohr says, that we must assume specific arrangements of the bodies surrounding an atomic object in order to be able to use words like "position" or "momentum" in the description of the phenomena described above. If we consider the experimental situation in which the diaphragm and the slits are forced by rigid constraints to be at rest relative to the framework, we can define the term "position of a particle at a certain instant of time," but we cannot define the term "momentum of a particle at a certain instant." On the other hand, if our experimental arrangement is a connection of the diaphragm to the frame by elastic springs, we can define the term "momentum of the particle in passing through the screen," but we cannot define the "position of the particle relative to the frame."

We have two experimental arrangements that exclude each other. In the first case, we can describe our phenomenon by using the term "position of a particle relative to the frame"; while in the second case, we can describe the phenomenon by the term "momentum of the particle in passing through the screen." In both cases, we mean by definition an "operational definition." The descriptions of our phenomena are of the following types: "If the particles have a certain position in passing through the diaphragm, the scintillations on the screen will follow a certain pattern (diffraction pattern)." In the second case, the description will be: "If an atomic object has a certain momentum as it passes through the slit, the diaphragm will start moving with a certain momentum and stretch the elastic spring by which it is attached to the frame." These two experimental arrangements exclude each other. In each case, we describe the passing of atomic objects from the source through the two slits to the screen in a different way.

Bohr says that these two descriptions "complement" each other. He has based his "principle of complementarity" on this state of affairs. It states, for simple cases of this type: the motion of an atomic object cannot be described by the trajectory of a particle that possesses at every instant of time a specific position and velocity. We can, however, consider "complementary" experimental arrangements which allow either a description in terms of positions or in terms of momenta of particles. In the first case, we can say— in a somewhat perfunctory way—that the atomic object can be regarded as a particle that possesses position but no momentum; in the second case, that it is a particle that possesses momentum but no position.

Very frequently, in popular presentations which are occasionally written by scientists, the laws governing the motions of atomic objects are formulated in a misleading way. Some authors have said that according to the contemporary laws of motion for atomic particles the position and velocity of a particle cannot be measured at the same instant. If we measure the coordinate (position), we "destroy" the possibility of measuring the momentum, and *vice versa*. This formulation is misleading because it gives the impression that before the measurement there was a "particle" that possessed both "position" and "velocity," and that the "measurement of its position" destroyed the possibility of "measuring its momentum." As a matter of fact, the atomic object itself cannot be described by the terms "position" or "velocity." Obviously, what does not "exist" cannot be "destroyed." Only if certain experimental arrangements surround the atomic object can the terms "position" or "momentum" be defined, but there is no arrangement in which both can be defined and measured.

3. "Position and Momentum of a Particle" Has No Operational Meaning

We have learned from Bohr's principle of complementarity that certain experimental arrangements allow us to define "position" and others to define "momentum" of an atomic object. These definitions are "operational definitions"; we can describe specific physical operations by which specific values are assigned to the coordinates or velocity components of such an object. If we know the "position" of an object in this sense, we can derive conclusions about the effect produced by the object in its environment, for example, scintillations on a screen or the observable motion of a diaphragm. Only if a definition can be used in this way shall we call it an "operational definition." This means that there must be physical laws in which the "operationally defined" term actually occurs.

We could, of course, relax the requirements that an "operational definition" has to meet, and drop the requirement that the defined term be actually employed in formulating a physical law. If we do so, we can certainly "define" the terms "position" and "velocity" of an atomic object by actual physical operations which are somewhat analogous to the operations by which these terms are defined in Newton's physics. We could, for example, consider an atomic particle (*e.g.*, an electron) passing through two parallel diaphragms in a direction perpendicular to both. There is one slit in each diaphragm, through which the particle passes. The distance between the slits may be D; the time in which the particle traverses this distance may be T. Therefore, the velocity with which the object passes through the slit of the second diaphragm may be defined by $v = D/T$. This definition seems to be natural because it is one of the possible definitions of velocity in Newton's mechanics; but while in this "classical theory" the knowledge of position and velocity at a certain instant allows us to compute the future motion, this is certainly not the case if we use the definition of velocity that we have just given ($v = D/T$). We well know that, according to the de Broglie theory, a particle that has just passed through the slit will not continue to move in the same direction, but will be deviated according to the law of diffraction. Therefore, position and velocity (so defined) do not determine the direction of the future motion. This "velocity" does not enter into any law of physics. It is not an "operational definition" as we have defined the term; this definition of velocity is not helpful in formulating the laws of atomic physics.

We could, of course, imagine a great diversity of ways of defining "velocity" if we did not require that this velocity occur in the laws of physics. If we consider this example, we realize that we can say more generally that there is no law of physics in which the term "position and velocity of an atomic object" occurs, whereas the laws of Newtonian physics employ the term "position and velocity of a medium-sized material particle." We can also say simply that the term "position and velocity of an observable body" has an operational meaning, but that "position and velocity of an atomic object" has no operational meaning. This formulation is much more general and instructive than the frequently used statement that the position and velocity of an atomic object can never be measured simultaneously. If we measure "position," we destroy the possibility of measuring the velocity of this particle. This formulation, which ascribes to measurement a "destructive effect," is misleading for the simple reason that it leads to the false concept that there "exists" a particle which "possesses position and veloccity" and which is somehow altered by the process of measurement. According to Bohr's formulation, the passing of an atomic object cannot be described adequately by the trajectory of a particle, and there are no laws about the motions of such small objects that contain the term "position and velocity of a particle." We must always keep in mind that arbitrary operational definitions cannot produce concepts that are helpful in physics. Any great advance consists in creating some operational definitions that allow us to formulate the laws of physics more adequately and practically than did the older definitions.

We shall now show, by means of a simple example given by Niels Bohr, how attempts at tracing the trajectory of a particle from the source to an individual point of scintillation on the screen lead to an experimental situation in which the pattern of scintillations disappears. Perhaps the most conspicuous proof of our inability to ascribe a trajectory to an atomic object is our inability to state through which slit the particle passed on its way from the source to

the screen.   We may assume that the source P is a point; the particles may diverge from it and pass through the two slits S_1 and S_2
in a diaphragm until they produce scintillations on a screen.   The
distances from the source to the diaphragm and from the diaphragm
to the screen may both be D.   The small angle between the directions from the source P to the two slits may be ω.   We have learned
that the pattern of scintillations on the screen cannot tell us anything
about through which slit the particle passed that produced the
scintillation at a certain point P'.   However, it seems that an
experimental arrangement could be envisaged in which it might be
possible to distinguish through which slit a certain particle passed.

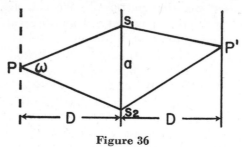

Figure 36

If we again make the arrangement that the diaphragm with the slits
S_1 and S_2 is connected with the frame by an elastic spring, the
passing of particles through these slits will impart momentum to the
diaphragm.   The direction of the momentum will be different
according to whether the particle passes through S_1 or S_2.   If the
momentum perpendicular to the diaphragm is again p, the difference
between momenta in the x-direction is ωp, if we compare passages
through S_1 and S_2 (Figure 36).

We can determine the slit through which the particle is passing if
we are able to measure differences of momenta which are of the same
order of magnitude as $\omega p = h\omega/\lambda$.   Since, as can be seen from the
figure, $D\omega = a$,   we   have $\omega p = ha/D\lambda = \Delta p_x$.   By   interaction
between the particle and the diaphragm, this momentum can be
transferred to the diaphragm.   We know that the two-body system
consisting of diaphragm and atomic object (particle) is itself an
atomic object.   If the "uncertainty" of its momentum is smaller
than or equal to $ha/D\lambda$ relative to the framework, there must be an
uncertainty Δx in the coordinate of the diaphragm relative to the

frame. According to the "uncertainty relation" $\Delta p_x \cdot \Delta x = h$, we conclude that Δx must be greater than or equal to $h/\Delta p_x$, or, to take the simplest case, $\Delta x = h/\Delta p_x$ or $\Delta x = D\lambda/a$; but, as we learned in Chapter 8, Section 4, this is just the distance between the two bright fringes in the diffraction pattern on the screen. If the uncertainty of the position of these fringes relative to the inertial frame is of the same magnitude as the distance between the fringes, the pattern is blurred out. We can see from these considerations that an experimental arrangement that permits us to find out through which slit the particle passes is an arrangement in which there is no sharp pattern of diffraction. Therefore, we would not be able to pursue a particle from the source through the slit to a scintillation point on the screen.

4. Facts, Words, and Atoms

As in every branch of physical science, and perhaps of every science, we work on two levels: the level of sense observation, and the level of description by a conceptual or, rather, verbal scheme. In the evolution of science, the discrepancy between these two levels has increased continually and has become very conspicuous in the realm of atomic objects, such as electrons, nuclei, and the like. A great many confused presentations concerning this field are due to the fact that the authors have not given sufficient attention to a clear distinction and to a well-defined correlation between these two levels. Our starting point must be the fact that all observable phenomena of atomic and nuclear physics can be described in everyday language. They belong in the realm of medium-sized bodies and can, therefore, be described in the language of Newtonian (classical) physics. From this fact we cannot, of course, conclude that such phenomena can be derived from the laws of Newtonian physics or, in other words, be "explained" by Newtonian physics. Niels Bohr said:[6]

> *However far the phenomena transcend the scope of classical physical explanation, the account of all evidence must be expressed in classical terms.* The argument is that simply by the word "experiment" we refer to a situation where we can tell others what we have done and what we have learned and that, therefore, the account of the experimental arrangement and of the results of the observations must be

expressed in unambiguous language with suitable application of the terminology of classical physics.

In our example given in the previous sections, the "experimental arrangements" consisted of the source of atomic objects, the diaphragm with the slits, and the screen. The "results" consisted in the pattern of bright and dark fringes or in the scintillation points on the screen. The problem that physical science has to face is that of inventing a system of principles from which statements can be derived by means of which we can predict or compute the "results" if we know the "experimental arrangement." As we said above, quoting Bohr, "the phenomena transcend the scope of classical explanation." The initial arrangements and the results can be described in the language of classical physics because we have to do with familiar mechanical and optical objects like medium-sized bodies, bright fringes, and point-like scintillations on a screen, etc. However, the principles from which we can derive the connection between initial arrangements and results are, as we have learned by now, not principles that contain concepts that are familiar from Newtonian physics; they cannot be expressed by describing trajectories of particles, propagation of waves in a medium, or similar concepts.

We may call a description of particles, waves, or similar concepts of classical physics a "picture." Bohr said:[7]

> Evidence obtained under different experimental conditions cannot be comprehended within a single picture, but must be regarded as *complementary* in the sense that only the totality of the phenomena exhausts the possible information about the objects.

A "single picture" would be provided, for example, by a description of all phenomena in terms of trajectories traversed by particles. Such a "picture" does not exist, but we can investigate for every experimental arrangement the phenomena occurring under these circumstances. The "objects" cannot be described as particles and their trajectories, but can only be described by describing all the phenomena that occur under different experimental conditions. "Under these circumstances," Bohr continues, "an essential element of ambiguity is introduced in ascribing conventional physical attributes to atomic objects." If I wish to describe the formation of scintillations on the screen in terms of particles with determined positions and velocities along trajectories, I ascribe to the atomic object "conventional physical attributes." According to Bohr, however, I cannot do so in an unambiguous way. If I want to ascribe position to the objects, I must use an experimental arrangement that is different from and even incompatible with the arrangement which allows me to ascribe momentum to the objects.

It is important, however, to understand that to "ascribe conventional attributes" even partially or "complementarily" is not necessary, if we are only interested in predicting the observable results of an experiment. If we want to understand how physicists actually doing research in atomic and nuclear physics proceed in predicting or producing observable results, we can do so by examining the example which has been used earlier in this chapter, the scintillation pattern produced by atomic objects after passing through slits in a diaphragm. If we are interested only in the observable results, the location of scintillations on the screen, or the motion of the diaphragm relative to the inertial frame, we can go back to the theory of the de Broglie waves. We shall present this theory here in a very elementary way, which is certainly an oversimplification but makes it understandable to readers who are not versed in mathematical physics. We must not forget that a great number of physics students and almost all chemistry students belong in this group.

If a stream of atomic objects (e.g., electrons) hit the diaphragm in which there are two slits, we must introduce de Broglie waves that pass through the two slits in a way similar to that in which light waves pass through. The waves passing through these slits (which are separated by the distance a) interfere with each other; the result consists in maxima and minima of the amplitude of the resulting wave. While in ordinary optics the intensity of illumination is given by the square of the amplitude, in atomic physics this square of the amplitude has an operational meaning that connects it with the number of scintillation points on a certain screen covered with zinc oxide that is placed in the path of the waves. The square of the amplitude is proportional to the number of scintillation points that are produced per unit of area on a screen that is at a certain distance from the diaphragm, or, in other words, is proportional to the probability that a scintillation occurs in a certain area on the screen. The conceptual scheme of the wave under a given experimental arrangement, for example, two slits in a diaphragm, is called the mathematical formulation of wave mechanics.   In this formulation the amplitude of the wave is uniquely determined if the initial experimental arrangement is given.

However, because of the operational meaning of the wave, we know only how many scintillations will appear on the average per unit of time in a certain region of the screen, but we can never predict at what exact location and at what precise instant of time a particular scintillation will occur.   Therefore, we must say that there is no causal law that allows us to compute from the initial arrangement the precise location of any single scintillation on the screen. We can compute the exact value of the wave amplitude at every point, but this amplitude is not observable; it is only connected with phenomena by the operational definition of the amplitude as proportional to the probability of a particle's being in a certain region. To be exact, we should not speak of a "particle's being in a certain region," because there are no particles.   We should speak of the probability that scintillations occur in some region of the screen or about the frequency of scintillations in this region.   Thus, the mathematical formulation, including the operational definitions of the symbols in this formulation, provides us with rules that connect the initial arrangements with "observable results."   In this example, we assumed that the diaphragm with the slits was rigidly connected to the inertial frame.   If it were connected by an elastic spring, no diffraction pattern with regular maxima and minima in the probability of scintillations would be observed; but we could instead observe motions of the diaphragm relative to the frame and compute the probability of a certain momentum if we knew the amplitude of the de Broglie wave.

In the above presentation, we have to do only with observable phenomena of the same type as those we observe in daily life, like the scintillations on a screen or the momenta of medium-sized bodies (for example, diaphragms).   These observations are coordinated with each other by a mathematical formulation, the de Broglie waves, and operational definitions.   In order to make it clear that in this presentation we do not make use of "conventional physical attributes," as Bohr calls them, we shall give a formulation that does not make use of any verbiage that reminds us of these conventional attributes.   We may say: If an "atomic object" passes from a source through a diaphragm to a screen, the space between the source and the screen is in a certain state which produces under some conditions events that are localized in a very small area around a geometrical point, like scintillations on a screen (point-events) or, under other circumstances, events that have a certain direction, like impulses imparted to a diaphragm (impulse-events). Both types of events are predicted by the mathematical formulation, not individually, but merely statistically; it is predicted how many point-events or momentum-events occur on the average in a certain region of space.

It is important always to keep in mind that this presentation is adequate to the aims of the physicist who is working in atomic research or on the technological applications of atomic theory. We may call this kind of presentation, with Reichenbach, a "restrictive interpretation" of atomic physics. We restrict ourselves from asking some questions which seem very natural and are suggested by some "verbal compulsion," to use an expression of P. W. Bridgman.[8] If a point-like scintillation occurs on a screen, we are very much tempted to say that the screen was hit at that point by a particle, and to ask along what path this particle has reached the screen. In the same way, if the diaphragm is moving, we are tempted to say that it received an impact from a moving particle that imparted a momentum to the diaphragm. If we actually attempt to find out what the path is of the particles that produce the scintillation, we bring up problems that do not deal with the phenomena. We ask questions that are not suggested to us by the task of improving our predictions of observable phenomena or of improving our techniques in developing helpful devices. If we search for the "path of nonexistent particles"—to put it a little flippantly—we are actually trying to describe the phenomena of atomic physics in a language that is as similar as possible to the language in which we describe the experiences of our daily life, or, in other words, we are looking for a theory that is as little remote as possible from a common-sense explanation.

5. Phenomena and Interphenomena

Those who have not been satisfied with predicting the experimental results by means of a mathematical formulation have tried to interpolate inobservable chains of events of a type that has become familiar to us in the traditional explanations of optical phenomena between initial arrangements and observable results. Such events are, above all, motion of particles and propagation of waves in a medium. Reichenbach introduced for these inobservable chains of events the name "interphenomena" in contrast to the observable phenomena, experimental arrangements and observable results.

Statements about these "interphenomena" are certainly not unambiguously determined by the description of the experiment: initial arrangements and results. It is therefore plausible that we can interpolate several chains of "interphenomena" without changing anything in the experimental arrangement and the observable results. On the other hand, it is not possible to interpolate into all experiments with atomic objects one and the same type of interphenomena. If this were possible, we could account for all phenomena of atomic and nuclear physics by the traditional laws of optics and mechanics. We know, however, that this is not the case, and that we need laws that are fundamentally different from Newton's classical laws. For this reason, it may be possible to interpolate interphenomena for each special experiment in atomic physics, but there cannot be a system of interphenomena which can be used for all possible experiments.

In order to permit an easier understanding of the role played by interphenomena, we shall explain it by means of a simple example which was used by Reichenbach.[9] We shall consider, as we did in previous sections, the passing of atomic objects through a slit in a diaphragm and the observable results of this experimental arrangement. We are considering here the diffraction produced by a single slit of the width b. The first and second maxima in the number of scintillations appear under the angles $\varphi = 0$ (center) and $\varphi = \lambda/b = h/mvb$. The only observable phenomenon is the distribution of scintillations on the screen; but, according to Reichenbach, we can interpolate an interphenomenon that is a "particle interpretation." We can assume that particles pass the slit and are deviated according to a probability law that determines a distribution which has its maxima at $\varphi = 0$, $\varphi = h/mvb$, $\varphi = 2h/mvb$, etc. This is a law concerning the motion of real particles although the law is very different from the traditional laws. The particles have at every instant of time a definite position and velocity as in Newtonian mechanics.

The "interphenomena" can be described separately without

regard to the experimental arrangements of which they are a part. Reichenbach calls such an interpolation of interphenomena a "normal" system. According to him, the "corpuscle interpretation" presented above is a "normal" system of interphenomena because it is a description of objective occurrences or, as Reichenbach puts it, the laws for these interphenomena are the same, whether or not the objects are observed. In the parlance of this book, we would say rather that the laws of these interphenomena (moving particles) can be formulated without knowing the whole experimental arrangement; we do not need to know, for example, whether the diaphragm is fixed or movable relative to the inertial frame. Reichenbach also stresses the point that the laws for the interphenomena are in agreement with the laws followed by the phenomena of atomic physics. The laws according to which the particles are deviated is a statistical law; the law according to which the observed source of electrons produces scintillations on a screen is also a statistical law, but it is a law about phenomena. Therefore, we can accept the "particle interpretation" as a legitimate chain of interphenomena.

If, however, we introduce a "wave interpretation," we shall have waves passing through the slit. If we want to derive the occurrence of a single scintillation, we must pursue the wave motion from the slit to the screen according to the laws of interference. We shall obtain maxima of the amplitude at $\varphi = 0$ and $\varphi = h/mvb$. However, in order to obtain the formula of a scintillation at a specific point on the screen, we shall have to assume that the waves are swallowed by the screen at that point and cannot be propagated further. We have here again an "objective" chain of events, "real" waves; but this "swallowing" of waves never occurs in the realm of observable phenomena. Hence, according to Reichenbach, this wave interpretation is not a legitimate interpolation of interphenomena.

If we have two slits at the mutual distance a, and again attempt a "particle interpretation," we must introduce a law of motion according to which the particles are deviated in accordance with a statistical law, with most of the particles deviated by $\varphi = 0$ and $\varphi = h/mav$. This would mean that a particle passing through the slit S_1 would be deviated by an angle φ that depends on the distance between the slit S_1 and another slit S_2, an action at a distance. This would be a law for the motion of particles that would be valid for interphenomena but would be different from all laws that are valid in the domain of phenomena. Therefore, Reichenbach rejects the particle interpretation in this case because it is not a "normal system." The wave interpretation has the same shortcoming as in the case of a single slit; it would involve the swallowing of a spherical wave by a small region on the screen. Reichenbach suggests, therefore, a modified wave interpretation in which the waves are not spherical but move in channels that start at S_1 and S_2, and converge at the scintillation point. Thus Reichenbach regards, in the case of a single slit, the particle interpretation, and in the case of two slits, the wave interpretation, as "normal systems" and, therefore, as the preferable systems of interphenomena to be interpolated between phenomena. We see that for every specific experiment we can interpolate a normal system of interphenomena, but we cannot find such a system that is valid for every experiment.

The presentation recommended by Bohr does not introduce any interphenomena. Neither "real" particles nor "real" waves, the state of which could be described independent of the surrounding experimental arrangement, are introduced.

It is perhaps helpful to emphasize two items in the discussion of interphenomena that have been frequently misinterpreted and misunderstood. They have played a great role in the "philosophical implications" of atomic physics that will be more thoroughly discussed in Chapter 10. At this point, we shall present them from the purely scientific angle in close connection with our presentation in previous sections. It has frequently been said that, whereas in traditional physics the old particle picture of optical phenomena has definitely been replaced by the wave picture, modern physics uses in some cases the particle picture and in other cases the wave picture. Some authors have even used the formulation that one and the same object of atomic physics can be regarded as a particle or as a wave, according to the specific experiment involved. Other authors have introduced the belief that this object is a kind of hybrid entity that offers two aspects, and some have even given to it a composite name like "wavicle." We have seen that actually all these formulations are misleading. The "wave interpretation" and the "particle interpretation" are two types of interphenomena that are interpolated between the observable phenomena of atomic physics. As we learned from the examples discussed by Reichenbach, both interpretations can be used in one and the same case, but one or the other will be preferable if we require that the interpolated chain be a "normal system." This requirement, as a matter of fact, does not provide an unambiguous criterion; the interphenomena never obey all the laws that are valid for phenomena. We must always introduce specific new laws, and it is a matter of taste in what case we should say that an interpolated system of interphenomena agrees essentially with the laws that hold for phenomena.

Alfred Landé,[10] in his lucid textbook, says that every experiment can be accounted for by the particle picture *and* by the wave picture. Reichenbach agrees in general with this view, but brings up certain criteria according to which in each case one of the pictures will be the preferable one. He writes:

> Given the world of phenomena, we can introduce the world of interphenomena in different ways; we then shall obtain a class of equivalent descriptions of interphenomena, each of which is equally true and all of which belong to the same world of phenomena. In other words, given the class of equivalent descriptions of the world the interphenomena vary with the descriptions, whereas the phenomena constitute the invariants of the class.

If we introduce with Reichenbach the concept of a normal system, every class of equivalent interphenomena is represented in the description of the world by the member that is a normal system. Wave interpretation and particle interpretation are just examples of such systems of interphenomena, and are as arbitrary and determined as such systems are in general.

The second point that has sometimes confused philosophers and others who are not experts in theoretical physics is the failure to distinguish between a wave interpretation by conventional three-dimensional waves and the de Broglie waves as a mathematical scheme. The latter comprise a mathematical formula which is used to compute observable results, like scintillations, from an observable experimental arrangement. They have nothing whatever to do with "interphenomena." They determine the location of scintillations according to the "operational meaning" of amplitude, and do not require us to make use of such physical laws as the swallowing of a wave by one point on a screen. All the observable facts of atomic physics can be derived by means of mathematical de Broglie waves, but from this fact we cannot conclude, as some authors have done, that the "wave picture" is a more adequate picture of atomic physics than the "particle picture."

6. The Variety of Formulations in Atomic Physics

If we are interested in atomic physics only as a system of principles from which the observable results can be derived, we can disregard the "interphenomena" completely. Then we have to do only with the observable facts, the mathematical formulation of atomic physics and the operational definitions. They form an unambiguous system of principles. As mentioned above, Reichenbach calls such a system a "restrictive interpretation" of atomic physics because it is restricted to the minimum of theory that is necessary to the scientist in his actual research. The presentation given (in Section 2) along the lines of Bohr's principle of complementarity is a "restrictive interpretation." If we describe the experiment by using wave pictures or particle pictures, we introduce interphenomena and achieve what Reichenbach calls an "exhaustive interpretation" of atomic physics. We gave (in Section 5) examples of exhaustive interpretations and we learned that there are for every experiment a great number, perhaps an infinite number, of exhaustive interpretations.

Some authors have taken the extreme point of view that "exhaustive interpretations" are "meaningless" because they contain statements about interphenomena that cannot be checked by observation. But we have to consider, as we learned in Chapter 3 (Geometry), that no single statement can be confirmed by observation, but only the system as a whole. In our case, what is checked by experience is the system consisting of statements about phenomena and interphenomena with the pertinent operational definitions. We can use different sets of interphenomena, different exhaustive interpretations, and account nonetheless for the same observable facts. We are at liberty to choose interphenomena to reach a system of principles that are as close as possible to our common-sense thinking and speaking.

Moreover, the dividing line between phenomena and interphenomena cannot be drawn sharply. If we consider, for example, "paths of particles" in a Wilson cloud chamber, do we really

observe "paths"?   Exactly speaking, we observe some dark interrupted lines; but Reichenbach suggested that we could just as well
say that we observe collisions between electrons and ions, or that we
observe a line-up of water drops.   He has stated:

> The logical difference between the physics of phenomena and the
> physics of interphenomena (including corpuscle and wave interpreta-
> tions) is therefore a matter of degree. . . . It is a question of volitional
> decision which of these systems we prefer; none can be said to be com-
> pletely restricted to observational data.

It is certainly true that the exhaustive interpretations are arbitrary because there are in every case several different interpretations that are admissible; but on the other hand, it would be wrong
to say that the exhaustive interpretations, wave and particle
pictures, say nothing about the objective physical world.   We
could also imagine a world in which it would be impossible to interpolate between the phenomena systems of the particle or wave
type.   Reichenbach says: "Nature allows us to construct, at least
partly, the world of interphenomena in agreement with the laws of
phenomena."
   We have learned in previous sections that the expression "position and momentum of a particle at a certain instant of time" does
not occur in any law about atomic phenomena, or, in other words,
the expression "simultaneous position and momentum of a particle"
has no operational meaning.   These statements belong to the
"restrictive interpretation" of atomic physics since they do not
deal with interphenomena.   There have, however, been several
authors who have formulated these statements in such a way that
they look like statements about interphenomena, particularly those
about moving particles.   These formulations speak about moving
atomic objects and say that before they are "observed" by a human
being it is not certain whether they have a position or a momentum.
If, however, an "observer" tries to measure its position (coordinate)
he interferes by his measuring instrument with the atomic object in
such a way that he makes the measurement of the momentum
impossible and *vice versa*.   If the observer tries to measure the
momentum, the object is influenced in such a way that the measurement of the position becomes impossible.

If we use the words in their usual sense, such a statement must be called "meaningless." Since the assumption that an atomic object behaves like a "real particle" is incompatible with the observable facts of atomic physics, neither position nor momentum is possessed by the object; they cannot be measured because they do not exist. For the same reason, the possibility of measuring them cannot be destroyed because it has never existed. According to Bohr's principle of complementarity, the "atomic object" by itself possesses neither position nor momentum; these words do not denote properties of an electron, but properties of an experimental arrangement as a whole. Inside a certain arrangement, an electron can have "position" and within a different (complementary) arrangement it can have "momentum." If we change the arrangement, we can change a situation in which the electron has a position into another situation in which it has momentum.

If we express this state of affairs by saying: "By observing position we destroy the possibility of measuring momentum," there is no objection to it, provided we keep in mind that the "observer" is only introduced as a figure of speech and can be eliminated without changing the meaning of the statements. The danger in introducing this figure of speech is only that it may induce us to forget that, in Bohr's words, no conventional physical properties (like position and momentum) can be ascribed to an atomic object. In his *Warsaw Lecture* in 1938[11] Niels Bohr, according to his own report:

> . . . warned especially against phrases, often found in the physical literature, such as "disturbing of phenomena by observation," or "creating physical attributes of atomic objects by measurements." Such phrases . . . are apt to cause confusion, since words like "phenomenon" and "observation," just as "attributes" and "measurements," are used in a way hardly compatible with common language and practical definitions.

We have learned in the previous sections that the initial conditions of an experiment in atomic physics do not allow us to predict the observable results with the same precision as in traditional physics. If we observe atomic objects passing through the slit in a diaphragm, we cannot predict at what precise location on the screen a scintillation will be produced.   On the other hand, if we consider the initial state of a de Broglie wave, we can compute unambiguously the final state of the wave.   This means that predictability and causality play in atomic physics a role that is somehow different from their role in Newtonian physics.   We shall discuss this point more elaborately in Chapters 11 and 12 (on causality).

10

Metaphysical Interpretations of the Atomic World

1. The "Spiritual Element" in Atomic Physics

Although from the purely scientific aspect the new theories used in subatomic physics, the quantum theory and quantum mechanics, are very different from the theory of relativity, the metaphysical interpretations of both types of theories, relativity and quantum, are in many respects very similar to one another. The claim has been made that a "mental element" is introduced into the physical world and that "materialism" is refuted. However, while relativity theory has been interpreted as supporting a belief in predestination, quantum theory has been regarded as a support for the doctrine of "free will."

In order to understand clearly the impact of quantum theory upon our general world picture, it is perhaps advisable not to ask philosophers or scientists, but writers, who have expressed the feeling of our twentieth century. George Bernard Shaw writes:[1]

> The universe of Isaac Newton which has been an impregnable citadel of modern civilization for three hundred years has crumbled like the walls of Jericho before the criticism of Einstein. Newton's universe was the stronghold of rational determinism: the stars in their orbits obeyed immutably fixed laws, and when we turned . . . to the atoms, there too were found the electrons in their orbits obeying the same universal laws. Every moment of time dictated the following

moment. . . . Everything was calculable: everything happened because it must: the commandments were erased from the table of the laws, and in their place came the current algebra: the equations of the mathematicians.

Shaw next describes how, for modern man, the belief in Newton's physics has become a kind of substitute for traditional religion. He continues:

Here was my faith. Here I found my dogma of infallibility. I, who scorned alike the Catholic with his vain dream of responsible free will and the Protestant with his pretense of private judgment.

He then describes how this substitute religion was shattered by twentieth-century atomic physics and quantum theory. "And now," he continues, "—now—what is left of it? The orbit of the electron obeys no law, it chooses one path and rejects another. . . . All is caprice, the calculable world has become incalculable." He accepts the widespread interpretation that the failure of Newtonian, mechanistic science has brought about a return to the pre-Newtonian organismic mechanics which was presented previously[2] as the mechanics of Aristotle and St. Thomas. "Purpose and design," Shaw continues, "the pretexts for all the vilest superstitions, have risen from the dead and cast down the mighty from their seats and put paper crowns on presumptuous fools."

The idea that the conceptions of organismic science have "risen from the dead," as Shaw puts it whimsically, has been taken very seriously by a great many philosophers. We may quote as an example the German philosopher and science writer, Bernard Bavink:[3]

There is today within the circles of natural scientists a willingness to restore honestly the threads from these sciences to all higher values of human life, to God and Soul, freedom of will, etc.; these threads had been temporarily all but disrupted and such a willingness had not existed for a century.

Bavink points out that this rebirth of organismic science had arisen from "purely scientific motives"; he mentions the remarkable coincidence that at the same period a type of political regime appeared that claimed to be hostile to materialism and to be based upon the organismic conception of science. These new regimes are obviously Italian Fascism and German Nazism. As a matter of fact, materialistic or antimaterialistic interpretations of science do not arise usually from "purely scientific motives" but generally have their origin in wishes to set up goals for desirable human conduct. These interpretations are connected with social, political, and religious trends.

This antimaterialistic interpretation of twentieth-century physics has appealed to men of action who were concerned about a scientific basis for their political goals. General Smuts,[4] former Prime Minister of South Africa, starts from the remark that it was difficult to introduce such words as "life" and "mind" into the mechanistic world picture that had prevailed since Newton. Einstein's and Minkowski's conception that space and time do not appear separately but only as a combination of both in the laws of science is interpreted by Smuts as follows: "The physical stuff of the Universe is therefore really and truly Action and nothing else." Smuts starts from the word "action" as a technical term in physics, where it means the product of energy and time. In this sense it has been used in the "principle of least action" and as the "element of action h" in quantum theory.

Later on, Smuts uses this word "action" as it is used in everyday language, where its meaning is rather vague and may denote physical movement as well as organismic growth and even mental activity. He continues:

> When we say that, when we make activity instead of matter the stuff or material of the universe, a new viewpoint is subtly introduced. For the associations of matter are different from those of Action, and the dethronement of matter as our fundamental physical conception of the universe must profoundly modify our general outlook and viewpoints. The New Physics has proved a solvent for some of the most ancient and hardest concepts of traditional human experience and brought a *rapprochement* and reconciliation between the material and organic or physical orders within measurable distance.[5]

In this metaphysical interpretation of modern physics, we can see clearly how "common-sense" analogies of recent theories are the decisive points of the argument; we have only to consider the way in which words like "action," "stuff," "material," "psychical," etc., are used. The British physicist James Jeans writes:[6]

Today there is a wide measure of agreement, which on the physical side of science approaches almost to unanimity, that the stream of knowledge is heading towards a non-mechanical reality. The universe begins to look more like a great thought than like a great machine. Mind no longer appears as an accidental intruder into the realm of matter. We ought rather to hail it as the creator and governor of the realm of matter.

All this amounts to the contention that twentieth-century atomic physics and quantum theory authorize the introduction of a "mental" or "spiritual" element into the physical world, while, allegedly, this was made impossible by Newton's theory of the physical world. This was certainly not the opinion of Newton, who introduced the sensorium of God into his interpretation of inertia. If we wish to learn precisely to what degree atomic physics has been interpreted as a support of spiritualism, in even the crudest sense of the word, we may consult the work of a contemporary German philosopher, Aloys Wenzel, who writes:[7]

This material world, in which there may be possibly also free and spontaneous happenings . . . this world cannot be called a dead one. This world is rather, if we are to make a statement about its essence, a world of elementary spirits; the relations among them are determined by some rules taken from the realm of spirits. These rules can be formulated mathematically. Or, in other words, this world is a world of lower spirits, the mutual relations between which can be expressed in mathematical form. We don't know what is the meaning of this form, but we know the form. Only this form itself or God could know what it means intrinsically.

In this interpretation, the quantum conditions that determine, for example, the energy levels in the hydrogen atom, are interpreted as forms in which "lower spirits" manifest themselves. The laws of quantum theory that cannot be expressed in common-sense language are interpreted by common-sense analogies like the "behavior of spirits," just as primitive tribes have interpreted sunrise and sunset as the behavior of organisms, superior to, but analogous to human organisms.

In order to understand more specifically the use that is made of common-sense analogies for metaphysical interpretations of atomic physics, we may study two examples taken from a pamphlet by Bernard Bavink.[8] The first example starts from the fact that in

Schroedinger's wave theory of matter the hydrogen atom is described by a special solution of the wave equation, a special superposition of de Broglie waves.   Bavink interprets this fact as follows:

> Matter and its worshippers, the materialists, simply laugh at us and tell us—"There is one single atom, the simplest one, the hydrogen atom.   Here, show me what you can do.   If you can show me how I can understand this atom as the product of merely spiritual process— I shall believe you."   It seems that spiritualism can today meet this test.

From the scientific aspect, it is hard to understand why the solutions of the Schroedinger equation are more "spiritual" than the solutions of the differential equations in Newtonian mechanics. But Bavink argues by way of analogies.   The solutions of Schroedinger's wave equation (ψ-functions) can be interpreted as probabilities; probabilities, however, are mental phenomena; hence the ψ-function is interpreted as a mental phenomenon that happens in a human mind; the hydrogen atom is described by ψ-functions; hence the hydrogen atom is a mental phenomenon and is a product of spiritual powers.   The case against materialism is proved.

Here again we see clearly that we have to do with a common-sense interpretation of physical theories.   The wave theory of matter, as it was presented previously,[9] cannot be stated in terms of our common-sense language.   However, by a metaphysical interpretation, the hydrogen atom is called a product of a spiritual power as, in our common-sense language, we ascribe the movements of our bodies to spiritual power and as, by a familiar analogy, the creation of matter is ascribed to the spiritual power of God, which is, again, conceived of as analogous to the spiritual power of man.   We could, of course, give a similar interpretation of Newton's "mechanistic" physics. "Gravitation" and "inertia" can easily be interpreted as analogous to spiritual powers.   Obviously, we have to acknowledge that old and new physics can be interpreted spiritually, but there is no argument to show that they must be.

The second example quoted by Bavink refers to the laws that govern the jumps of an electron from one orbit around a hydrogen nucleus to another orbit.   The laws of quantum mechanics tell us which are the possible orbits for electrons around the nucleus; but if a specific electron circles around, there is no law that would tell us precisely, for every instant of time, what this electron will do in the next instant—whether it will jump or not jump to another orbit. The theory can only predict the average number of electrons that will jump in the next second, but not when an individual electron will jump. Bavink, in his above mentioned work, gives the following interpretation of this state of affairs:

> We must remember firstly, that the individual elementary act (of jumping) as such is not calculable, but left free; secondly, that the real essence of this freedom is perhaps or probably a psychical event. . . . In other words, the "free" choice of the elementary act, which is left undetermined by physics, exists actually only as a part of an embracing "plan" or "form," exactly speaking of a "hierarchy" of "forms"; the superior form always absorbs the inferior one and performs a higher synthesis. . . . What is new is only the fact that physics itself suggests trying out this idea.

In this case the analogical character of this interpretation is obvious. Since the rules of wave mechanics cannot be formulated in common-sense language, the author compares the behavior of the electron with the behavior of a living being which is "free" to choose what to do in the next instant. The word "free" is used here in the vague common-sense parlance according to which we call the action of organisms "free" because we do not know the rules by which we can predict their behavior in the next instant of time. After the existence of "freedom" in the physical world has been established, one makes use of this "fact" in order to make it plausible that human decisions may be "free." Man certainly couldn't be less free than an inanimate physical object. The justification of the doctrine of free will by atomic physics has been one of the reasons why it has been announced, solemnly and repeatedly, that physics is now more compatible with traditional religion than it has been for centuries.

We must, of course, keep in mind the fact that statements like "recent advances in physics have introduced mental factors into science" or "modern physics justifies the doctrine of free will" do not speak of physics from the "scientific aspect." They deal actually with the metaphysical interpretations of recent physical theories. To give to these statements a precise meaning, we should say: Contemporary physics lends itself to a metaphysical interpretation, according to which the hydrogen atom is a product of spiritual powers and the electron jumping from one orbit to another carries out acts of free will. Thus, we have to ask whether Newtonian mechanics cannot be accompanied by a metaphysical interpretation that authorizes the intervention of spiritual powers and free will in physics. Since all such interpretations are essentially presentations of common-sense analogies to physical theories, we can only ask whether it is more "natural" or "common-sensical" to interpret quantum mechanics by spiritual powers than to do so with Newtonian mechanics.

2. Popular Interpretations of Atomic Physics

The interpretation of atomic physics that has been current among philosophers, educators, and ministers, and also among many scientists and laymen with scientific interests, is presented in books like *The Limitations of Science* by J. W. N. Sullivan.[10] The author presents very lucidly the argument according to which twentieth-century physics has brought back into the universe the role of the spirit which had been all but banned by the Newtonian type of physics. Since this argument has occurred recently in an immense number of books, papers, and lectures, it is perhaps instructive to present it at this point. The "mechanistic" science, prevalent since the seventeenth century, assumed, as Sullivan writes, that:

> Of all elements of our total experience, only those elements which acquaint us with the gravitation aspects of material phenomena (*e.g.*, masses and velocities of particles) are concerned with the real world. None of the other elements of our experience, our perceptions of color, etc., our response to beauty, our sense of mystic communion with God, have an objective counterpart.

Sullivan stresses the point that twentieth-century physics does not speak of a "reality" as mechanistic physics does in speaking of "matter and motion." In relativity and quantum theory, Sullivan claims, "we do not require to know the nature of the entities we discuss, but only their mathematical structure. Which, in truth, is all we do know." We have Einstein's differential equation of the gravitational field or, perhaps, of the unified gravitational and electromagnetic field; we have Schroedinger's or Dirac's wave equation. Both, if we include operational definitions, give us advice on how to predict future observations, but they do not tell us at all what "physical reality" is behind this mathematical structure, while Newtonian physics told us that "behind" the equations is a reality consisting of "matter" in motion. "The fact," Sullivan argues, "that science is confined to a knowledge of structure is obviously of great 'humanistic' importance. For it means that the problem of the nature of reality is not prejudged." Since we are no longer compelled to believe that only "matter in motion" is real, "we are no longer required to believe that our response to beauty, or the mystic's sense of communion with God, have no objective counterpart. It is perfectly possible that they are what they have so often been taken to be, clues to the nature of reality."

This argument amounts, briefly, to stressing one sharp cleavage in the evolution of physics; in the period of "mechanistic (Newtonian)" physics, only material masses and their motions were regarded as realities, while twentieth-century physics does not say anything about what reality is. Therefore, Newtonian physics objected to regarding "beauty and religious faith" as realities, while relativity theory and quantum theory are compatible with the belief in "beauty and faith" as realities. This argument is certainly a gross oversimplification. The assertion that material masses are the only reality is in no way inseparably connected with Newtonian physics. There was, for example, in the second half of the nineteenth century the school of energetics who maintained that material masses are not real at all, while "energy" is the only reality in physics. As a matter of fact, this school, the leaders of which were Rankine in England, Ostwald in Germany, and Duhem in France, had already claimed that by this interpretation entities like beauty and faith are no longer excluded from "reality." We find in the energetic school, particularly among the followers of Ostwald, a certain attempt to regard mental entities like fortune, beauty, enthusiasm, etc., as kinds of "energy."

On the other hand, there are many scientists who have made statements about the "realities" underlying the theories of quantum mechanics. Some authors assert, for example, that the only reality in subatomic physics consists of the de Broglie waves. Then we have, of course, the same difficulty as in mechanistic physics; it is hardly more plausible to regard beauty and mystical communion with God as de Broglie waves than to regard them as material masses.   All the mental entities, beauty, religious experience, etc., are no more a part of quantum mechanics than they are of Newtonian physics.   They are added as metaphysical interpretations, which can be done just as well on the basis of Newtonian physics.

In other words, these mental entities enter science as analogies to common-sense experience.   The only real cleavage consists in the fact that the principles of twentieth-century physics are much more remote from the common-sense propositions than the principles of Newtonian physics.   If we interpret, for example, the principle of indeterminacy[11] by common-sense analogies, we easily arrive at statements about mental phenomena of a "higher," or rather, more complex type.   We say that the world is freed from the iron chain of causality; that, as Sullivan says,[12] the electron is at present a very hazy entity.   It is nothing like so definite a thing as the "hard, substantial little atom of the Victorians," hinting that the loosening of the Victorian rigid code of behavior may have had something to do with the new atomic physics.   The prominent British physicist, Jeans, formulates his common-sense analogies by comparing the universe with a prison.   He writes:[13]

> The classical physics seemed to bolt and bar the door leading to all freedom of will; the new physics hardly does this; it almost seems to suggest that the door may be unlocked if we could only find the handle. The old physics showed us a universe which looked more like a prison than a dwelling place.   The new physics shows us a universe which looks as though it might conceivably form a suitable dwelling place for free man, and not a mere shelter for him—a home in which it may at least be possible for us to mould events to our desires and live lives of endeavor and achievement.

It is instructive to learn what these interpretations look like when they penetrate from the physics books by way of the philosophy books into periodicals that serve the purpose of enlightening the educated layman.   The prominent writer on education and politics, Erwin D. Canham, writes:[14]

> Throughout the nineteenth century and part of the twentieth, we lived in an atmosphere of self-confident materialism. . . . It was a mechanistic world and we were sitting on top of it.   We were laying hold of the apparent forces of matter, and matter was our God.   The natural scientist challenged the spiritual revelations of the Bible with doctrines

of atheism and rationalism. One might almost say that this era of materialism lasted until the day the atomic bomb burst over Hiroshima. . . . I think it is correct to say in a very real sense that matter has now committed suicide—matter has been revealed as no longer able to protect or to serve anybody anywhere unless the thinking behind it is oriented to new concepts of mutual interdependence.

This argument very correctly stresses the point that not the material that is exploded in the bomb is responsible for the good and evil resulting from the explosion, but the men who are threatening each other in such a way that research, manufacturing, and military operations follow as the result of these human stresses. This was the situation at the time of the "orthodox bombs" that exploded according to the rules of Newton's physics just as well as the Hiroshima bomb exploded according to the rules of twentieth-century physics. In the above mentioned article Canham writes:

The new concept of a cosmos in which matter has ceased to have its old reality and substantiality is now generally recognized by most natural scientists. In the nineteenth century they felt they knew all the answers, and that the universe was wrapped up in a neat mechanistic package. Today their explanations rotate around Heisenberg's self-described "Principle of Uncertainty." This is a very hopeful change.

This formulation is a very good example of the way in which "common-sense analogies" used in many presentations of modern physics can be misleading if they are taken literally. The older physicists who "knew all the answers" are contrasted with modern physics where the "principle of uncertainty" is the dominating principle. From our presentation of the "scientific aspect" of subatomic physics, one can clearly see that the word "uncertainty" in Heisenberg's principle does not mean "uncertainty about the truth of a scientific theory." It refers rather to the vagueness which we necessarily introduce into the description of a physical system if we attempt to make use only of the old Newtonian concepts like "particles." We must remember, above all, that the word "reality" itself belongs to those "common-sense analogies" if it is applied beyond our common-sense language. One could devote a more elaborate investigation to the way in which the words "real" and "reality" have been used and to the importance of these words for

the relation between physical science in the strict sense and the use
of this science for influencing human behavior.

3. Science and Metaphysics in the Principle of "Indeterminacy"

We have learned[15] the "scientific aspect" of subatomic physics;
we have learned how to predict by computation the future statistical
distribution of point-events and impulse-events if the initial experimental conditions are given.   The method used for this prediction
consists in integrating Schroedinger's wave equation, computing the
de Broglie wave described by the ψ-function, and applying the
operational definition according to which the square of ψ is the
average number of point-events taking place in a certain region.

All practical work in subatomic physics is more or less of this
kind.   For example, we wish to predict the number of uranium
nuclei that are induced to fission by the impact of neutrons because
this prediction is of great relevance for the application of quantum
theory to the production of nuclear energy.   This kind of prediction
is mathematically of the same type as every prediction of the number
of point-events in a certain region.   Therefore, all practical work
can be done without paying much attention to the principle of inde-
terminancy.   This principle only plays a role when we ask the
question of to what degree we can formulate our problem by using
the traditional concept of particles that traverse trajectories "in
space and time" or, in other words, the rectangular coordinates
x, y, z which we can describe as continuous functions of the time t.
It was revealed by the work of men like Heisenberg, Schroedinger,
and de Broglie that the point-event cannot be predicted by introducing trajectories that pass through the points in space where the
events are to occur.   The prediction can only be made by computing the ψ-function and by applying its operational meaning.

However, this describes the method that is actually used in
atomic physics.   We use concepts that are very remote from the
concepts by which we describe the world in our traditional common-
sense language.   The "principle of indeterminacy" is an attempt
to introduce common-sense concepts like particles and trajectories,
which imply, of course, that a particle moving on a trajectory has
at every instant of time definite coordinates and definite velocity
components.   The attempt has been made to invent such laws for

the motion of particles that the observable results are identical with the results found by the application of wave mechanics, *i.e.*, the computation of the ψ-function from Schroedinger's wave equation. In order to formulate such laws for the behavior of particles, we must, of course, abandon the conception of trajectories traversed by those particles; to assume the existence of trajectories would lead to an attempt to obtain the results of quantum mechanics from Newtonian mechanics. Therefore, completely new laws of motion have to be introduced.

In order to obtain the results of wave mechanics by describing the behavior of particles, we must introduce concepts like "indeterminacy of coordinates," or "momenta of a particle," or "average number of hits that a particle produces in a certain region of a screen," etc. These laws for the behavior of particles are, of course, very different from the Newtonian laws and very different from our common-sense ideas about particles. As Bohr pointed out, we must avoid ascribing to an atomic object (such as an electron) traditional properties of a particle. As we have learned,[16] "position and velocity of a particle" is an expression without operational meaning if applied to small particles. In order to "humanize" these particles, which do not have the traditional properties of particles, we must ask the question of how a particle would behave if it had the traditional properties but was very small. Then we would obtain the famous result that was first announced by Heisenberg: If one tries to measure the position of a particle, one frustrates the measurement of the momentum and vice versa.

This way of speaking retains a certain analogy to common-sense experience; it preserves the habit of speaking of particles that traverse real trajectories; they "have" position and momentum but they cannot be observed simultaneously. This way of speaking is correct if we consider it as an analogy taken from common-sense experience. If we claimed that this is what really happens, this way of speaking would become a "metaphysical interpretation" of how atomic objects actually behave. Alfred Landé,[17] in his excellent textbook on quantum mechanics, writes very correctly: "It is unphysical to accept the idea that there are particles possessing definite positions and momenta at any given time, and then to concede that these data can never be confirmed experimentally, as though by a malicious whim of nature."

Landé also points out correctly and lucidly that we could avoid this metaphysical interpretation by following the way in which Niels Bohr described the situation in which the conditions of indeterminism have to be applied.   Bohr said very clearly:

> When an experimental arrangement or state can be interpreted in terms of particles whose positions are defined with a margin Δx, then the same arrangement or state cannot be interpreted in terms of particles with momenta defined more exactly than $\Delta p_x = h/\Delta x$, and vice versa.

The salient point is that according to Bohr every state or arrangement can be "interpreted" by particles; but he does not claim that particles "exist."   If we restrict ourselves to the way in which Bohr presented the laws of subatomic phenomena, the physical theory of this type of phenomena will be, in principle, no different from any physical theory.   It will have in general the same logical structure that we described before by the discussion of special cases. The philosophical interpretations which are introduced by Heisenberg's principle of indeterminacy and Bohr's principle of complementarity do not introduce any element of vagueness or irrationality if we keep strictly in mind the conception that these principles are "interpretations" of the subatomic phenomena in the sense that we have described in this chapter as well as in Chapter 9 on subatomic phenomena from the strictly scientific aspect.

The "interpretations" lead us into "deep water," however, if we take them too seriously, which means if we regard them as statements about reality.   We invite trouble if we ask the question, what are the "real" physical objects in subatomic physics.   Are the particles "real" or are the de Broglie waves (described by the ψ-function) "real?"   If we say that "particles are real," what sense does it make to say that a "real particle" has at a certain instant of time an indeterminate position?   If we say, on the other hand, that the de Broglie waves are "real," we must remark that the operational meaning of ψ is connected with the probability that point-events take place in a certain region of space.   If we say that these "waves of probability" are "real," we use the word "wave" in the same sense that it is used in expressions like "wave of suicides," "wave of disease," etc.   To speak of a "wave of flu" as a "real wave" would be an unusual use of the word "real."

Henry Margenau,[18] a prominent author in physics as well as in the philosophy of science, discusses elaborately the problem of reality in subatomic physics and quantum mechanics. We shall restrict ourselves to the special problem of whether, in subatomic physics, particles (electrons, neutrons, etc.) can be regarded as "real things" or are only "constructs," while the de Broglie waves constitute "physical reality." Margenau argues that:

> The use of probabilities as essential tools in the description of nature has brought about a separation of our experience into two domains: one composed of immediacies (observation, measurement) that are not all predictable in detail, and another, refined and rational, which is the locus of laws and regularities, of permanent substances, of conservation principles and the like.

Margenau suggests calling the first domain *historical*, the latter *physical reality*. If we apply this dichotomy to the phenomena and interpretations of subatomic physics, it seems clear that the dark spots produced on the screen by the hits of electrons or photons are elements of *historical reality*; they are not individually determined by the experimental arrangement. However, the pattern itself, or, in mathematical terms, the solution of the wave equation (Schroedinger's ψ-function) is unambiguously determined by the experimental arrangement; the ψ-function can, moreover, be computed for all time if it is known from the whole space at one instant of time. Margenau therefore regards the ψ-function as a part of *physical reality*. He feels justified in this by "the philosophical view which identifies the real with the elements of experience that are causally connected in time and space. . . . Hence it is safe to say . . . that physical science would lose its hold on reality if an appeal to law and order were interdicted as a major claim." If we look at the individual events, *e.g.*, individual hits on a screen and appearance of individual spots, we note that there are no regularities. If we consider, however, the pattern of hits or spots on the screen as a whole, it follows a very simple and obvious law. The ψ-function that determines the distribution or probability of these spots obeys a causal law. "Regularity," Margenau writes, "is found primarily in aggregates or, when assigned to individual events, in the *probabilities* which inhere in these events. Laws govern these probabilities, they do not govern simple occurrences." Since, according to Margenau, "reality" is to be ascribed to the quantities which follow a causal law, he concludes correctly:

> To be in harmony with the spirit of physical science we must therefore accept a conclusion unpalatable to many thinkers of the past, the conclusion that *probabilities are endowed with a measure of reality.* . . . Thus as physical reality, the probability extended through all of space, like a continuous medium, devoid of matter. It formed in fact a field.

Since the probability of events is measured by the square of the ψ-function, we can also say that the wave described by the ψ-function is a physical reality. We shall discuss, in the section on causality in atomic physics,[19] this role of the ψ-function from the purely scientific aspect. We shall see that the law of causality is valid if we describe the state of a system by the ψ-function. Up to this point, no philosophical or metaphysical interpretations come into play. The interpretation becomes more and more remote from statements of physical science, it becomes more and more "metaphysical," to use this expression, the more we speak of "reality," particularly of "physical reality." To say that the ψ-function obeys a causal law is merely physics. But to say, as Margenau suggests, that the ψ-function describes physical reality *because* it obeys a causal law is a metaphysical interpretation. To say that only that which obeys a causal law is real is to state an analogy with common-sense experience. The ordinary bodies of our common-sense experience, stones, or planets, or bodies of animals, follow a causal law. Therefore, we call analogously all objects "real" if they obey causal laws too. In this sense, Margenau calls the probability or the ψ-function part of physical reality.

This statement is certainly true if, and only if, we keep in mind that we call these objects "real" because they obey causal law, and do not forget that this connection between "reality" and causality is only based on an analogy with common-sense experience. We can see this easily if we direct our attention to the fact that other authors deny the "reality" of the "probabilities" and claim that particles (like electrons, neutrons, etc.) are part of physical reality. We may mention among them, as one of the most prominent, William H. Werkmeister,[20] who has written two textbooks and many papers on the philosophy of science. He starts from the statement that every conception of "physical reality" has to start from "ordinary things" like rocks, and planets that are "real" according to the parlance of common sense.   Then, he argues, one has to add those objects that are in direct interaction with "ordinary things."   In the usual parlance of physics we say that electrons or other subatomic particles interact with macroscopic pieces of metal and this interaction (*e.g.*, scattering) has been described by the formulae derived in mathematical physics.   Therefore, Werkmeister argues, these subatomic particles have to be included in "physical reality" along with ordinary things.   But probabilities or "ψ-functions" do not interact with ordinary things if we choose our words as close to common-sense usage as possible.

According to Margenau, de Broglie's "waves of probabilities" belong to physical reality and particles do not, while according to Werkmeister material particles are the only "real objects" in subatomic physics.   According to the conception of philosophical interpretation that is presented in this book, there is no contradiction between the statements of Margenau and those of Werkmeister on "physical reality."   Both statements are based on the same scientific doctrine, de Broglie's and Bohr's quantum theory, but they interpret the same scientific theory by different analogies taken from common-sense experience.   The two statements become contradictory only if one believes that metaphysical statements are results of "seeing with the intellect" or "perceiving the ultimate reality behind the observable facts."   In this case, Margenau believes that the "ultimate reality" consists of "waves of probabilities" which are certainly not material, but rather mental or spiritual entities, while in Werkmeister's statement the "ultimate reality" consists of material particles or, as in the language of some philosophers, of clumsy parcels of matter.

In the first case, subatomic physics becomes a support of idealistic or spiritualistic world views, while in the second case, twentieth-century atomic physics cannot be used in this way to bolster up "idealism" and to refute "materialism."   Such metaphysical interpretations have very frequently the purpose of supporting some desirable human conduct, some cherished way of life.

There is scarcely any doubt that Margenau regards the statement that the "waves of probability" are part of physical reality, not as an interpretation by analogies, but as a "true statement" about reality.   This becomes particularly clear if we compare Margenau's attitude on the question of reality with Bohr's.   According to Bohr's principle of complementarity, there is no unique description of isolated atomic objects.   Different experimental arrangements, excluding each other, yield different descriptions of the same atomic objects.   Since we cannot describe, in conventional terms, the atomic object itself, we have to start from the object within a specific frame of experimental arrangement which we may also call a specific operation of observation.   Given this specific situation, the application of the mathematical formulation of quantum theory or, in other words, of the differential equation of the ψ-function permits us to predict the phenomena we may expect as the result of our experiment.   According to Bohr, the word "phenomena" is to be understood as in common-sense language.   As Bohr puts it:[21] "However far the phenomena transcend the scope of classical physical explanation, the account of all evidence must be expressed in classical terms."

Bohr would refuse to give to the "phenomenon" or to the ψ-function the attribute "real."   Margenau, however, describes the attitude taken by Bohr as follows:

> Physics *has* a choice between describing nature in terms of classical observables (positions of particles, etc.) and in terms of abstract states, such as ψ-functions.   The first choice permits visualization (phenomena) but requires that causality be renounced; the second forbids visualization but allows causality to be retained.   And these alternatives can never be reconciled. . . . Bohr does not ask science to *make* a choice—he asks science to resign itself to an eternal dilemma. He wants the scientist to learn to live while impaled on the horns of that dilemma and that is not philosophically healthy advice.

Bohr himself does not see a dilemma in this situation.   According to him, there are two descriptions of the same atomic object which serve different purposes and are not contradictory to one another.

To quote Margenau, "Science has in fact made its choice and its choice was the second alternative (description by ψ-functions)."[22] As a matter of fact, the choice of science has been that the use of the ψ-function is the best method of computing the result of a certain experiment.   If this is meant by "reality," Margenau is right in asserting the "reality" of the "probability wave."   According to

Bohr, the presentation by the ψ-function and by "observables" are not "rival theories," as Margenau calls them, between which a decision can and will be found, but they are two descriptions within one and the same theory. Margenau calls Bohr's view "agnosticism," which is dangerous because it encourages resignation and discourages research in other fields. But for Bohr, his view is not agnosticism and lack of decision, but it is a particularly profound type of decision, the decision to combine in one and the same world picture the phenomena that appear under all possible circumstances. It is, if we speak on the level of philosophical interpretation, the decision to integrate all possible analogies with common-sense experience into one world picture.

4. Physics and "Free Will"

It has been repeated over and over that our twentieth-century mechanics of subatomic bodies solves the conflict which had existed previously between the doctrine that there are "free decisions" of the human will and the doctrines of Newtonian mechanics. The position and velocity of every material particle can, according to Newton's mechanics, be computed if we know the state of motion at any previous instant of time and the force acting upon all masses, from equations of the type $m \times a = f$, where m stands for mass, a for acceleration, and f for the force acting upon the particle with the mass m. These equations can be solved only if we know the force f. All our physics is based in practice on the assumption that there are only three types of forces, gravitational forces, electromagnetic forces, and, in the most recent physics, nuclear forces. If we take $m \times a = f$ in its purely physical sense (see Chapter 4), we cannot substitute for f in the Newtonian equation any "spiritual power" or "will power." If all observed motions can be computed without taking "will power" as a component of f, "will" cannot influence the motion of material masses. Since any human action leads to some motion of masses, will power cannot bring about any action if Newtonian mechanics is right in its strictly physical sense. If we assume, however, that will power could be substituted for f in the Newtonian equation, no conflict between Newton's mechanics and "free will" would exist.

During the period when Newtonian mechanics was recognized as the undoubted basis of physics, many attempts were made to find "gaps" in the predictions of mechanics and to use them as "loopholes" for the introduction of free will.   Many authors, for example, pointed out that a particle that moves perpendicular to the direction of the force acting upon it would do no work and would therefore not consume any energy.   Hence the "force of will" could produce such a motion without violating the law of conservation of energy derived from Newtonian mechanics.   A more sophisticated way of finding a "gap" was to point to the "singular" points in the differential equations of mechanics.   At such a point, the field of force is infinite or indeterminate, and the acceleration a cannot be computed unambiguously from the equations.   There, some philosophers have argued, will power could take over and determine the motion of the material masses.

If, however, the "force of will" can be substituted for f in Newton's equations, no "gaps" are necessary to make "freedom of will" compatible with mechanics.   If, on the contrary, the "force of will" cannot move a material mass, the "gaps" are of no help.   If, for example, the masses are to be moved perpendicular to the mechanical force, we need a "force" to move it in any particular direction.   Although none of the motion needs energy, it needs a "force" in order to determine the direction.   One needs, for example, an increase of momentum to choose one particular motion.   In the same way, if there is a "singular" point, the motion beyond it can only be determined by introducing an additional law of motion.   But then the problem is again the same as it was before.   If a "force of will" can determine the motion of a particle at a "singular" point of the differential equations, there is no reason to declare that the substitution of this force for f in the Newtonian equation is illegitimate.   In this case, however, the motion of a particle at a "regular" point would also depend on "will power" in addition to gravitational and electromagnetic forces.   "Free will" would be understandable without introducing any "gaps."

Hence, all justification of "free will" by gaps is irrelevant; if will power can replace a physical force, the gaps are unnecessary; if such a replacement is illegitimate, the gaps are of no avail because the need is for an additional law that determines the motion of a mass starting from the gap.   This additional law is certainly a physical law because it determines the motion of material masses.   If will power can play a role in this additional law, it could just as well play it in the original Newtonian laws. For all these reasons, the introduction of "gaps" into mechanical laws does not provide any help in the problem of "free will." If we approve this argument, it follows easily that the replacement of Newtonian mechanics by Bohr's or Heisenberg's quantum mechanics or, in other words, by the mechanics of subatomic particles, cannot be of any help for the problem of "free will" or "free arbitration."

To make this clear, it is perhaps best to explain it by the example from subatomic physics that we used in the treatment of the "scientific aspect."[23] We may choose a beam of electrons that pass through a hole in a diaphragm and hit a screen. According to Newtonian mechanics, the point where a particle hits the screen can be predicted from the state of the particle at the time when it passes through the hole in the diaphragm. This state consists in Newtonian mechanics of the position and momentum of the particle at the time of passing through the diaphragm.

In subatomic mechanics, such a state does not exist. We know only the position of the hole in the diaphragm and the way in which the beam of electrons is launched. From these conditions, the theory of subatomic mechanics allows us to compute the statistical distribution of the hits on the screen; we can find out how many hits will occur, on the average, each second at a square inch of screen. This distribution is uniquely determined by the laws of wave mechanics plus operational definitions. These laws replace the causal laws of Newtonian mechanics. If no Newtonian (gravitational or electromagnetic) force is acting upon the electrons passing through the diaphragm, they will, according to Newtonian mechanics, follow the law of inertia and, according to wave mechanics, produce on the screen a pattern of spots around the central spot that would be produced according to the law of inertia.

If we apply a distributing force, e.g., an electromagnetic field, the single spot produced according to the law of inertia would be displaced into another single spot. If we assume that wave mechanics is valid, the pattern of spots appearing on the screen if no forces are acting will be changed when an electrostatic or magnetic or other physical force is applied. In other words, every physical force produces a different statistical distribution of hits on the screen. Since the physical forces determine only the statistical distribution of the spots, but do not determine where, at a particular instant of time, a spot will be produced, we have again a "gap" situation. Since the appearance of a spot at a particular instant of time is not determined, "will power" may come in and determine the appearance of a spot at a definite instant at a definite place on the screen without violating the laws of physics.

However, this argument is as irrelevant as in the case of other types of gap. If we assume the "scientific aspect" of subatomic physics, the statistical distribution at the end of the experiment is completely determined by the experimental arrangement and the applied forces. This means that no physical force can restrict the statistical distribution in such a way that we could predict when a specific point of the screen will be hit. Hence, no addition of a physical force (gravitational or electromagnetic) can convert the statistical law into a causal law where single events are uniquely determined. Thus, we must again assume that a "force of will," a "spiritual force," may produce a choice between the possibilities left open by the statistical law. But if we assume that a "spiritual force" can produce a motion of material masses, we can just as well assume that a spiritual force can be substituted for f in the Newtonian law $m \times a = f$. If this were legitimate, we would need no gaps in the equation of motion, and, accordingly, no statistical laws of motion.

From all these considerations, it seems to follow that by changing the theories of physics we cannot make a contribution to the understanding of what has been called the problem of "free will," or "free arbitration"; in other words, the problem of physical determinism has very little to do with the problem of free will. There is only one place in physics where the words "free" or "freedom" can be usefully applied. Everyone who has taken up mechanics, even in an elementary way, will know the distinction between "free vibrations" and "forced vibrations." For example, if we consider a pendulum which is abandoned to the force of gravity, it will perform vibrations with a frequency that is dependent only upon the length of the pendulum (L) and the acceleration of gravity (g). This frequency n is, as it is computed from Newton's laws, $n = \sqrt{g/L}$.

We call n the characteristic frequency of the pendulum or the frequency of free vibration because it does not depend upon a frequency which one would try to impose on the pendulum from outside. One can, however, submit the pendulum to periodical pushes that are repeated with a frequency N which depends, of course, only upon the external influences and not upon the internal frequency of the pendulum. If the pendulum is abandoned to itself, it will perform a "free vibration" of the frequency n; if it receives periodical pushes of the frequency N, it will perform an "enforced vibration" which depends upon the simultaneous effect of n and N. The actual vibration will be a superposition of vibrations with the frequencies n and N. It will become more intensive the nearer N is to n, and will exhibit the phenomenon of resonance, a vibration with large amplitude, if N comes very close to n.

We can say that in the first case the behavior of the pendulum is only dependent upon itself; the motion is "free." In the second case it is determined by external influences. The behavior of human beings has also been analyzed in the same way. The mode of behavior is dependent partly upon "internal factors" which would still exist if the human being were not acted upon by the environment at all. An example of a "free action" of a human being would be thinking that comes completely from a person's own mind and is not influenced by reading books or listening to other people. We understand that this characterization of "freedom" has a meaning only on a common-sense level of description, but becomes very vague if we attempt a scientific understanding of the terms. A "free action" of a human being would be a movement of his legs or hands which is caused only by internal stimuli.

It is clear that, strictly speaking, such a movement does not exist. On the other hand, it is also true that, speaking in everyday language, we can say correctly that some actions are due to external influences and some mostly to internal stimuli. The external cases are easy to recognize. If a strong man gives a weak man a push, the latter will move; this action is certainly not "free." If a man at rest feels no external push that would move him, but gets up because he wants to do so, we can describe this phenomenon by saying "he got up freely." But everybody feels that this distinction is very vague. As a matter of fact, the distinction between "free vibration" and "forced vibration" in mechanics is only sharp if we speak in an oversimplified way. The statement that the vibrations of a pendulum that is abandoned to the force of gravity have a frequency that does not depend upon any external influence and is "free" is only true if we neglect the problem of producing the pendulum from raw material and of providing the raw material by the economic cooperation of a great many people. Only in this vague way can the word "freedom" be used in physics and be taken over from physics into the problem of human actions.

If we want to have a judgment about what physics can do for the problem of "free will," we have to investigate what the people think who are actually interested in the existence of "free will," and regard this existence as very important for the understanding of the world and for the desirable conduct of human beings. We shall see that all these people admit gladly that in the field of physical phenomena there is strict determinism. Their point is that there are events that are not physical but spiritual that follow laws that are completely different from the physical laws.

We found in discussing the difference between Newtonian mechanics and twentieth-century subatomic mechanics that the difference is completely irrelevant for the problem of free will. However, the prominent British astronomer and mathematical physicist, Sir Arthur Eddington actually made use of twentieth-century subatomic physics to support the popular belief in "free will" that has seemed to a great many people to be a necessary presupposition for a belief in traditional religion and ethics. He starts[24] from the statement that "the future is a combination of causal influences of the past together with unpredictable elements" because of the basically statistical character of laws in subatomic physics. "Science thereby withdraws its moral opposition to free will." This is, of course, only valid if volition can fill the gaps left by statistical laws. In the conclusion of his book, Eddington makes (p. 350) the historical statement:

> It will perhaps be said that the conclusion which can be drawn from these arguments from modern science is that religion first became possible for a reasonable scientific man about the year 1927.

It was in 1927 that Heisenberg advanced his principle of indeterminacy. In a later book,[25] Eddington gives a more thorough analysis of his previous argument in favor of free will. He states flatly that the hypothesis that volition may act by superposition upon statistical law or, as we have said, by infiltration through the gaps left by those laws is just "nonsense." He argues again in favor of free will by advancing the hypothesis of "spiritual force," but denies that one can support the belief in free will by Heisenberg's principle of indeterminacy. However, a great many authors have adhered to Eddington's older, rather perfunctory argument, and have paid no attention to the much more profound judgment in his later book. This attitude of a great many philosophers and scientists has been caused by the long tradition of justifying "free will" by "gaps" in the "iron causality" of Newton's physics.

The philosophers and theologians who advocate this belief are not interested in advocating indeterminism in physics, but in claiming that there are events and phenomena that follow laws that are different from the physical laws. What is advocated by metaphysicians and theologians is the doctrine that there is determinism in the realm of physics but "freedom" in the realm of the spirit. It is instructive to learn what the most popular Catholic philosopher and theologian in this country, Bishop Fulton J. Sheen, writes about this point:[26]

> St. Thomas asserts that a changed conception of empirical science does not involve a change in the metaphysics which ruled this science. Philosophy is independent of science. . . . [page 148] There is absolutely nothing in quantum theory and the principle of indeterminacy to show that any physical event is uncaused. There is therefore no basis in physics for the freedom of will. . . . The problem of the freedom of the will is not a problem of physics but a problem of philosophy.

We may add a quotation from a contemporary Hindu philosopher, a fervent believer in metaphysics as seeing with the intellect and a strong believer in "free will." Nalim Kanta Brahma writes about the attempts of Eddington and other scientists to "prove" the "freedom of will" by the advances in physical science:[27]

> If future experiments reveal to us that the indeterminism supposed to exist in the movements of the electron is really non-existent, philosophy would find itself helpless to prove its position if it now accepts the argument of Professor Eddington. . . . Freedom and other metaphysical truths cannot be proved in the sphere of phenomena, where space, time and causality are the only categories that rule.

In order to learn how the doctrine of freedom of will is formulated and proved by those groups who advocate it for moral reasons, we shall look into a French philosophical periodical published in 1953.

It contains a paper on "free arbitration" advancing the usual formulation of those who advocate this doctrine.   The author writes:[28]

> Two classical demonstrations establish, according to our opinion, in a satisfactory way, the existence of free arbitration; the first one is moral and has to induce us to believe in liberty; the second one is psychological and has to confirm this belief.   Our moral conscience confronts us with duties which are real when we become aware of them. A real duty makes sense only if it can be obeyed or disobeyed at will. This amounts to the statement that duty presupposes liberty or, in other words, if I have the belief something is my duty this implies the belief that I am free.

This argument which says briefly that there can be no belief in duty unless there is a belief in free arbitration is certainly convincing if one uses the words "duty" and "free will" as they are used in the common-sense parlance in which we have been trained from our childhood.   If we attempt to apply the more profound analysis of science, the argument becomes certainly much more complex, and eventually becomes very similar to the psychological argument to which we now turn.   According to our author:

> The psychological argument does not need to be presented elaborately, because it consists in the testimony of our interior observation: the invitation to become aware of this observation is sufficient. . . . Does not everyone know by experience what it means to assume the moral responsibility of an action?   To assume such a responsibility is identical with perceiving that one is free.

Whatever one may think about the cogency of these demonstrations, it seems, as our author puts it, that the essential question is: How can the will bring about a free decision between duty and pleasure?   The answer to this question, however, becomes in no way easier if Newtonian mechanics are replaced by subatomic mechanics.   It is a purely psychological question and the answer which we give depends completely upon the psychological theory which we hold.   In a way that is quite satisfactory from the point of view of older psychological theories, the answer was given as early as 1673 by Spinoza in his *Ethics*.   He wrote:[29] "In the mind there is no absolute or free will, but the mind is determined by this or that volition, by a cause, which is also determined by another cause, and this again by another and so on ad infinitum." Then Spinoza gave the following demonstration:

> The will is a certain and indeterminate mode of thought, and therefore, it cannot be the free cause of its own actions, or have an absolute faculty of willing or not willing, but must be determined to this or that volition by a cause which is also determined by another cause and this again by another and so on ad infinitum.

According to Spinoza, the mental states are a part of the causal chain of physical states, and the question that we can ask is not whether there are "gaps" in this chain, but rather how it comes about that our interior observation seems to tell us that we can make "free" decisions. Spinoza gave a good answer to this question also, in the appendix to Part I of his *Ethics*:

> It will be sufficient if I here take as an axiom that which no one ought [to] dispute, namely that man is born ignorant of the causes of things, and that he has a desire, of which he is conscious, to seek that which is profitable to him. From this it follows, firstly, that he thinks himself free because he is conscious of his wishes and appetites, while at the same time he is ignorant of the causes by which he is led to wish and desire, not dreaming what they are. Secondly, it follows that man does everything for an end, namely for that which is profitable to him, which is what he seeks. Hence it happens that he attempts to discover merely the final causes of that which has happened, and when he has heard them he is satisfied, because there is no longer any cause for further uncertainty.

If a stone falls from a roof, we no longer say that it fell purposively, in order to hit a target and perhaps to destroy it; we "explain" the fall of the stone by Galileo's and Newton's laws of gravitation. Spinoza points out that the behavior of a human being is just as determined by causes as the fall of a stone; because the human organism is a very complex system, we do not know the causes of its motions and substitute "final causes" or "goals." Such a goal is what we call in everyday language "will." According to Spinoza, "will" is a mental phenomenon like imagination or thought that accompanies our actions, but is never the cause of our actions. We experience mental phenomena that we describe in common-sense prescientific language as "free choice" or "free arbitration." If a falling stone could think and speak, it would also say that it has a "free choice."

The actual scientific problem is to investigate how that feeling of free choice arises and what the use of this feeling is to the human organism. Common-sense psychology describes, as we have learned, this situation as a conflict between "duty" and "pleasure." These terms are certainly understandable and meaningful on the level of everyday experience. Every child understands that attending school is a duty and attending the "movies" a pleasure. But situations may arise when attending school may become a pleasure and attending boring film performances for the purpose of "social studies" may become duty. Scientific psychology has replaced the common-sense concepts of duty and pleasure by a more complicated system of concepts, as physical science has replaced the common-sense concepts of "rest" and "motion" by the conceptual pattern of Einstein's theory of relativity.

As an example of such a scientific psychology, we may consider contemporary "depth psychology" based upon the theories of Sigmund Freud, known under the name of "psychoanalysis." Freud studies the "anatomy" of human personality structure. Besides the "ego" that cares for pleasure and tries to achieve it in a reasonable way, there is hidden in the subconscious region of human personality the "id" and the "superego." The former consists of the remnants of elementary animal instincts in human personality, while the "superego" consists of the personality traits that are acquired under the influence of the parents, the schools, the churches, the military and civic indoctrination. A part of what Freud called the "superego" corresponds vaguely to what is called in everyday parlance "conscience."

The conflict between "pleasure" and "duty" which is, according to prescientific formulations, decided by "free arbitration" was described by Freud[30] scientifically in the following way:

> The proverb tells us that one cannot serve two masters at once. The poor ego has a still harder time of it; it has to serve three harsh masters, and has to do its best to reconcile the claims and demands of all three. These demands are always divergent and often seem quite incompatible; no wonder that the ego so frequently gives way under its task. The three tyrants are the external world, the superego and the id. When one watches the efforts of the ego to satisfy them all, or rather, to obey them all simultaneously, one cannot regret having personified the ego, and established it as a separate being. It feels itself hemmed

in on three sides and threatened by three kinds of danger, towards which it reacts by developing anxiety when it is too hard pressed. . . . In this way, goaded on by the "id," hemmed in by the "superego," and rebuffed by reality, the ego struggles to cope with its economic task of reducing the many influences which work in it and upon it to some kind of harmony.

We do not see one ego that makes "decisions" or "choices," but the "ego" is one part within the personality structure struggling with other parts and with the external world. We can ask the questions, under what conditions have we the feeling of "free choice," what is the function of this feeling in human life, and what is its value for formulating the laws of human behavior. The problem of "free choice" belongs in this context, and has nothing to do with physical determinism or indeterminism.

If we want to reach a well-balanced judgment about whether the indeterminacy of subatomic physics has made a contribution toward supporting the doctrine of "free will" as advocated by traditional religion, we must look into the writings of recognized religious leaders. We may quote, for example, Thomas Merton, who has been one of the most impressive Catholic writers of our days. He writes:[31]

> Freedom does not consist in an equal balance between good and evil choices but in the perfect love and acceptance of what is really good and the perfect hatred and rejection of what is evil, so that everything you do is good and makes you happy, and you refuse and deny and ignore every possibility that might lead to unhappiness and self-deception and grief . . . only the man who has rejected all evil so completely that he is unable to desire at all is truly free. . . . God, in Whom there is absolutely no shadow or possibility of evil or of sin, is infinitely free. In fact, He is Freedom.

If we read such statements, we understand clearly that the term "free" is used in a sense that has nothing to do with the difference between Newtonian mechanics and twentieth-century subatomic physics. There is no chain of reasoning that would lead from the statistical character of physical laws to the statement that freedom consists in "love of what is really good . . . and hatred of what is evil."

11

Causal Laws

1. The Meaning of "Predetermination"

All philosophers, of every school, imagine that causation is one of the fundamental axioms of science, yet, oddly enough, in advanced science, such as gravitational astronomy, the word "cause" never occurs. . . . The Law of Causality, I believe, like much that passes among philosophers, is a relic of a bygone age, surviving like the monarchy, only because it is erroneously supposed to do no harm.

Thus writes Bertrand Russell.[1] In the language of our daily life, we understand very well what is meant by saying that an event A is the "cause" of an event B or that B is the effect of A. When a boxer punches his opponent's nose, "the punch is the cause of the fracture of the nose," and "the fracture is the effect of the punch." Everyone understands what is meant by the statement that "the drop in temperature is the cause of the contraction of mercury in the thermometer."

If, however, an attempt is made to formulate this statement in the language of theoretical physics, it will be apparent that the clear distinction between "cause" and "effect" becomes a little blurred. When a motor accident occurs, we might say that "the darkness was the cause of the accident," or the "negligence of the driver," or the negligence of the pedestrian involved," or the "slipperiness of the road," or the "rumours of a threatening war," or the "wrath of the gods." It is uncertain which of these statements is correct. A common-sense solution has been to say that every one of these assertions gives a "partial cause." But if the number of "partial causes" becomes greater and greater, it will eventually include all phenomena in the universe, and then the assertion is only that something in the universe is the cause. The assertion becomes tautological and conveys no information. Hence, if we attempt to formulate the state of affairs scientifically, we note that it becomes very difficult and complicated to give a satisfactory formulation of the principle of causality.

On the other hand, when an affair goes to court, e.g., a motor car accident, the judge or the jury must decide what the "cause" of the accident is in order to fix the compensation to be paid to the victim. In such cases, an attempt is made to fix the "responsibility" for the accident—thus interjecting an expression of "moral judgment" into the problem. Some authors, e.g., Hans Kelsen,[2] even believe that the concept of "cause" itself has its origin in legal or moral language. As a matter of fact, even in speaking of inanimate objects the expression "responsible" is sometimes used—for example, the bad weather is "responsible" for the accident. Legal procedure requires the judge or the jury to fix the "responsibility" or, in other words, to establish the cause of an event, although scientific analysis shows that the concept of "cause" or "causality" is either very complicated or very vague. In this chapter we shall deal exclusively with the role of these concepts in science itself and disregard its role in ethical, political, or religious language.

Perhaps the best way to understand the difficulties involved in a scientific formulation of causality is to start with a formulation that a great many people, philosophers, scientists, theologians, and laymen alike will regard as adequate. They wish to say that if we believe in causality as a general law, the future of the universe is unambiguously "determined" by its past and present. At first glance, this seems to be a statement containing only expressions in our common-sense language; but if we try to understand such an assertion about "predetermination" exactly, we shall experience tremendous difficulties. We shall soon see that the statement about the "determination of the future" is tautological and gives no information about the empirical world. The statement that "the future is predetermined" seems to us to belong to the language of common sense because we are, from our religious—Judaeo-Christian —tradition, accustomed to the idea of an omniscient Intelligence in whose mind this predetermination takes place. To the pagans, since their gods were imagined as more human, this predetermination took place, not in the minds of the gods, but in the mind of a "Fate" above the gods, which has become popular in our culture through the Wagnerian opera, *Twilight of the Gods*.

If science does not care to include an omniscient Intelligence in its conceptual scheme, by the statement that "the future is determined" it can only mean that it is determined by law. If nothing specific is said about this law, it is easily seen that the mere "existence" of such a law, unless we mean "existence in an omniscient Intelligence," is a tautological statement about the world that does not restrict any possibility. Bertrand Russell[3] presented the argument for the tautological character of any statement about the "future's being determined" very lucidly. In order to simplify matters, let us assume that an event in the world consists in the motion of a single mass point. Whatever the future may hold in store, it will be described by giving the coordinates x, y, z of the mass point as functions of the time (t). In other words, the future is determined by the three equations: $x = f_1(t)$; $y = f_2(t)$; $z = f_3(t)$. If such functions, $f_1(t)$, $f_2(t)$, $f_3(t)$, exist, the "future of the world is determined." Since the world exists only once, these functions are determined by the course of events in the world. "It is true," says Russell, "that the formulae involved (f_1, f_2, f_3) may be of almost infinite complexity, and therefore not practically capable of being written down or apprehended." But this only means that we are not capable of knowing these formulae; their "existence" follows from the statement that "there is only one world." Thus the material universe must be subject to formulae and the future must be determined.

It may be, however, that human knowledge cannot grasp these formulae. For this reason, an "omniscient Intelligence" has been introduced in order to interpret the "existence of the world formula" as a phenomenon in the mind of a being that, despite its higher abilities, can only be understood by an analogy to the human mind. In order to describe what the principle of causality means in our actual science, we have worked out formulations that are much less general than the statement that "the future is determined." What matters is not *that* the future is determined but *how* it is determined.

2. Laplace, Newton, and the Omniscient Intelligence

We have learned that the goal of science is to set up a system of relations between symbols and operational definitions of these symbols in such a way that the logical conclusions drawn from these statements become statements about observable facts that are confirmed by actual sense observations. We must, therefore, ask what is the place of "causality" in such a system of relations and definitions. If we start to investigate this, we soon note that it is very difficult to assign to the law of causality its proper place among the principles of science. Perhaps the best thing to do is to start from the science in which logical analysis has advanced the most. We omit geometry, in which the concept of time does not enter and only static phenomena are treated (Chapter 3) and turn to the laws of motion in their traditional, Newtonian form (Chapter 4). There was a period in the evolution of science, and a very important and successful period at that, when scientists and philosophers believed that these laws were the fundamental laws of all phenomena of nature or, at least, the fundamental laws of physical science.

At the end of the eighteenth century, the great French mathematician and astronomer, Laplace, made a statement that may perhaps be regarded as the sharpest formulation of what has been regarded as the "law of causality" as it is used in science. Laplace wrote, in the introduction to his *Theory of Probability*:[4]

> Let us imagine an Intelligence who would know at a given instant of time all forces acting in nature and the position of all things of which the world consists; let us assume, further, that this Intelligence would be capable of subjecting all these data to mathematical analysis. Then it could derive a result that would embrace in one and the same formula the motion of the largest bodies in the universe and of the lightest atoms. Nothing would be uncertain for this Intelligence. The past and the future would be present to its eyes.

We can easily see in what way Laplace imagined this Intelligence to operate. He continued:

> The human mind provides in the perfection which he has succeeded in giving to geometry a weak outline of this Intelligence. . . . All efforts of men in the search for truth have the aim of approaching closer and closer to that Intelligence, but man will remain forever infinitely far removed from it.

Looking at the astronomy of that period as it was presented in Laplace's *System of the World*, we can easily describe the structure of the world formula that should be produced by the Superior Intelligence invoked by Laplace. He imagined the work of this Intelligence as being similar to the work of the astronomer who observes the present positions of the celestial bodies and computes from these their positions at any time t. The Superior Intelligence does more than the astronomer by assuming an arbitrary number of bodies, arbitrary initial conditions, and forces between the bodies that may not follow Newton's law of gravitation. However, Laplace assumed one restriction on these forces which has been, since the decline of Aristotelian and the rise of Newtonian mechanics, almost taken for granted. He said: "the orbit performed by a single molecule of air or vapor is determined with exactly the same certainty as the orbits of planets. The differences between them are only due to our ignorance."

We shall now describe the way in which the astronomer computes the future positions of celestial bodies from the knowledge of their present positions and velocities. We shall call every law a *causal law* which allows us to infer from information about one region of space and time some information about another region of space and time. Newtonian mechanics, by means of which the future positions of celestial bodies are computed, certainly contains causal laws. The "principle of causality," however, makes certain claims about the range over which causal laws work or can be put to work for the prediction of the future. Laplace's Superior Intelligence would be in possession of causal laws that would allow the prediction of the whole future of the world from knowledge of its present state. The causal laws by which the orbits of celestial bodies have been predicted can be derived from Newton's laws of motion.

We have learned in Chapters 3 and 4 that physical laws do not consist of relations among symbols only. Such relations would be Euclid's axioms of geometry or Newton's laws of motion. In order to draw conclusions about observable facts, we must add operational definitions of the symbols. Hence, we must keep in mind that the following statements about predictions refer merely to the future values of the symbols. Their bearing upon the prediction of observable facts depends upon our choice of operational definitions.

Let us consider N mass points with the masses $m_1, m_2 \cdots m_N$.

The Cartesian coordinates of the mass-point m_k are x_k, y_k, z_k. The force acting upon m_k may have the components X_k, Y_k, Z_k that are given functions of the coordinates x_1, y_1, z_1, $\cdots$ x_N, y_N, z_N. The equations of motion are then: $m_k \dfrac{d^2x_k}{dt^2} = X_k$, $m_k \dfrac{d^2y_k}{dt^2} = Y_k$, $m_k \dfrac{d^2z_k}{dt^2} = Z_k$, where X_k, Y_k, Z_k are given functions of x_1, y_1, z_1, $\cdots$ x_N, y_N, z_N. If we introduce the velocity components $u_k = \dfrac{dx_k}{dt}$, $v_k = \dfrac{dy_k}{dt}$, $w_k = \dfrac{dz_k}{dt}$, we can write the equations of motion in such a way that we can recognize that they are causal laws. We must remember that at every instant of time every particle has a position and velocity given by x_1, y_1, z_1, $\cdots$ x_N, y_N, z_N, u_1, v_1, w_1, $\cdots$ u_N, v_N, w_N. If these quantities are given at a certain instant of time, the laws of motion allow us to compute the values of these quantities at any past or future instant t. This is obvious from the fact that what the laws of motion actually do is as follows: the increases of the quantities x_1, y_1, z_1 $\cdots$ x_N, y_N, z_N, u_1, v_1, w_1, $\cdots$ u_N, v_N, w_N are given as functions of those quantities themselves.

$$m_k \frac{du_k}{dt} = X_k(x_1 \cdots w_N) \qquad\qquad m_k \frac{dv_k}{dt} = Y_k(x_1 \cdots w_N)$$

$$m_k \frac{dw_k}{dt} = Z_k(x_1 \cdots w_N)$$

$$\frac{dx_k}{dt} = u_k \qquad \frac{dy_k}{dt} = v_k \qquad \frac{dz_k}{dt} = w_k$$

In order to make the scheme of these equations of motion simpler, we may introduce for the components of the coordinates of all particles in all directions x_1, x_2, $\cdots$ $x_n (n = 3N)$ with the corresponding velocity components u_1, u_2, $\cdots$ u_n. Then the equations of motion look like this: $m_k \dfrac{du_k}{dt} = X_k(x_1, \cdots u_n)$, etc., $\dfrac{dx_k}{dt} = u_k \cdots$ etc. This is a special case of a more general type of system. If we call x_1, x_2, x_3, $\cdots$ x_n, u_1, u_2, u_3, $\cdots$ u_n the "state variables" of our mechanical system and denote them all by the same symbol with subscript ξ_1, ξ_2, $\cdots$ ξ_{3N}, the equations of motion have the form $\dfrac{d\xi_k}{dt} = F_k(\xi_1, \xi_2, \cdots) \; k = 1, 2, 3N$. This means that if the "state of a system" is described by $3N = n$ state variables, the "incre-ments of these variables" $\dfrac{d\xi_k}{dt}$ are given as functions F_k of the present values (initial values) of these state variables.

The mathematical theory of differential equations teaches us to "integrate" a system of the form $\dfrac{d\xi_k}{dt} = F_k(\xi_1, \xi_2, \cdots)$, $(k = 1, 2, \cdots n)$. This means that if the values of the state variables for one instant of time (*e.g.*, $t = 0$) are given, we can find the values of $\xi_1 \cdots \xi_N$ at any time t if the value for $t = 0$, and the differential equations (in other words, the functions F_k) are given. The differential equations are an instrument that allows us to compute the values of the state variables at all times t if the values for one instant $t = 0$ are given. We can call this a "prediction" because we compute the future values from the present values. We must keep in mind, however, that we can just as well compute the values of the state variables for a time $t < 0$, *i.e.*, for the past.

3. The Mathematical Form of a Causal Law

We have learned that the Newtonian laws of motion allow a pre-diction of the future based on knowledge of the present because these laws are of the form $\dfrac{d\xi_k}{dt} = F_k(\xi_1, \xi_2, \cdots)(k = 1, 2, \cdots n)$. According to the mathematical theory of differential equations, if the values of the "state variables" are known for the present instant of time $t = 0$, one can "predict" their values for any past or future time t. All laws of this type are called "causal laws." The general "principle of causality" would claim that all phenomena are governed by causal laws which have the form $\dfrac{d\xi_k}{dt} = F_k(\xi_1, \cdots \xi_N)$ where $\xi_1, \cdots \xi_N$ are any variables that determine the "state" of a physical system at the time t.

We shall postpone, at this point, the discussion of the general principle of causality and only stress the point that belief in this general principle is supported by the special case of astronomy where the ξ_k are the coordinates and velocities of mass-points and the functions F_k are known to be simple mathematical formulae derived from Newton's laws of gravitation. In addition, in all applications that actually can be carried out, the number n is a small one; in other words, the initial conditions at the present time $t = 0$ are given in a simple way. From these simple assumptions, the immense complexity of the actual orbits of celestial bodies could be computed by using the rules of ordinary mathematical analysis. What caused the success was the simplicity of the laws in comparison with the complexity of the observed facts.

However, if we regard the F_k as arbitrary functions of the ξ_k and admit complicated initial conditions, the causal law $\dfrac{d\xi_k}{dt} = F_k(\xi_1 \cdot \cdot \cdot \xi_N)$ may be "valid" but will not guarantee the same kind of success. It may be that the law is as complex as the observed facts. Then there is no advantage in substituting for the direct description of observation another "indirect" description called "law" that is in no way simpler than the direct one.

Our problem now is to investigate the role of the general law of causality in the case that we no longer have to do with the special case of astronomy, but allow that the F_k and the initial conditions are of arbitrary complexity. To overcome this difficulty, Laplace introduced his "Superior Intelligence" for which, because of its overwhelming abilities, the general case was supposed to be as simple as ordinary astronomy is for human beings. The principle of causality says then that there exists a Superior Intelligence that knows all initial values of the state variables, knows all the functions F_k, and is such an accomplished mathematician that it can predict all future values of the state variables by solving the differential equations.

We must keep in mind that in the formulation of causality a "Superior Intelligence" was explicitly introduced by Laplace. He did not, of course, believe that this Intelligence was essential in the formulation of causality. There is a famous anecdote about Laplace's submitting to the Emperor Napoleon[5] a copy of his *System of the World*. Napoleon asked him what was the place of God in the system. Laplace answered: "Sir, I do not need this hypothesis." It is interesting to note that Laplace, who did not need the hypothesis of God in his book on astronomy, needed a Superior Intelligence in his formulation of the principle of causality. He believed, of course, that this is only a convenient way of speaking, and that this Superior Intelligence could be eliminated and the principle of causality could be formulated by reference to human ability alone.

This, however, is not so easy as it looks. We must say: *there are* functions F_k of the $\xi_1 \cdots \xi_N$ which have the property that $\dfrac{d\xi_k}{dt} = F_k$ allows us to predict the future values from the present values of $\xi_1 \cdots \xi_N$, but if we do not know specifically what these functions are, the mere existence of such functions means only that the future values of the ξ_k are "determined" somehow by the initial values or, in other words, the increments of the ξ_k $\left(\text{the } \dfrac{d\xi_k}{dt}\right)$ are determined by the ξ_k themselves. If we do not introduce a Superior Intelligence which "knows" the functions F_k or present explicitly a formula for the F_k, the word "determined" can only mean that we assume some property of the F_k that restricts the possibilities. This is similar to what we learned in Chapter 4 (Laws of Motion), that Newton's laws become practically useful only when we add the assumption that the "forces" are "simple" functions of the coordinates. If we interpret, as we do in this section, Newton's mechanics as a system of causal laws, this means that the F_k are simple functions of the $\xi_1 \cdots \xi_N$. If we admitted an arbitrary complexity of the functions $F_k(\xi_1 \cdots \xi_N)$, the mere statement of "existence" would be no statement about facts, but a tautological statement that could not be refuted by any experiment.

In either case, we can always regard the future values of the ξ_k as functions of t and the initial values; if we assign any arbitrary values of the ξ_k to the values of t, the dependence can always be described by a formula. If a causal law is to be of actual value in the prediction of the future, we must introduce the vague quality of "simplicity," which is, of course, dependent on the psychological and sociological status of the scientists of a certain period. A "simple" formula means, in this context, a "workable" formula. Since "prediction of the future by ordinary human beings" is an activity within the whole range of human activities, it is understandable that the criterion of "simplicity" can be applied to a causal law although it depends on psychological and sociological considerations whether a certain law is "simple."

4. Relevant and Irrelevant Variables

If we merely assume that the formulae $F_k(\xi_1 \cdots \xi_N)$ are "simple" and that the initial conditions are simple, we are not yet certain that the Principle of Causality is a statement about observable facts. If we assume, however, that there are operational definitions which allow us to assign numerical values to the state variables $\xi_1 \cdots \xi_N$ by an operation of measurement, the Principle says that all observable facts are governed by causal laws that allow us to predict measurable values of the $\xi_1 \cdots \xi_N$ from their measured values at the present. We do not bother, at this point, to specify the operations by which we assign these values. Is it possible, then, to test the principle of causality? This could only be done if we could imagine a world in which the principle of causality was wrong. In such a world we could not predict the motions of planets by a law that was comparable in simplicity to Newton's laws; but we could not, of course, prove by any experiment that it was impossible to find such a causal law. The principle of causality in this form could only be dropped in despair, by recognizing that success could not be realized by starting from this assumption.

There is, however, one conclusion that can be drawn from the principle of causality that is independent of the special form of the functions F_k and which perhaps could be tested by experiment. Whatever the specific form of F_k may be, one thing is certain: The changes of the variables in time $\dfrac{d\xi_k}{dt}$ are dependent upon the present values of $\xi_1 \cdots \xi_N$ only. Whenever these variables reassume their original values, the changes in time reassume their values too. In other words, if a state of the system repeats itself, all following states repeat themselves too. If we call a value system of the ξ_k the "state A" of our system, we can say: If a state A of our system is followed by a state B, every time that A returns it is followed by B. This is a formulation of the principle of causality which does not make use of expressions like "simple formula." We must keep in mind that saying that the system "has a state A" or "a state B" means only that the ξ_k have certain numerical values. The statement "A returns" or "B returns" refers only to the numerical values of the state variables, not to observable facts.

Now, how must we proceed in order to find an observation that could refute the statement that "if A returns B will return too?" We must observe an actual recurrence of the state A and observe that B does not return the second time, although it had followed A

when it occurred the first time. We must take it for granted that
we know a procedure of measurement by which we can assign
numerical values of the ξ_k. If the principle of causality is to be
right, we must understand by "state A" and "state B" states of
the whole universe. This means a state is defined if the coordinates
and velocities of all masses in the world have specific numerical
values. Since the number of these masses is enormous, of the
order of billions and billions, the "return of a state A" means that
the billions and billions of variables should reassume their original
values. Such an event can, of course, not be checked by any
observation. This means that the validity of the principle of
causality cannot actually be checked by the return of a state A.
In order to make a test possible, we must not require more than
that the approximate return of A brings about the approximate
return of B.

We can bring the principle of causality into a "testable" form,
by formulating its contents as follows: We define as "state A"
a state of the world in which a certain group of the ξ_k have
specific values while the values of the other ξ_k are "irrelevant."
Then the principle of causality would claim that if B followed A
once, a return of A would bring about a return of B, whatever the
initial values of the "irrelevant" ξ_k might be. Obviously, this
principle cannot be refuted by experiment. If we observe that B
does not follow A in a certain case, this proves only that the observed
return of A was not a "real" return. We can only observe that a
number, say n, of the variables ξ_k reassume their initial values; some
other "irrelevant" variables may not. This would disprove the
principle of causality only if we knew exactly which variables are
relevant. But it may just as well be that there are, besides the
variables that we know to be relevant, others which must also
reassume their original values in order to make it certain that B
will follow. Theoretically, in every case in which the state B does
not follow A we can always assume that we were wrong in taking it
for granted that all the other variables that did not assume their
original values were "irrelevant."

What we actually can confirm is only the fact that, practically, in
a great many cases one can single out a relatively small set of
"relevant" variables. Then the principle of causality says that in

every situation in the physical world we can introduce a small number of "relevant" variables that have the following property: The return of a "small" number of variables to their original values would imply that a "state A" has returned. This would imply, again, that the "state B" will also return. This "principle of causality" can, of course, be "confirmed" by experience or observation. This confirmation has a certain vagueness because it is vague to say that the return of a "small number of variables" to their original values (A) is sufficient to bring about the return of the state B.[6]

5. Causal Laws in Field Theory

We have so far overlooked entirely the question of the relationship between the Newtonian mass-points and the phenomena that are actually observed. For Newton and his immediate followers, this relationship was regarded as a very simple one which did not demand much discussion. However, in the application of mechanics to actual, technical problems, we do not regard a solid or fluid body as a system of mass-points but as a continuum. We do not describe the state of a fluid, for example, by describing the positions and velocities of its mass-points, but by regarding it as a continuum and describing each point of it by its coordinates x, y, z. The "state of the fluid" is then described by the velocities of all masses at every instant of time. The mass located at a point x, y, z at a time t may have three components of velocity u, v, w. The state of the fluid at a certain instant of time t is known if we know the u, v, w as functions of x, y, z and the time t. If we know the functions $u(t,x,y,z)$, $v(t,x,y,z)$, $w(t,x,y,z)$, we know the state of our fluid in the present $(t = 0)$, the future $(t > 0)$, and the past $(t < 0)$.

A causal law would enable us to compute the "state variables u, v, w" for all future times if they are given for the present $t = 0$. If we denote, as usual, the increments of u, v, w per unit of time by $\dfrac{\partial u}{\partial t}$, $\dfrac{\partial v}{\partial t}$, $\dfrac{\partial w}{\partial t}$, a causal law would give these increments as functions of the present values u, v, w and their spatial derivatives $\dfrac{\partial u}{\partial x}$, $\dfrac{\partial u}{\partial y}$, $\cdots \dfrac{\partial w}{\partial z}$, $\dfrac{\partial^2 u}{\partial x^2}$, $\dfrac{\partial^2 u}{\partial x\,\partial y}$, $\cdots\cdots$. A causal law has, therefore, the form

$$\frac{\partial u}{\partial t} = F\left(u, v, w, \frac{\partial u}{\partial x}, \cdots \frac{\partial^2 u}{\partial x^2}, \cdots \right) \text{ plus the analogous equations}$$

for $\frac{\partial v}{\partial t}, \frac{\partial w}{\partial t}$. The form of the function F depends upon the cohesion between the parts of our fluid or solid body. The principle of causality means then that all motions of solid or fluid bodies follow laws of the form $\frac{\partial u}{\partial t} = F(u, v, \cdots)$, where the functions F are dependent upon the nature of the body, the viscosity of a fluid, the brittleness or elasticity of a solid body.

It is clear, then, that we can never find an observable phenomenon that could falsify the principle of causality, for we can never confirm that there is *no* function which could be used to describe the phenomena in a fluid or solid body. On the other hand, the confirmation of the principle of causality is based on the experience that for a great many bodies we can assign functions F of the properties of the bodies and can predict correctly the future values of the u, v, w if the present values are given. The existence of such functions is the real physical meaning of the principle of causality. In other words, this principle expresses the belief, or at least the hope, that for every type of body, liquid or solid, plastic or elastic, such a function can be found. This principle would hold for all physical phenomena if we believed that all physical phenomena could be reduced to the laws of traditional mechanics. We have learned, however, that the principle of causality can also be extended to much more general physical hypotheses.

Toward the last quarter of the nineteenth century, it was accepted more and more that the phenomena of electromagnetism were not to be reduced to Newtonian mechanics, but were to be deduced from a separate system of principles, of which, in turn, the Newtonian laws were a special case. The "state of a system" is no longer described by the velocity at a certain point x, y, z and a time t, but by the electric and magnetic field strengths at x, y, z and a time t. A causal law in the theory of the electromagnetic field is now an equation that allows us to compute from the present distribution of field strengths the future values of field strengths. Mathematically, the causal laws look exactly like those in mechanics except that the velocities u, v, w are replaced by the field strengths. This theory of the electromagnetic field has been generalized into a "general field theory." We may assume that there are, besides the electromagnetic field, other fields like the gravitational field or the nuclear field. If we denote by u_1, u_2, u_3, $\cdots$, u_n the components of all these fields, a causal law has the form:

$$\frac{\partial u_k}{\partial t} = F_k\left(u_1, u_2, \cdots u_n, \frac{\partial u_1}{\partial x_1}, \cdots \frac{\partial u_n}{\partial x_n}, \frac{\partial^2 u_1}{\partial x_1{}^2}, \cdots\right)$$

$$(k = 1, 2 \cdots n)$$

The principle of causality would claim then that all physical phenomena can be described by equations of this type. Then if the u_1, $\cdots$ u_n are given at the present time $t = 0$ for every x, y, z, we can compute the future values from the equations.

The principle of causality in the field theory of physics is obviously much more vague than in mechanistic physics, where all phenomena are regarded as deducible from Newton's laws of motion. Not only are the functions F_k indeterminate as the "forces" were in Newtonian physics, but the state variables u_k themselves are indeterminate, whereas in Newtonian physics all state variables were known to be positions and velocities. Hence, all difficulties experienced in Newtonian physics in formulating the general principle of causality survive in field physics. The assertion that the future is determined by the present state has a palpable sense only if we introduce either an omniscient Intelligence or give specific equations by which the increment of the state variables per unit of time is determined, *i.e.*, the functions F_k.

The principle of causality in field physics obviously has less factual content than in Newtonian physics. In the latter case, it would be possible to say: If all positions and velocities of masses reassume at $t = t_1$ values which they possessed at $t = t_0$, they will after t_1 traverse the same values as after t_0. In the case of field physics, however, the meaning of the statement that all state variables u_1, u_2, $\cdots$ reassume their original values except if we can actually enumerate all state variables is indeterminate; otherwise, the principle of causality says only that if a state A is followed by a state B once, A will, whenever it returns, be followed by B. This statement, however, is clearly tautological. We would only recognize that a state A did return if it were again followed by B. To give factual meaning to the principle of causality we must assume, at least, that whenever a small number of state variables reassume their original values the "state A" has returned and the "state B" will also return. The greater the number of state variables u_1, u_2, · · · the smaller is the factual content of the principle of causality. If the number is very great, we can never know when a state A has actually returned; we can always surmise that there is still a variable left which we have overlooked and which did not reassume its original value. Then we cannot expect that B will again follow; we are not certain that A has actually returned. This is certainly the case if the number of state variables becomes infinite. Then the principle of causality becomes a tautological one; it is not a statement about physical reality at all.

If we want to prevent the general principle of causality from slipping into a tautological one, we must formulate it as follows: By the introduction of a small number of state variables we can make sure that the return of this small number of variables to their original values is followed by the return of the state B that had followed the first time. This means, in the usual language of physical science, that there are only a few kinds of forces that determine the changes of states—gravitation, electromagnetism, etc. If we were not certain that the number of forces is small, we could never be certain that a "state A" can return, for if all known forces had the same values, there could always be another unknown force that would make the new state different from the original one. Then the return of B could not be expected. In order to prevent the principle of causality from becoming tautological, we must introduce vague terms like, "small number of state variables" and "simple laws of forces (F_k)." We have a choice between making the principle of causality precise and tautological or vague and factual.[7]

6. "Gaps" in Causal Laws

In the philosophical and religious attitudes toward causal laws, a great deal of attention has been directed toward the possible "gaps" in causal chains. Is the future motion of a system determined at every "state A," or are there some, perhaps exceptional, states which do not determine future states B unambiguously? If we start from Newton's classical formulation, "mass times increment of velocity is equal to force," this increment is certainly determined everywhere when "force" is determined. If we mean by "force" simply Newton's attraction between mass-points, inversely proportionate to the square of distance $(1/r^2)$, this expression becomes undefined when the distance disappears $(r = 0)$, where mass-points collide mathematically. This means that the future becomes indeterminate when the present state is a "singular" point in the differential equation of motion. We know from the mathematical theory of differential equations that the solution is not uniquely determined if we give the values of the coordinates at a singular point. Starting from a singular point, the solution can be continued in various ways. This situation becomes even more obvious if we apply these considerations to the world of atoms and molecules. The mechanical theory of heat regards gases as great numbers of colliding molecules. At each collision there is a singular point, and the future motion is not always determined by the colliding molecules, *e.g.*, when two equal molecules collide head-on with equal speed. Thereafter there are various motions of the molecules that would be in agreement with the motion of the molecules before the collision.

So far we have spoken of causal laws only as statements about mathematical deductions from the equations of motion. The "state of a system" has been defined by the set of values ascribed to the state variables, *e.g.*, to the coordinates and velocities of mass-points. By "values" we have meant the "real numbers," in the mathematical sense, ascribed to the variables. However, the situation changes when we ask: Is it possible to predict future observations from the present observed state of a system or the whole world? The result of a measurement is never a number in the mathematical sense, but always a certain interval. For example, we cannot distinguish by observational measurement whether the length of a certain body is a rational or an irrational number. Hence, the initial state A of a system is not given by numbers assigned to the state variables $u_1 \cdots u_n$ but by intervals within which these numbers are included. A great many "mathematical states" of a system A_1, $A_2 \cdots$ may correspond to one and the same observed state. All these states A_1, $A_2 \cdots$ are very near to each other, and we can choose an average value A as the approximate value of all the A_1, $A_2 \cdots$.

If the observed value A is followed by another observed value B, are we, on the grounds of our mathematical causal laws, certain that a return of A will be followed by a return of B? To put it precisely, can we conclude from the mathematical causal laws that an analo-

gous causal law will also be valid for the observed states of a system?
Obviously, it will not. From the mathematical causal law we can
derive that after a certain time t the state A_1 becomes B_1, A_2
becomes B_2, A_3 becomes B_3, etc. If the states B_1, B_2, B_3, $\cdots$ are
so near each other that they correspond to one and the same observa-
ble state, we can certainly conclude that when an observable state
A recurs, the observable state B will recur also, and we shall have a
causal law for observable phenomena. If, however, B_1, B_2, $\cdots$
are not near each other, we cannot predict that an observable state
A will always be followed by the observable state B. "The
observable state A recurs" may mean in some cases that A_1 recurs
and in other cases that A_2 recurs. In one case, it would be followed
by B_1, in the other case by B_2. However, B_1 and B_2 may be very
remote from each other and may not correspond to one and the
same observable state.

We note that the validity of a causal law for observable states is
based upon one assumption about the mathematical law, namely:
If two states A_1 and A_2 of the system at the present instant ($t = 0$)
are very near each other, the states B_1 and B_2 reached by the system
at a later instant ($t = T$) will also be very near each other, whatever
the value of T may be. In other words, a small change in the
initial state ($t = 0$) cannot result in a large change in the final state
($t = T$). The technical expression in mechanics for this requirement is that the motion starting at A must be "stable." Hence, a
mathematical causal law leads to a causal law for observable phenomena only if solutions of the mathematical equations are "stable"
solutions.

We can easily give examples of states A which lead to solutions
that are not stable and do not lead to causal laws for observable
phenomena. Let us consider the ridge of a chain of mountains and
assume for the sake of simplicity that the ridge is a horizontal
straight line. A solution of the mathematical equation of motion
may be found for the initial condition that a mass-point is situated
on the ridge and has an initial speed c; the force that is acting may
be gravity. An initial mathematical state A_1 may be described by
imputing to the mass a velocity in the horizontal direction. The
initial states A_1, A_2, $\cdots$ may be given by velocities with slightly
different directions but all with the same speed c. After a time T
the state A_1 will have become a state B_1; the mass-point will then
be located on the horizontal ridge at a distance cT from its starting

point.   The state A_2, however, will have become a state B_2 which is very remote from B_1.   Since the velocity is different from the horizontal, the mass will fall down along the mountain and will reach a depth of $1/2gT^2$ below the position B_1 on the ridge.   If we do not know whether our observed initial state A corresponds to the mathematical state A_1 or A_2, we cannot predict whether after the time T the mass will be on the ridge or at a distance of $1/2gT^2$ below, where g is the acceleration of gravity.   Since the slightest departure from the horizontal direction of velocity produces the downward motion, and there is no possible observation by which we can distinguish between A_1 and A_2, we cannot predict by observing our mass at $t = 0$ whether after the time T it will be on the ridge (B_1) or at a distance $1/2gT^2$ below the ridge.   Hence, the future state cannot be predicted from the observation of the present state.

The motion of a mass along a ridge is an "unstable" motion.   The observation of the initial state cannot lead to a prediction as to where the mass will move.   We have here a case in which the observation of the initial state will not allow us to predict the subsequent state; there is no causal law for the observable facts.   Such a situation also arises, for example, within a gaseous body.   If we assume that the molecules are mass-points that attract and repel each other according to Newton's laws of motion, every head-on collision is a state of motion that is not stable.   If we were able to observe the states of these molecules, we would find that a great many of these states were in the neighborhood of unstable motions and that the future was as unpredictable as the future of a mass moving along a ridge.

We can conclude from these remarks that there are deep and broad "gaps" in the applicability of causal laws to mechanical systems.   Even if we assume the strict validity of Newtonian mechanics for all physical phenomena, we cannot conclude that from an arbitrary initial state which we observe the future state can be unambiguously determined.   From this remark, it becomes even less plausible that Laplace's omniscient Intelligence could be replaced by a human agent.   The idea of general predetermination seems to be tied up with the existence of a "superhuman" or "supernatural" agent.   From the scientific point of view, as remarked above, predetermination of the future is either a tautological concept, or it assumes the existence of causal laws connecting few variables by simple relations.[8]

12

The Principle of Causality

1. Discussion of How to Formulate the General Principle of Causality

Humean and Kantean Causality. We have, by now, tried to formulate the principle of causality by starting from a special theory like Newtonian attraction or field theory. But we must remember that this principle should be applicable not only in physics, but in every field of knowledge, in biology as well as in psychology, in the social sciences as well as in the natural sciences. Only then we can speak of a "general principle of causality." The question has been raised again and again of whether the principle of causality is as valid in biology and sociology as it is in physics and chemistry, but the question itself does not make sense unless we know how to formulate the principle in such a way that it is applicable to all these different fields. If we take our cue from the formulation we used in mechanics and in field theory, we may tentatively say: By the state A_0 of the world at the time t_0, the state A_1 at every subsequent instant t_1 is uniquely determined. We already know the difficulties that are hidden in the term "determined"; we can interpret it in two ways. On the one hand, we may say: Every time the world is in the state A_0, it will be after the time $(t_1 - t_0)$ in the state A_1. On the other hand, we may say: "There is a law which allows us to compute from every state A_0 at a time t_0 the state A_1 at the time t_1, if $t_1 > t_0$.

We have already learned that even in the special cases discussed

278

in the previous section it is not easy to distinguish between the definition of a law and that of the return of a state. These difficulties are, of course, much more serious when we approach the general case where the laws are no longer the known laws of mechanics or field physics. It becomes increasingly difficult to define "the return of the same state" without making use of the causality concept in the definition. But if this is done, the principle of causality becomes tautological. We would define "equal states" as such states A_0 that "have the same effects A_1." Then it obviously follows that if a state A_0 recurs, its subsequent states A_1 recur also. This statement, however, is only a definition of the term "equal states" and gives no information about the physical world.[1]

If we assume that we can solve this difficulty and give a definition of "equal states" that does not make use of the causality concept, there still remains another difficulty that is perhaps more serious. This supreme difficulty arises from the fact that the state variables that occur in the causal laws that are used in science cannot be uniquely correlated with the actual observations of men; but without this correlation the causal laws would only contain symbolic quantities; they would be definitions of these terms. We learned, for example, in Chapter 3 that the axioms of geometry are definitions of the terms "straight line," "intersection," etc., that occur in the axioms of Euclid's or, to speak more exactly, Hilbert's geometry. The layman in science and, in many cases, even the scientist may be inclined to disregard the distinction between the symbols and the observable quantities that are correlated with each other by "operational definitions" or "physical interpretations," or, more generally speaking, by "semantic rules."

The difficulty in such a correlation has existed as long as there has been a physical science. It has its ultimate root in the simple fact that the figures resulting from physical experiments and observations are always the average of a great many actual observations, whereas the figures assigned to the symbols are sharp mathematical quantities. They may, for example, be irrational numbers which can never occur as the result of observation. This difficulty has existed in the mechanics of fluids, where no "velocity" of a mass-point can be observed, although these velocities occur nonetheless in the equations of hydrodynamics. Similarly, the electric field strength within an electron can never be observed. If we wish to speak precisely, we must say that even in the formulation of Newton's laws the "velocity of a mass-point" is merely a symbol. To correlate it with observation, we must remember that "velocity at a certain instant of time" is the first derivative of a function of time, and is computed as the limit of a great many observations.

If we attempt to formulate the principle of causality as a statement about observable facts, it is obvious that we are faced with a very complex and difficult task. It is sufficiently clear by now that statements like "the future is determined by the present" or, more simply, "the future is predetermined" are not statements that can be checked by experiment and observation. In the special cases where we assumed that the "states" of the world could be described either by the positions and velocities of particles or by the field quantities, we indicated the two ways in which the general principle of causality can be formulated without slipping into tautology. We can either require that equal states be followed by equal states, or we can postulate that all phenomena take place according to law. In order to have a short denotation for these two types of formulation, we shall follow the suggestion of Margenau.[2] He takes his cue from the way in which two of the "founding fathers" of modern philosophy formulated the principle of causality in the eighteenth century. David Hume formulated causality as the repetition of sequences; when A recurs, the sequence AB recurs. Immanuel Kant, on the other hand, defined causality as the existence of laws according to which states follow one another. Accordingly, Margenau distinguishes between Humean and Kantean causality.

Hume[3] writes:

> The only immediate utility of all science is to teach us how to control and regulate future events by their causes. . . . Similar events are always conjoined with similar, of this we have experience; therefore, we may define a cause to be *an object followed by another and where all the objects similar to the first are followed by objects similar to the second.* Or, in other words, *where, if the first object had not been, the second never had existed.* The appearance of the cause always conveys the mind by a customary transition to the idea of the effect. Of this also we have experience. We may, therefore, suitably to this experience, form another definition of cause and call it *an object followed by another and whose appearance always conveys the thought to that other.*

In the Humean sense, the principle of causality is regarded as a device for practical purposes. It deals with facts that are directly observable. The principle of causality tells us that there is always a suitable cause by which we can produce a desired effect.

If we introduce values systems A and B of state variables instead of sequences of observable states, we take it for granted that there is a simple operational definition that assigns values of state variables to observable facts. The Humean formulation of causality, the recurrence of sequences AB, can be precisely understood only if we mean by A a set of observations made at a certain instance of time t_0, and by B a set of observations made at the time t_1. There is nothing in the conception of state A that refers to observations that occur at a later or earlier instant than t_0. However, as we have seen repeatedly, the causal laws that are actually used in physics or other sciences define a state A in such a way that observations occurring during a certain interval of time are used. It is obvious, for example, that the operational definition of "velocity at t_0" requires observations of positions during an interval of time.

Exactly speaking, a state A can only be defined by the effect that a "body in the state A" has upon other bodies, our instruments of measurement. Immanuel Kant made the point, therefore, that the law of causality intervenes and plays a role in the definition of a state A or B of a system. While Hume stressed the point that sequences of arbitrarily defined states AB recur, Kant put the emphasis on the existence of general rules. We can choose such definitions of state, A or B, that they follow each other according to those general rules. To quote a characteristic passage, Kant says:[4]

> We cannot study the nature of things *a priori* otherwise than by investigating the conditions and the universal (though subjective) laws, under which alone such a cognition as experience is possible. . . . A judgment of observation can never rank as experience, without the law "whenever an event is observed, it is always referred to some antecedent, which it follows according to a universal rule."

This formulation has a certain metaphysical tinge. However, we shall here understand by "Kantean causality," in the sense of Margenau, a conception that is as scientific as Humean causality. The latter puts the emphasis on the fact that states A are defined by

observations and the causal law derived from them by inductive
inference, while the Kantean causality directs our attention to the
fact that we define the states A in such a way that universal laws
for the sequences can be set up.

2. Causality as a Recurrence of Sequences

If we argue on the basis of a conceptual scheme in which a state
of the world is defined by a finite number of given state variables,
the meaning of the statement that a state A returns is clear. If,
however, we start from the whole or part of the empirical world, it is
difficult to make clear the meaning of the statement that a state A
returns. Obviously we cannot mean that the same observable
properties return. If we take this definition at its face value, since
a magnetic piece of iron may look like an ordinary piece of iron, the
replacement of one by the other would be the return of a state A;
but the next state B would certainly not be the same in both cases.
If we want to define precisely the meaning of "return of a state A,"
we must state precisely how we describe this state. If we accept
the formulation: "If a state A of the whole world returns, the next
state B will also return," this statement has factual meaning only
if the course of the world consists in an infinite number of cycles,
in the eternal recurrence of the same events. If this is not the case,
our formulation becomes tautological, empty of any factual content.

If there is no recurrence of the same state A, the principle of
causality—"If A recurs, B recurs also"—is valid whatever may
happen in the world. If we exclude the recurrence of a state A of
the universe, we must restrict ourselves to "incomplete cycles."
Let us consider, for example, a body falling to the ground from rest:
When this phenomenon occurs, we certainly do not have a complete
cycle. The initial position and velocity relative to the earth
recurs, but the position relative to the sun, and even to the environment on the earth, is certainly different in each case of recurrence.
If we neglect the environment, we can say that the state A recurs
and is always followed by the same state, if we mean by this state
simply the position and velocity relative to the earth. If we wished
to be very precise, even this recurrence would not be exactly true
since an influence is exerted by all surrounding bodies. The applicability of the principle of causality (recurrence of sequences) is
based on the fact that it can be applied to incomplete cycles. We

may say that, practically, the principle of causality claims that all phenomena in the world can be described by resolving them into incomplete or approximate cycles.

If we really want to understand the meaning of causality in actual science, we must keep in mind that by "recurrence of a state" we can mean very different things if we attempt to carry out the resolution into approximate cycles in situations which actually present themselves in nature. We can point out these differences by using an example that is familiar to everyone: the physical state of our atmosphere that is called "weather." We may speak of the recurrence of a weather situation A if we again have the same temperature, atmospheric pressure, direction and intensity of wind, density of electric charges, etc. If we define "recurrence of A" by the recurrence of a weather situation in the sense described above, the law of causality would allow us to set up a system of weather prediction that would assert: If a weather situation A is followed by B, every time that A recurs, it will be followed by B. This method of weather prediction has the advantage that it makes use of quantities that are very near to observable facts and, therefore, easy to handle. This method of weather prediction has been used through the ages in practical meteorology and even in popular "farmer's almanacs." It presupposes the existence of cycles in weather situations. The belief in these cycles is occasionally based on superstition, for example, on the recurrence of a weather situation every hundred years. The weather situation has been usually described by temperature and pressure in the atmosphere.

By "temperature" or "pressure" we mean here the values that are registered in meteorological tables. They define the "meteorological state" of an atmosphere. In these tables, "temperature" or "pressure" is the average value of these quantities in a major area, such as temperature in Boston or pressure in Worcester, Massachusetts. This is certainly a rough description of the weather; temperature and pressure are actually changing within much smaller areas. We could, for instance, mean by "temperature" or "pressure" the average value within a cubic inch or even in a lesser volume. These values are the state variables that describe the state of a fluid (like air) in the differential equations of aerodynamics. We could then regard the state A of the atmosphere at an instant $t = t_0$ as the initial conditions required for a solution of these differential equations. In this case, the description of the state A of an area consists of a very large number of values and is highly complex. If we assume that the differential equations can be integrated for arbitrary initial conditions, we can, mathematically speaking, compute the values of temperature, pressure, etc., for any instant t if we know them for $t = t_0$. Predictions obtained in this way are as reliable as the equations of aerodynamics, but the solutions are so complex that they are practically without value.

However, there are situations in which even the "aerodynamical states" of our atmosphere will not follow strict causal laws; such phenomena are rapid fluctuations or turbulence. In such situations, we must state the positions and velocities of single molecules as state variables. The number of variables is then increased into the millions of millions. The prediction would be as reliable as the Newtonian mechanics of particles, but the practical usefulness would be almost nil. If we wish to refine the "molecular state" even more, we must go into subatomic parts of the molecule. According to what we learned in Chapter 8 (Atomic Physics), in this domain positions and velocities of atomic objects are no longer possible state variables and no causal law can be set up in terms of these quantities. We must use the amplitudes of de Broglie waves as state variables. They are connected with observable phenomena in a rather complicated statistical way.

Generalizing these remarks, we can say that the meaning of "recurrence of sequences" depends upon what kind of state is assumed to recur. In our example, causality can mean very different things, depending upon whether we define it by the recurrence of meteorological, aerodynamical, molecular, or subatomic states.

The question of whether the principle of causality is valid in the historical and social sciences as well as it is in physics or chemistry has been much discussed. The argument has been advanced that history investigates events that happen only once, while physics investigates recurrent sequences of events. This argument was used by the German philosopher, Rickert,[5] in a famous book. It has become a kind of banner in the fight of the representatives of the humanities against the "expansion" of the scientific method into their field.

If we exclude the possibility that the whole universe moves eternally in cycles, repeating its states again and again, it is obvious that the world process has happened only once. If we regard causality as the recurrence of sequences, it makes no difference whether we say that the world process as a whole obeys or disobeys the principle of causality. Whatever we may think about the relation between physical and biological phenomena, one thing is certain: The cycles of physical facts that are interpreted as examples of causal laws are small cycles within the whole world process that as a whole is probably no cycle. The motions of heavy bodies toward the ground are regarded as a cycle within which the same series of events recurs. We know, of course, that exactly the same event does not recur. The starting points in time and space are different, the size of the falling body, the season of the year, the environment, etc., are all different. However, the relevant characteristics of the cycle repeat themselves. If we know how position and velocities succeed each other in one part of the cycle, we can conclude how they will follow in another part. As a matter of fact, all causal laws are found by dissecting the world process into such incomplete cycles or, in other words, by finding out what state variables can be and must be disregarded in order to see in the world process a great many incomplete cycles.

If we understand that in physical science the concept of causal law is entirely founded on the existence of such incomplete "subcycles" in the world process, we can easily understand how to look for causal laws among historical and social events. It is certainly true that there is no complete recurrence of historical events, but neither is there any complete recurrence of physical facts. The causal laws in physics are discovered by finding out what state variables we can disregard in the definition of "recurrence." The more variables we can disregard and the fewer of them we keep, the more frequently recurrences take place and the closer we come to the causal laws of physics in which, as we have learned, the essential point is the recurrence of states that are defined by a small number of variables.

The reduction in the number of variables can be achieved in various ways. Two typical ways can be easily described by using the example of weather prediction. If we make use of the "meteorological" states, large areas are described by one temperature and one pressure. The small number of variables is achieved by averaging. If we choose, on the other hand, the "molecular" description of state, we have an immense number of variables if we include position and velocity of every single molecule. If, however, we disregard the space and time locations of the molecules, each individual one is described by very few variables, in the simplest case position and velocity only. The causal laws for the single molecule are, then, very simple. In the case of the "meteorological description of state" we would get causal laws as they are actually used in practical weather forecasting. The following might be an example: If there is in November a small pressure difference between certain points on the North American continent, there will be a very cold winter. Such causal laws are very practical in long range forecasting, but their accuracy is not very great. If, on the other hand, we consider the laws determined by the motion of a single molecule, essentially Newton's laws of motion, they are valid with great accuracy if the isolated conditions are produced under which they can be applied. However, their application in practical weather forecasting is very involved and often not really feasible.

If we change the emphasis a little, we can also say: The recurrences in terms of meteorological variables are useful within a very limited range of phenomena, the weather situations. The aerodynamic description would be helpful in predicting all phenomena taking place in gaseous bodies, while the molecular description would be helpful in tackling all kinds of material bodies. While the range of phenomena to be handled increases, however, the distance between theory and observable phenomena becomes greater and greater. The operational definitions become more and more involved. For these reasons, the formulation that there is "recurrence of sequences" is incomplete unless we give a specific definition of what is meant by "recurrence." In our example, it may mean recurrence of meteorological states, or of aerodynamical states, or of molecular states, or even of subatomic states.

3. Causality as the Existence of Laws

We have learned by now that the Humean formulation of causality as the recurrence of sequences fizzles out if we attempt to make it very precise. It seemed, at first glance, very clear and even common-sensical. If we mean by states A or B sets of actual observations, it is clear what "recurrence" means, but if any one conclusion can be drawn from all our considerations, it is certainly the conclusion that the law of causality does not hold if we mean by A and B sets of actual observations. If we compare a magnetized piece of iron with an ordinary piece, they do not look different from one another when we observe them in the ordinary sense of the word. We say, of course, that the molecular structure is different—in one case, the magnetic axes of the molecules are ordered and in the other they are not. We can set up experiments by which this order can be produced or destroyed, for example, by exposing the iron to heat in a magnetic field. Thus we can learn that if a certain group of experiments performed with a piece of iron yields certain specific effects, the iron will also have the effect of a magnet. If the group of effects is produced by a group of experiments, we ascribe to the iron a state variable which we call "magnetization." This state variable is combined with others like density, temperature, and pressure to define a state A of the iron. "Recurrence" in the formulation of causality then means "simultaneous recurrence of the values of these state variables." This statement, however, is very far from the original Humean formulation that sets of observable phenomena ("objects" in Hume's language) are followed by the same phenomena whenever the previous phenomena recur. Our new formulation states that if we make a great many types of experiments with a certain group of bodies, we can ascribe to them state variables u, v, and w in such a way that there are laws that determine the values of these variables that follow a given set.

Our reformulation of Hume reminds somehow of the formulation given by Kant (Section 1) according to which the state of a body cannot be defined unless we have observed the results of experiments performed on this body. Kant, in his way of speaking, presents this situation by stating that we must take the validity of the causality principle for granted if we wish to be entitled to say that a certain experiment reveals to us a certain property of a body; we must assume that the result of our experiment is the "effect" of this property. However, in the language that is used in the science of today, we would say rather that we attribute to a body such properties that the results of our experiments with this body can be expressed in the form of causal laws. We can, for example, introduce the property or state variable "magnetization" in such a way that we can describe experiments in the field of magnetism by statements such as: If a body has a certain magnetization, a certain temperature, etc., the laws of thermomagnetism tell us how this body will act upon other bodies under given conditions. Hence, the general principle of causality can be expressed as follows: We can attribute state variables to bodies in such a way that a small number of such variables will suffice to enable us to express the results of experiments made on these bodies in the form of causal laws.

If we start from this conception of causality, we must keep in mind that the "properties" or state variables assigned to bodies may be very far from observable properties if we use the word "observable" in the sense in which it is used in everyday life or with reference to pointer readings of physical instruments. Let us examine such a very simple property as the "length of an iron rod." In the formulation of laws, it is denoted by a symbol, the letter L. We say, for example, that a lever is in equilibrium if between two weights W_1 and W_2 at the distances L_1 and L_2 from the fulcrum, the relation $L_1W_1 = L_2W_2$ is valid; but what does this mean in terms of observable facts? L_1 stands for a "real" number that may be rational or irrational. Every observation provides us with a number that has a finite number of digits: They are never irrational. In the best case, a single observation may tell us that L_1 is between two numbers like 1.001 and 1.002. In order to find the number L_1 that is to be introduced into the physical law, we make a great number of measurements and take the average. This average is the result of a manipulation and not of a direct measurement. This result is introduced into the law $L_1W_1 = L_2W_2$, and the validity is checked by examining whether or not the expressions L_1W_1 and L_2W_2, both results of a great many observations, are equal. We often use the expression that the length of the rod "possesses" a value L, but we must not forget, however we may express this fact, that the value L is the result of a computation that uses a great number of observations. Hence, the symbols between which a causal law establishes a connection do not stand for single observations. The "operation" that defines the "operational meaning" of a symbol like L_1 or L_2 consists in a great number of pointer readings combined with mathematical computations like formations of averages. If we wish to use expressions like the "length possesses a value L," we must realize that the procedure of establishing possession is a fairly involved one.

Quite a few writers in the fields of science and philosophy would take exception to calling the law of the lever, "$L_1W_1 = L_2W_2$," a "causal law." Such a law establishes in every case a sequence of events in time. However, we can easily see that the law, $L_1W_1 = L_2W_2$, actually does establish a sequence in time. It says: If for a lever at the instant $t = t_0$ the relation $L_1W_1 - L_2W_2 = 0$ holds, and the angular velocity around the fulcrum is zero, the lever will at any future time $t = t_1$ be in the same position as at $t = t_0$. The relation $L_1W_1 = L_2W_2$ is a prediction of the future under special initial conditions. We can easily see that $L_1W_1 = L_2W_2$ is a special case of a more general "causal law" that determines the rotation of a lever around the fulcrum. We may denote by ϕ the angle included by the rod of the lever and a fixed direction, by α the angular acceleration of the lever, by I the moment of inertia of the lever around the fulcrum, and by M the momentum of the external force around the fulcrum. The equation of motion for a rotation around the fulcrum is analogous to the Newtonian equation for the motion of a mass-point, with the mass m, upon which a force F is acting: $ma = F$, where a is the linear acceleration. For the rotational motion, we replace the mass by the moment of inertia, the linear acceleration by the angular acceleration, and the force by the moment of force. In this way we obtain $I\alpha = M$. If we apply this equation to the case of the lever, we have $M = W_1L_1 - W_2L_2$, and the equation of motion becomes: $I\alpha = W_1L_1 - W_2L_2$. If we apply this equation to the case in which the lever remains at rest, we have $\alpha = 0$ and $W_1L_1 - W_2L_2 = 0$, which is again the law of equilibrium for the lever.

We learn from these considerations that all laws of equilibrium are special cases of causal laws. They state the conditions under which we can predict that there will be no motion (or, at least, no accelerated motion) in the future. Even the laws of geometry can be interpreted in this way. If we assume the validity of the Euclidean axioms[6] for triangles made of a certain material, and we find by measurement that the sum of the angles is equal to two right angles, we can predict that these triangles will remain at rest if the initial velocities are zero. This geometrical law is not mentioned in mechanics because it is taken for granted. But, to speak precisely, the geometrical laws are pertinent in every prediction of future motions. If the sum of the angles were not equal to two right angles, internal tensions would have to be considered in any prediction of future motion of a triangle.

Many authors have distinguished between laws that contain time and laws that connect states at the same instant of time like the geometrical laws, but the distinction is not an essential one. We have learned that all laws of equilibrium are special cases of more general causal laws. This becomes even more obvious if we take into account the theory of relativity (Chapter 5). According to this theory, it depends upon the arbitrary system of reference whether two or more events happen at the same instant of time or at different instants. The law of inertia is just as much a causal law as Newton's second law, according to which force produces an increase of momentum.

4. Causal Law and Statistical Law

If we launch a bullet with a definite speed in a definite direction, we can derive from Newton's laws of motion which specific point of a target will be hit; but if we toss a coin and observe it striking the table, we cannot predict whether "heads" or "tails" will come up. We can, however, predict that among a thousand throws about half will turn up "heads." In the first case, we speak of a "causal law," in the second of a "statistical law." Let us attempt to define the difference between these two types of laws as exactly as possible. As in all our argument, in geometry, in mechanics, in atomic physics, etc., we have learned that we must distinguish sharply between the description of what is actually observed or observable and the symbolic pattern by which the scientist describes the phenomena. For example, we must distinguish between the distance D between two points that obey the axioms of Euclidean geometry and the physical operation by which such a distance is measured and by which one can check whether or not the two points are actually separated by a well-defined distance. Bearing this in mind, we shall analyze the difference between "causal" and "statistical" law, using the example of a mass-point launched with a definite initial position and velocity toward a target that is hit at a definite point.

If we know the initial conditions (the "cause"), can we predict the point on the target which will be hit (the "effect")? What, exactly speaking, is the "causal law" that connects the cause with the effect? The initial position may be a point P, and we may launch our mass-point in a direction that is perpendicular to the plane of the target. If the mass-point passes precisely through P and is exactly perpendicular to the target, it will hit at the center C of this target. If we launch a great number of mass-points under the same conditions, they will all hit the target at C. This result is computed by the application of Newton's laws, in particular the law of inertia. In order to make the argument as simple as possible, we shall neglect the effect of gravity and assume that the paths of the particles are in the horizontal direction. We can hit the center of the target with certainty if we are certain to launch the mass-point under exactly the right initial conditions. This means we must start at exactly the point P and launch it in precisely the horizontal direction.

If we try to carry out this experiment in practice, we become aware of the technical difficulties in launching a mass in exactly this way. We repeat the experiment under the same practical conditions; this means that in every case we make the same type of technical arrangement in order to achieve the intended conditions. Then we observe that the target is actually not hit in each case at the center C, but that the points that are hit form a certain pattern around the center C. If we investigate the pattern, we see that the pattern of hits exhibits rotational symmetry around C and that the frequency of hits decreases with the distance r from the center. Mathematically speaking, the frequency of hits decreases as a Gaussian distribution function ($e^{r^2/2D}$) of the distance r from the center. Every such distribution is characterized by the constant D, the "dispersion" of the pattern of hits. The smaller D is, the more the hits are crowded around the center; the greater D is, the more these hits are dispersed. By a simple computation, we can conclude that $D^2 = \Sigma r^2/N$, where Σr^2 means the sum of the squared distances of all the hits from the center and N the number of hits that we consider. The "dispersion" D is a measure of the failure of the launched masses to hit the center. It can disappear only if every mass hits the exact center since from $D = 0$ it follows that every single r must disappear.

If we describe the initial conditions (the "cause") in terms of a feasible technical operation, we cannot predict exactly the point of the target that will be hit, but only the pattern of hits and the "dispersion" D. The "effect" is a pattern and a "dispersion."

This state of affairs is described in science as follows: If we could give the mass the exact initial position P and a velocity that was exactly perpendicular to the target, it would strike the center C exactly. But in practice, because of the insufficiency of the devices by which we establish position and direction, the actual initial position has a certain distance from P and the velocity has a certain component that is parallel to the target. We may call the projection of the distance from P upon the target Δq, while the component of momentum (mass $\times$ velocity) parallel to the target may be called Δp. Then we can say that not all masses hit the target at the center because not all the Δq and Δp are equal to zero. Then there is a dispersion of Δq: $D_q{}^2 = \Sigma(\Delta q)^2/N$ and a dispersion of Δp: $D_p{}^2 = \Sigma(\Delta p)^2/N$, and we can say that the center would only be hit by all the masses if both dispersions D_p and D_q disappeared. Unless D_p and D_q both disappear simultaneously, we cannot predict the "effect" of launching each single mass, but only the pattern produced by launching the whole "swarm" of masses. We have a "statistical" law.

However, it follows from the laws of motion, by purely mathematical argument, that exact hits at the center can be predicted if we can produce "swarms" with disappearing dispersions of position and momentum. If, as in this case, by suppressing the dispersions in the initial conditions, one can also eliminate the dispersion in the effect, our law is according to the axioms of Newtonian mechanics a "causal law." It is also a "causal law" in terms of observable phenomena if we are certain that we can invent improvements in practical devices by which we can make both dispersions D_q and D_p as small as we please. Then we may be certain that by the application of this device we can predict the "effect," the hit at the center, as exactly as we please.

We shall now try to characterize laws that are not "causal" in the sense that we have just described. A "causal law" had to meet two requirements, a mathematical and a physical (empirical) one. According to the first one, the dispersion of the "effect" would disappear if the dispersion of the initial conditions disappeared; according to the second, there are physical devices by which the dispersion of the initial conditions (D_q and D_p) can be nullified simultaneously. This is certainly in agreement with Newton's laws of motion if we assume that these laws, including the operational definitions of position and momentum, are valid for values of p and q no matter how small we choose to make them. Only from this assumption can we derive the possibility of nullifying D_q and D_p simultaneously and, hence, of predicting the hit at the center precisely. We have learned that in terms of observable phenomena even the law of inertia that allows for hitting the center of a target precisely is a law for a pattern of hits, a statistical law. We call it "causal" only because by contracting the dispersion of the initial conditions simultaneously we can predict the precise location of the hit. Therefore, we obtain different types of laws if we turn to motions which also allow for the prediction of a pattern of hits, but for which the dispersion of the "effect" is not nullified by suppressing the dispersion in the initial conditions. This is obviously the case if we describe the game "heads or tails" as a mechanical phenomenon produced by launching a coin. As everybody knows, the result of this mechanical experiment is that among a great number of throws about half of them turn up "heads" and the other half "tails."

This experiment has frequently been verbalized by saying that we know no reason why either "heads" or "tails" should turn up more frequently than the other. However, the motion of the coin is guided by the same mechanical laws as the launching of a mass-point toward the center of a target. Our ignorance of what will happen cannot replace the Newtonian laws of motion. We must realize that in our coin experiment we have a "cause," the way in which we launch the coin by some device, and an "effect," the frequency with which "heads" or "tails" turns up on top. This is analogous to the case in which aiming at a target is the "cause" which produces as an "effect" the frequency of hits at different distances from the center. If we toss a coin N times, where N is a great number, we can regard all these tosses as one experiment. Each time, the coin is tossed under the same initial experimental conditions. The "effect" can be described as a pattern by denoting each toss by H or T according to whether "heads" or "tails" appears on top. The "effect" is then described by $HTTHHTHTTTHHTH. \ldots$ The farther we go, the clearer it becomes that there are approximately the same number of H's and T's in this pattern.

However, there is one essential difference between the "effects" determined by the initial conditions of aiming at a target and of tossing a coin. In the first case, by narrowing more and more the dispersions of the initial conditions, we could also narrow the effect until we could predict almost precisely the values of q and p in hitting the target. But in the case of tossing a coin, we can predict almost with certainty a pattern HTT . . . in which the frequency of appearance of "heads" is approximately $\frac{1}{2}N$ among N throws. This pattern is independent of the initial position and velocity of the coin. By narrowing the dispersion of the initial conditions we could not alter the "effect"—the "frequency" of H and T occurring in a long series of tosses.

If we observe a small number of throws, we cannot predict anything; the pattern seems to be capricious and unpredictable. The larger the series becomes, the more a dependable frequency appears. We speak of a "frequency $\frac{1}{2}$,"meaning by "frequency" of the event "heads" the ratio between the number of cases in which heads turn up and the number of all tosses. In this case, we say that the operation of tossing a coin as "cause" brings about almost with certainty an "effect," a series of tosses in which "heads" has the frequency $\frac{1}{2}$. We can also make a similar statement about aiming at a target. If we launch a very small number of mass-points, we observe a small number of hits on the target, and we cannot predict anything about the distance from the center C at which the hits will occur; but if we launch a great many masses, we can predict almost with certainty that a definite pattern will be formed. If we draw circles with increasing radii r around C, the ratio of the number of hits within a circle of radius r to the whole number of hits will be a definite function of r if the number of hits becomes very great.

We see from these two examples that we can derive from the Newtonian laws of motion two types of laws that are valid for observable phenomena. We must keep in mind that Newton's laws themselves are not statements about observable facts but only a conceptual scheme. There are, however, two main types of conveniently testable laws that can be derived. On the one hand, we have the "causal laws." Aiming at a target by launching a mass-point is an example of this type. In this case, the situation is a simple one. The field of force is simple, and the initial conditions are describable by giving the values of a few variables. We can narrow down the dispersion in the prediction of the final values of the variables (values at the target) with great certainty in this case.

However, we can also draw consequences of a completely different kind from Newton's laws. In this second type, the situation is a very complex one. The initial condition of a coin cannot be described by a few variables. In its path through the air, the coin collides with an immense number of air particles with irregular velocities. In this case, we cannot predict anything about the values of the state variables at a certain time, but only about a certain average behavior; we record only whether after striking the ground "heads" or "tails" is on top. However, this pattern can be described by a simple law, in our case by the frequency $\frac{1}{2}$. A law of this type is called a "statistical law." In this case, the predicted frequency is not altered by any narrowing of the dispersion in the initial state variables of the coin.

All applications of Newton's mechanics to observable phenomena are based upon these two types of law. Both are equally results of Newton's equations of motion; they result from two different kinds of approximate solutions that can be derived under different circumstances. In the first case, a narrowing dispersion of initial values allows us to predict with a high degree of accuracy the occurrence of specific values of the state variables. Then we speak of "causal laws." In the second case, we can predict from the approximated initial conditions with great certainty a certain pattern in the values of the state variables or, in other words, the "frequency" with which different values of these variables occur in a certain time. We speak of "statistical laws."

The advance of science has made it clear that not only in mechanics but in all fields of science we have to do with these two types of law—the "causal laws" and "statistical laws." The difference is, of course, that in domains that are believed to be governed by Newtonian mechanics we can derive both types of law from Newtonian mechanics. The derivation of statistical laws from Newton's laws of motion can ultimately be traced back to the "Ergodic Theorem."[7] If we assume that in a mechanical system, which may be of any complexity, all masses remain always within fixed finite limits and the energy remains constant, we can derive from the differential equations of motion the result that our system passes again and again near all the positions and velocities that are in agreement with the prescribed limits and energy. If we pursue this system over a long period of time, every state of the system (position and momentum) is traversed with a certain frequency that depends on this state only. This "Ergodic Theorem" states for a mechanical system a "statistical law," and we are led to assume that any statistical law is based ultimately on a theorem that is similar to the "Ergodic Theorem."

In other fields, we just assume that the same two types of law exist. We study them directly without demanding proof that they are of common origin. However, the fact that in the case of mechanics both types can be traced back to Newton's laws makes it sufficiently plausible that causal and statistical laws are not two types of law that are irreconcilable with one another. If we speak in terms of observable phenomena, all laws are statistical. From observing the procedure of aiming at a target we can predict a pattern of hits; in the same way, from observing the procedure of tossing a coin we can predict the frequency of the appearance of "heads" or "tails." From our mathematical laws of motion we can, however, introduce a "causal law," in the first case by narrowing the dispersion in the initial conditions. Having computed this limit, we introduce the following manner of speaking: If a mass were launched in exact agreement with some mathematical condition, it would hit the target in the exact center. But to say that "a mass-point has a certain figure as its velocity" has no other meaning than the statement about the existence of a limit. An observed velocity is always a statistical average of observations and cannot occur in a precise causal law except as a limit. To speak precisely, there are statistical laws that allow limits; these are called "causal laws." On the other hand, there are statistical laws, like the laws for tossing coins or drawing lots, which admit no limits.

In these considerations, we must keep in mind that in all these arguments we have taken it for granted that the phenomena in question follow laws that can be expressed in terms of some special variables, positions, and velocities. If we consider more general phenomena like sociological, biological, or psychological phenomena, we can only speak of "laws" if we also specify for what variables there should be causal or statistical laws. If we describe the game of tossing coins by regarding the frequency of "heads" as a variable, we have a "causal law." From the procedure of tossing it follows with certainty that the frequency of "heads" will be $\frac{1}{2}$.

13

The Science of Science

1. The Place of Induction in Ancient and Modern Science

The great British philosopher, John Stuart Mill,[1] wrote in a letter in 1831: "If there is any science that I am capable of promoting, I think it is the science of science itself, the science of investigation, of method."

It has been a custom of long standing to characterize the difference between modern and medieval science by stressing the changed roles of induction and deduction. Medieval science, following the line of Aristotelian philosophy, proceeded by deduction, by conclusions from general principles, to single facts, while modern science (after 1600) starts from observed single facts and proceeds to general principles by the method of "induction." The natural sciences, physics, chemistry, biology, have been called the "inductive sciences." One of the most important books on the history of natural science is William Whewell's[2] *History of the Inductive Sciences*. What is actually meant by distinguishing modern science in this way from older science? We mean to stress the point that the modern scientist collects single facts by observation and experiment and proceeds from these facts by the method of "induction" to general principles, whereas a man like Aristotle, in his book on Physics, started from general principles and derived from them by logical conclusion (by "deduction") individual events which could be observed. Actually, when a modern scientist has set up general principles by induction he has to draw logical conclusions from them in order to obtain individual facts that can be checked by experiment. On the other hand, a scientist like Aristotle did not find his general principles in his dreams, but advanced them on the basis of experience that consisted in the sum of the individual facts that had been observed.

Hence, in actual practice science has always made use of both induction and deduction, but there has certainly been a difference in the way the general principles have been set up on the basis of observed facts. The contemporaries of Plato and Aristotle certainly knew from observation that the celestial bodies performed orbits in the heavens that could be identified vaguely as being circular. The principle that all motions of celestial bodies were circular certainly had its source in those observations; hence, we would say that the principle of circular orbits was formed by induction if we used the word in the general sense of deriving general propositions from propositions about single facts. However, we described elaborately[3] that the principle of circular orbits for celestial bodies was "believed in" much more firmly than this "inductive inference" from observation would warrant. Men believed in it as an "intelligible principle"; it seemed to be very plausible that perfect divine beings like the celestial bodies should also move in "perfect orbits," and the perfect curve is the circle. If one derives from the principle of circular orbits results about the position of a planet on the sphere, the principle is approximately confirmed; however, if one derived a great number of positions and measured them precisely, one would find that the orbit would not be exactly circular but, as Kepler found, elliptic. Originally, the argument taken from "perfection" carried so much weight that not much attention was paid to the fact that the positions of planets derived from the circularity principle did not agree precisely with the observed positions. But, in time this attention had increased.

The difference between ancient and modern science was not the use of induction—ancient science was based on induction as is modern science—but the criteria by which a discovered principle was recognized to be valid. The method of "verification" is different now; more weight is given to the agreement of the results with observed facts than to the agreement of the principles with a world picture that has been accepted for what we called (in Chapters 1 and 2) "philosophical" reasons. Many authors have said that

modern science is characterized by its firm resolution to recognize
as "verification" of general principles only and exclusively the
agreement of their results with "observed facts." However, this
resolution must be taken with a "grain of salt" even in our own
century. Practically, this resolution has meant that a verification
of astronomical principles must consider only verification by the
observation of astronomical facts. To consider the agreement or
nonagreement of the Copernican system with theological and philosophical doctrines had been declared "out of order" from the point
of view of "modern science."

However, it was an "observed fact," too, that the Copernican
system violated the habits of common-sense thinking; it produced
psychological pain: these were observable facts the reality of
which cannot be doubted. On the other hand, modern science has
not accepted its principles of physical science exclusively on the
basis of the physical facts that can be deduced from these principles
and observed by our senses. Certainly, the Copernican system
would never have been accepted before the belief in the literal
interpretation of the scriptures had lost power. Einstein's principle
of relativity would never have been accepted if the metaphysical
belief in absolute space and time had not been shaken by the
empirical philosophy of men like Ernst Mach. Therefore, we
cannot draw a strict dividing line between the type of verification
used by Aristotle and the type used in modern science. We can
only say that modern science prefers criteria of verification that can
be comprehended and checked by everybody who has had sufficient training. It is understood that every person with normal
senses and normal mental ability can achieve such a training. We
must, of course, admit the unavoidable vagueness of terms like
"normal" and "comprehend." In order to be sure that a method
of verification is "sharable," science prefers "observations" that
consist, e.g., in seeing at a certain point of space at a certain instant
of time a red spot, or feeling a warm rounded surface or experiencing
similar "sense observations." It is important to understand that
the scientist does not hold metaphysical views like "the real world
consists in sense observations"; he prefers sense data only because
they can be repeated and checked by everybody.

We can also formulate the difference between ancient and modern
science in another way: Aristotle and medieval science proceeded

very quickly from the observed sense data to very general principles that seemed to be intelligible.   They set up from few vague observations of the positions on the sphere the general law of perfect circular orbits.   One could enjoy the beauty of these general principles, but there was no way back to the laws of lower generality from which facts could be deduced which were nearer to precise observation than the facts from which one started.   As is well known, Francis Bacon[4] was a philosopher who regarded induction as the basis of science.   He described adequately the difference between the roles of induction in ancient and modern science.   Although he is the man who is usually credited with the turn from Aristotelian philosophy to modern science, he never maintained that Aristotelian philosophy and medieval science were not based on induction; but he characterized very aptly the difference between ancient and modern science.   In his book, *Novum Organum*, Bacon pointed out the new approach that science must adopt in order to be useful in mastering nature.

Francis Bacon wrote:[5]

> There are and can be only two ways of searching into and discovering truth.   The one flies from the senses and particulars to most general axioms and from these principles, the truth of which it takes for settled and immovable, proceeds to judgment and to the discovery of middle axioms, and this way is now in fashion.   The other derives axioms from the senses and particulars, rising by a gradual and unbroken ascent so that it arrives at the most general axioms last of all.   This is the true way but, as yet, untried.

We see clearly that Bacon conceives of both ancient and modern science as starting with induction; but what is the difference between the uses of induction in ancient and modern science? Bacon describes the difference clearly:[6]

> Both ways set out from the senses and particulars and rest in the highest generalities; but the difference between them is infinite.   For the one just glances at experiment and particulars in passing, the other dwelling duly and orderly among them.   The one, again, begins at once by establishing certain abstract and useless generalities, the other rises by gradual steps to that which is prior and better known in the order of nature.

The essential point in Bacon's formulation is what he says about the role of the "middle axioms" or what we called (in Chapter 1) the principles of "intermediate generality." These are our physical laws, like the law of gravitation or the equivalence of heat and mechanical work. Statements about observable phenomena can be exactly derived from these principles. In modern science, these laws are constructed from our observations and experiments by "induction," while in medieval science or philosophy the process of induction from observations passed over the head of these "middle axioms" and produced the most general philosophical principles like "perfection," "search for a natural place," etc., from which the "middle axioms" were derived by deduction. Hence, there was never a precise agreement between the most general principles and the observed facts.

2. Induction, General Laws, and Single Facts

If we review the structure of the sciences that we discussed in our presentation of geometry, Newton's theory of motion, theory of relativity, and motion of atomic objects,[7] we note that everywhere we started from a system of axioms and derived from it theorems. The chief problem with which we are now faced is how to arrive at these axioms or general principles. The raw material that is given to us, and from which we must build up these principles, consists first in the results that we have obtained from physical observations and experiments. We may call it, briefly, our "observational material." On the other hand, the direct building material of the principles are words or mathematical formulae, together with rules by which these words and formulae are connected among themselves, the rules of syntax or deductive logic. We may speak briefly of "linguistic material." The task of science has been to infer from the observational material the general principles that are made of symbols and connected by logical operations. We may say that on the basis of "observational material" science has to build a structure of "linguistic material." In the widest sense, the procedure of building such a structure is called "induction."

Obviously, both the observational and the linguistic material develop in the course of human history. Their growth depends upon sociological and psychological factors. Using a more familiar language, men have often described these two elements as "facts" and "ideas," as did Whewell in his fundamental books on the history and philosophy of the inductive sciences.[2]   C. J. Ducasse[8] points out[9] that "many pairs of terms exist which all refer to one or another aspect of this fundamental antithesis." It is perhaps instructive to quote several of these "pairs" in order to free our thinking from a too rigid adherence to some verbal pattern.   Whewell speaks, for example, about the antithesis of "Thoughts" and "Things."   Thus, in the knowledge that a year consists in 365 days, there are included on the one hand the sum as given, and on the other the mental act of counting.   "Without Thoughts there could be no connections; without Things there could be no reality." Another familiar way of speaking is to contrast "Necessary" and "Experiential Truths."   "Necessary Truths," writes Whewell, "are derived from our own Thoughts; Experiential Truths are derived from observations of Things about us."   Even the opposition of Deduction and Induction can be regarded as another aspect of the same fundamental antithesis.   Whewell points out that "Deduction" starts from statements that are supplied by our "Thoughts" while "Induction" starts from the observation of external "Things."   Another antithesis of a similar kind is between "Theory" and "Fact."   The pair of terms that, according to Whewell, separates the members of this antithesis most distinctly is "Ideas" and "Sensations."   Whewell says:

> I term space, time, cause, etc., "Ideas" . . . These relations involve something beyond what the sense observations can furnish. . . . We use the word "Ideas" to express that element, supplied by the mind itself, which must be combined with "Sensation" in order to produce knowledge.[9]

If we use, with Whewell, expressions which are not very precise but which are near to the language of common sense, we could say that "induction" starts from sensations, facts, things . . . connects these elements by ideas, theories, thoughts of necessity . . . and leads to general principles from which by "deduction" new things, facts . . . can be derived.   Ultimately, as we see, "induction" leads from observed facts to the discovery of new facts.   To give a simple example: We observe the positions of planets on the sphere; these are facts.   Kepler added the idea of elliptic orbits by which he connected the observed facts; he arrived in this way at

Kepler's laws of motion. From these laws one could find by deduction positions of the planets which had not been observed before. In this way one could, on the basis of observed facts, predict new facts. The question has been raised whether the process of induction leads from observed facts to general laws or from observed facts to new facts that had not been observed. As we have learned by now, it is obvious that from known facts there cannot be a way that leads to new unknown facts by logical conclusions; such a way can only go through general principles from which facts can be derived. The general principles, on the other hand, can only be found by induction from the given facts. Nevertheless, there has been quite a dispute about whether "induction" is a method that leads from given facts to new facts or from given facts to general laws.

It may be instructive to take a look at the discussion[10] between John Stuart Mill and William Whewell on this point, since both authors gave, a hundred years ago, more thought to the problem of "induction" than any of their contemporaries. Even today, their writings should be studied as background material. Indeed, Whewell was the first author who formulated the structure of science in the way in which it is conceived today. Mill, on the other hand, presents induction in a way that is nearer to common-sense ideas and, therefore, from our present scientific point of view, slightly obsolete. Therefore, the discussion between them gives a clear picture of how today's science is constructed. Whewell dissents, for example, from Mill, who includes in his notion of induction the process by which we arrive "at individual facts" from other facts "of the same order of particularity" . . . Such inference is, at any rate, not induction alone; if it be induction at all, it is induction applied to an example.

If a ball strikes another ball, we can find by induction the law of conservation of momentum; but, according to Whewell, the term "induction" cannot apply if an ordinary billiard player by his skill, without thinking of momentum, can knock a ball in a desired direction. Mill stresses the point that this is a type of induction of which even brutes are capable.

This action by "conditioned reflexes," the fact that "a burnt child or even a burnt dog dreads the fire," according to Mill, is an action guided by "Induction." Whewell, however, objects " . . .

though action may be modified by habit and habit by experience, in animals as well as in man, such experience so long as it retains that merely practical form, is no part of the material of science." One may perhaps think that it is only a question of words whether or not we call the reactions of children or animals "induction." However, Whewell directs our attention to a point that has been very important in science and philosophy. When we prefer one definition of a term like "induction" to another, we do it for a certain purpose. Definitions are used to formulate propositions about facts as briefly and simply as possible. In our case, we like to use the proposition: "Speaking of Induction we mean the kind of procedure by which the sciences now existing among us have been constructed." As Whewell emphasizes, science is not constructed of practical habits and tendencies but of general principles. Hence, to denote practical habits or conditioned reflexes as induction would make the statement that the sciences grow by induction a fallacy. The correct statement would then be that the sciences grow by the special type of induction that uses general principles, whereas induction by practical habits plays a minor role. This formulation would probably be correct, but would be complicated and would exclude the simple statement that "science grows mainly by induction."

3. Induction by New Concepts

If we stick to this formulation as a basic statement about the "science of science," we must investigate how science actually grows in order to learn what is meant by induction. If we take our cue from our old example, we may consider Kepler's[11] law, according to which the planets move in elliptic orbits. He started from the observed positions of the planet Mars on the sphere and derived his law from this "observational material" by induction. It is again interesting to examine how Kepler's inductive inference was described by John Stuart Mill and William Whewell.[10] According to Mill, Kepler added nothing to his observed positions of the planet Mars; he saw only that all these positions were situated upon an ellipse. Kepler "asserted," he says, "a fact that the planet moved in an ellipse. But this fact, which Kepler did not add to, but found in the motion of the planet . . . was the very fact, the separate parts of which had been separately observed; it was the sum of the different observations." In contrast to this, Whewell emphasized strongly that it "was not the sum of the observations *merely*; it was the sum of the observations, *seen from a new point of view*, which point of view Kepler's mind supplied."

Whewell illustrated the difference by a familiar analogy. Kepler's laws are contained in his books, but unless one knows Latin, one will not find them.

> We must learn Latin in order to find the laws in the book. In like manner, a discoverer must know the language of science, as well as look at the book of nature, in order to find scientific truth.

The book of nature consists of what we have called the "observational material"; but in order to discover and formulate a law, the discoverer must be sufficiently in control of the linguistic material in order to read the book of nature. In a very impressive and lucid way, Whewell writes words which come very near to the conception of science that is the predominant one in twentieth-century science. "Man is the *interpreter* of Nature; not the Spectator merely but the Interpreter. The study of language, as well as the mere sight of observation, is requisite in order that we may read the inscriptions which are written on the face of the world."

According to Whewell, the essential point in every successful induction is a new concept, a new order that is formed by the author out of his linguistic or logical material. This concept was, in Kepler's case, the ellipse; in Galilean mechanics, the concept of acceleration; in Newton's thinking, the concepts of acceleration and gravitation; in modern optics, the concept of waves, etc. Mill had denied that these concepts are different from the observed facts themselves; the conception is a copy of the facts. "The conception is not furnished *by* the mind," says Mill, "until it has been furnished *to* the mind." Whewell, however, stressed the point that the "concepts" that lead to new inductions are not forced upon us by the observed facts, but are built up by an activity of our minds which constructs these new conceptual schemes, using as building material the linguistic material that has either existed within our minds for a time or has been built up just for the purpose of securing an adequate system of concepts. While Mill insisted again and again that the general law is in the facts and has only to be noticed and read, Whewell insists, on the contrary, that the general law is a product of human activity. "If," writes Mill, "the facts are rightly classed under the conceptions, it is because there is in the facts themselves something of which the conception is a copy." Whewell, however, objects, "But it is a copy which cannot be made but by a person with peculiar endowment; just as a person cannot copy an ill-written inscription so as to make it convey sense except he understand the language."

The salient point in every induction on the basis of observed facts is the invention of a new concept that ties the facts together, that "colligates" them, according to Whewell's terminology. The introduction of a new concept whenever observed facts are colligated manifests itself by the fact that every induction is accompanied by the introduction of some new verbal expression or new technical term. Whewell says:

> At least some term or phrase is henceforth steadily applied to the facts, which had not been applied before. Kepler asserted that Mars moved around the sun in an "elliptic orbit"; Newton asserted that the planets *gravitated* toward the sun. These new terms *elliptic orbit* and *gravitate* mark the new conceptions on which induction depends.

Whewell points out that the history of physics consists not only in the discovery of new facts but, what is not less important, in the formulation of new concepts. He says bluntly that:

> The history of the inductive sciences is the history of discoveries, mainly so far as concerns the *facts* which were brought together to form science. The philosophy of the inductive sciences is the history of ideas and conceptions by which the facts are connected.

It is certainly, to a large extent, a pure question of words, whether an induction like Kepler's law of elliptic orbits is called a summarizing of facts or an addition of concepts made by our mind. It is very difficult to draw a boundary between theories and facts that can be applied in every case. Whewell himself tells us occasionally that *facts* are nothing but theories that have been very well confirmed and have become very familiar. In his emphasis upon the role of concepts that are products of our minds, Whewell was certainly under the strong influence of Kantian philosophy. We have already mentioned[12] that, according to Kant, science is produced by fitting the observed facts into a frame produced by our minds. Kant believed this frame to be of eternal validity and not subject to change by any progress of science. Whewell believed with Kant in the great importance of the "linguistic material" produced by our minds for the advance of science and contributed in this way a great deal to a better understanding of what the structure of science is, and of how science is growing; but he did not believe with Kant that the conceptual frames produced by our mind were unchangeable. He assumed rather that they change along with the increasing "observational material," and was in this way perhaps the first author who conceived the growth and the structure of science in the sense in which it is understood in twentieth-century science.

Whatever may be the origin of this view, it is instructive to consider the great importance which the introduction of new concepts and terms has had for the advance of science into unknown territory. Let us start again from the example of Kepler's law. We observe the consecutive positions of a planet on the sphere and represent them by a series of points on a piece of paper. If we look at these points simultaneously and describe them by saying that "they can all be connected by a curve called an 'ellipse,'" have we said more than we said by enumerating the positions? Can we infer all positions of the planet from the observed ones without introducing the concept "ellipse"? If we mean by "induction" the process by which we infer from the observed positions all the other positions between the observed ones, we can introduce "induction by enumeration," a kind of interpolation. If we mean by "induction" the creation of the equation of an ellipse in our minds and the computation of all positions from this equation, we can introduce "induction by creation of concepts." Mill favored, to a certain extent, the first type of induction, while Whewell strongly favored the second.

If we formulate the alternatives in this way, we can create a dispute that would practically never be settled; but we can easily show the great usefulness of the "induction by new concepts" in the advance of science. If we consider only the motion of a planet under the influence of one body (the sun), the orbit is strictly elliptic, and it makes little difference whether we say that it is a fact that all the positions are included on an ellipse or that the concept of ellipse is added to the positions that are the only facts. Let us, however, consider a third body (*e.g.*, a second planet) and ask what the orbit will be of the first planet under the gravitational attraction of two other bodies. If we consider the positions traversed under these circumstances, under a "perturbation" as the astronomer says, we see no decent curve on which they could be situated. In the theory of "perturbation," it is shown that we can describe these positions by assuming an elliptic orbit which is not at rest but is slowly rotating and moving in space. The motion under the influence of perturbation can only be described and computed if we start from the concept of an ellipse and ask how an "ellipse" must move in order to represent the more complicated motion. The concept "ellipse" that was, in the simple planetary motion, only an instrument to make the description of facts simple and convenient becomes, in the theory of perturbation, the indispensable instrument for the solution of the problem of "perturbed motion" that is of an immensely higher complication than the simple planetary motion covered by Kepler's law.

This example leads us to a general path of induction by which we can proceed from simpler to more complicated problems of physics. The great British (to speak precisely, Scottish) physicist, James Clerk Maxwell,[13] made the exact point that the first stage into the land of the unknown is to work out mathematical concepts that describe the known territory as simply as possible. Maxwell says:[14]

> The first process therefore in the effectual study of science must be one of simplification and reduction of results of previous investigation to a form in which the mind can grasp them. The results of this simplification may take the form of a purely mathematical formula or of a physical hypothesis.

Maxwell emphasizes very strongly that the work done in "pure mathematics," by producing concise and elegant analytical expressions, is an essential part of the advance in physical science. If we should be restricted to the formulae produced by elementary mathematics or even elementary calculus, the conceptions needed for further advance would be expressed in a clumsy and circumlocutory way.

As an example of the dynamic role of fitting mathematical or logical concepts, we can quote the role of the concept "curl or rotation of a vector field." If we have electric charges at rest, they produce an electric field that obeys Coulomb's law that has the same form as Newton's law of gravitation; it has at a distance r from a charge E the intensity E/r^2. Such a field is derivable from an electric potential V; it is the gradient of this potential. Every vector field that is the gradient of a potential has the special property that its curl or rotation is zero. Hence the equation: "curl of a field = 0" characterizes an electrostatic field, a field produced by electric charges at rest. The disappearance of the curl of a field is mathematically identical with the existence of a "potential energy" from which the field can be computed as its gradient. The introduction of the concept of "curl" for an electrostatic field is a purely mathematical device which permits the formulation of laws in a very concise way, but adds nothing to our physical knowledge about such a field that is not contained in Coulomb's law for the interaction of two charges. A great many people would say that it is a superfluous whim of mathematicians to introduce such a complex mathematical concept as a "curl" to describe such a simple thing as an electrostatic field. However, when Maxwell stated his generalization from the electrostatic field to the general electromagnetic field, he found that his chief instrument of generalization was the concept of "curl." He assumed that in a general field the "curl" no longer disappears as in the static field but changes in time. Maxwell's hypothesis could be formulated in a simple way by stating that the curl of the electric field is proportional to the time increment of the magnetic field. This is the application of the general idea that Maxwell formulated in the passage quoted above. The same procedure for the discovery of generalizations has been repeated in science again and again.

One of the most conspicuous examples is Einstein's theory of gravitation, a generalization of Newton's "classical" theory of gravitation. Minkowski formulated in 1908 the "special theory of relativity," which Einstein had advanced in 1905, by using the "four-dimensional space" and a tensor calculus in this space. At that time, the introduction of the "four-dimensional world" seemed to be only a mathematical trick to give to Einstein's theory a shape that was to the mathematician elegant and even thrilling, while to the physicist it seemed to be rather obscure, too remote from common-sense conceptions and containing superfluous mathematical difficulties. At a certain time, this was even the opinion of Einstein himself. As we learned[15] in the study of the theory of relativity, the special theory of relativity was only concerned with systems that moved along a straight line with constant speed. Accelerated and rotational motion were regarded by Newton, and originally even by Einstein, as "absolute" motion. However, Einstein had always attempted to generalize the principle of relativity in order to make it applicable to motions that were not uniform. He very soon found that the original formulation of the special theory, which made use only of the traditional presentation in three-space coordinates and one time coordinate, was so complicated that it was difficult to visualize how a generalization could be achieved. Einstein, however, noticed that in Minkowski's four-dimensional presentation the special relativity theory was mathematically so simple that it could easily be generalized to embrace accelerated and rotational motions. The "induction" that provided us with the general theory of relativity had been made possible because of the elegant and simple shape that the "special theory" had been given by Minkowski's four-dimensional space-time.

We do not, however, have to go so far into the difficult theories of modern physics; the greatest and most dramatic example of the usefulness of an elegant mathematical pattern is the Copernican theory of the planetary system. When Copernicus advanced his system of concentric orbits around the sun, everyone recognized that this was a mathematical scheme much superior to the Ptolemaic scheme of circles and epicycles around the earth. But one could and did answer that a theory cannot be judged according to mathematical simplicity alone; it must be judged also by its approximation to "truth." If we disregard "philosophical truth" as a control of "scientific truth" and restrict ourselves to the latter, we would judge the comparative truth of scientific theories according to their comparative usefulness for the advancement of knowledge. We would prefer the theory that lends itself better to generalizations, which would, in turn, lead us to a theory that embraces more observable facts. While the Ptolemaic system was in close agreement with the motion of the planets if one neglected their mutual interaction, the Copernican system turned out to be a good starting place for investigation about how the circular orbits, *e.g.*, of the earth, would be influenced by gravitational force exerted by other planets, *e.g.*, Jupiter. When the effect of the interaction is computed from the Copernican theory, the result can, of course, also be interpreted in the Ptolemaic theory by computing the orbits of all the planets concerned relative to the earth; but these orbits would be so complicated that they would practically never have been found by starting from the Ptolemaic epicycles. We see that the superiority of the Copernican system is based upon its special fitness for generalization. We have learned that this great fitness is based upon its great mathematical simplicity and elegance. We understand now that the great mathematical simplicity made the Copernican system superior not only because of its aesthetic properties like elegance, but also by its "dynamic quality," by its fitness for generalization.

We learned from Geometry[16] that if Euclidean geometry is valid, the sum of the angles in any plane triangle is $2R$ (or 180°), or if the angles are α, β, γ, the "defect" $\Delta = 180° - (\alpha + \beta + \gamma) = 0$. We also learned the meaning of the "curvature of space" (C) which is the defect of a triangle per unit of area (a): $C = \Delta/a$. The mathematicians succeeded in deriving a very elegant formula for C. The physicists did not ascribe any great importance to this formula because in the Euclidean space—and every space was regarded as Euclidean—C was equal to zero anyway. However, the formula that was in Euclidean geometry nothing but a mathematical symbol that was interesting because of its elegance and simplicity became a chief instrument for discovering and presenting the general theory of relativity. One could now formulate the basic hypothesis of Einstein's theory of gravitation by assuming that the curvature of the four-dimensional space-time continuum should be proportional to the gravitational masses present in this space. Such a hypothesis would never have been formulated if the mathematicians had not put the formula for C at the disposal of the physicists. We see here again the great importance that convenient and simple mathematical formulations of known facts have for finding new and more general facts. The "deduction" of adequate mathematical formulae is an indispensable basis of the "induction" which teaches us new generalizations and therefore new facts.

4. Concepts and Operational Definitions

We have seen—from the analysis of geometry, the laws of motion, the motion of atomic objects, the new language of the atomic world, the causal laws (Chapters 3 through 10)—that results from induction can only be tested by experience if the concepts occurring in it have "operational meaning."[17] A concept (*e.g.*, "length") has an operational meaning if we can give an "operational definition" of that concept. This means that we have to describe a set of physical operations, which we must carry out, in order to assign in every individual case a uniquely determinate value to the concept (*e.g.*, to the length of an individual piece of iron). We know that the "length" depends on the temperature, pressure, electric charge, and other physical properties. Since Einstein's theory of relativity, we know that the length of a body will "alter" with its speed. Hence, the description of the operation by which we measure a length contains also the operation by which we keep temperature, pressure, speed, etc., constant. Or, in other words, the operational definition of length contains, strictly speaking, also the operational definitions of temperature, pressure, speed, etc. In order to know how to measure a length while keeping other factors, such as pressure, temperature, speed, etc., constant we must know a great many physical laws. Hence, every operational definition of an individual quantity like "length" must be taken "with a grain of salt," and is to be understood as an approximate definition. In other words, only under "favorable" circumstances can a set of operations be described which would provide unambiguously an operational definition of a single quantity like "length" or "time distance."

P. W. Bridgman[18] made an elaborate and astute study of the conditions under which the fundamental concepts of thermodynamics, like "temperature, supply of heat, supply of mechanical energy, etc.," can be defined by a set of operations of the physical or "paper and pencil" kind. It is obvious that no measurement of temperature can be performed if there are great changes within small space and time intervals. If we want to measure the heat Q supplied to a body through a certain surface, we may apply the methods of traditional calorimetry. Then we must assume that heat flows "smoothly" through the surface without considerable motion of matter within the body. On the other hand, we define the mechanical work which is done by the product of pressure and the increase of volume. This relation is only valid, of course, if there is no impulse or turbulent motion of masses within the body. Bridgman writes:[19] "It is only under highly exceptional conditions in practice that a clear-cut analysis can be made into heat flux and flux of mechanical energy." As an example, he considers Joule's original setup for describing the mechanical equivalent of heat by the elevation of temperature of a pail of water stirred by paddles. The conversion of mechanical energy of the paddles into thermal energy of the water is to a certain extent a degradation phenomenon, involving the transformation of large-scale motion into turbulence of an ultimately molecular scale. Bridgman stresses the point that it would be hard to analyze what is observed into heat or mechanical work; the result would certainly depend on the scale of the measuring instruments. "A study of the most general conditions under which heat flow and work have meaning would doubtless be of great interest, and so far as I am aware have never been attempted." Although concepts like "heat flow" and "work" have operational meaning merely under special "smooth" circumstances, they can be used as the basic concepts in our handling of physical experiments; the laws governing actual experiments are relations between measurements and are valid only under those "smooth" conditions under which concepts like "heat" and "work" have their own operational meaning. If, for example, we apply the first law of thermodynamics to the paddle situation, we draw a boundary surface around the region of turbulence "at such a distance that heat and work are there defined as can obviously be done." Then the first law tells us that for a cyclic process the input of heat is equal to the output of work. We are faced by a similar situation in all fields of physics and, as a matter of fact, in all fields of science. All "operational definitions" are limited to certain "smooth" or "simplified" conditions. We can even go a step further. We can easily see that, practically, operational definitions cannot be constructed in a domain of experience for which we don't know physical laws.

One of the main objections that has been raised against the concept of "operational meaning" since Bridgman proposed it is the assertion that we cannot form, for instance, an operational definition of "length" unless we already have some conception of length in our minds. This objection seems to have a certain validity unless we examine the actual construction of operational definitions more thoroughly. We may start from the example of the temporal difference between two instants of time. For example, we may consider the stretch of time between the beginning and the end of a lecture scheduled for a duration of one hour. The operational definition of this hour is the reading of the angle traversed by the hands of the clock fixed on the wall.   The angle traversed by the big hand is equal to four right angles (360°).   This is not, however, an arbitrary definition.   We must be certain that the clocks in all other classrooms show the same time, that the pocket watches of the teacher and students show the same time too.   Whether the clocks and watches are moved by falling weights or by elastic springs should make no difference.   This shows that the operational definition of an hour by the angle traversed by the hands is only of practical value if the clocks of all sizes and blueprints show one and the same length of time.   This is only the case, however, if there is a physical law that connects the oscillations of the pendulum under the influence of gravity with the oscillations of the hairspring under the influence of elasticity.   This is a law of mechanics.   Moreover, all the mechanical clocks provide a definition of temporal distance which allows us to formulate the laws for the propagation of light or of electromagnetic waves in a very simple way.   Hence, the spatial distances traversed by light can also be used as operational definitions of time distance.   The usefulness of all these operational definitions of time is based upon the fact that equal stretches of time remain equal whatever operational definition of time is used. It would, therefore, be wrong to say that we have the original concept of time and invent operational definitions for the purpose of measuring this time.   The basic facts are actually the identical results of the measurements by different operational definitions, which allows us to conceive a stretch of time not as defined by specific operational definition but by a large class of definitions that contain operations of very different types.

However, many philosophers, and even scientists, would object that there is, besides all these definitions by physical operations, an immediate feeling of time; we are able to estimate by a certain mental process how much time has elapsed during a lecture.   We can compare, subjectively, this time which is estimated by our consciousness with the duration ascribed to the same lecture by spring watches, pendulum clocks, or the propagation of light.   Some authors are inclined to say that the length of a time interval defined by direct observation is the "natural" length of it, while the operational definitions by clocks of different constructions are "artificial." If we give the matter more penetrating thought, however, we soon recognize that the definition of a time interval by our "subjective sensation of time" is actually one among several possible operational definitions. If we regard a human being as a measuring instrument, his estimation of a time interval, his reaction to the experience of a one-hour lecture, corresponds exactly to the pointer-reading on the dial of a mechanical clock. There are different methods of subjective estimations, as there are different types of mechanical clocks. The estimation of the duration of a lecture can be based on the degree of boredom produced in the audience, the degree of physical exhaustion, of hunger, of thirst, or of longing for pleasant company. The usefulness of mechanical clocks is derived from their agreement with "subjective" estimation of time. If, on the average, the students would not be exhausted, to a similar degree, by a lecture which takes an hour according to a mechanical clock, the operational definition by these clocks would be of no practical value. Hence, all operational definitions of time intervals, objective and "subjective," make sense only if two intervals that are equal according to one definition are also approximately equal according to another of their definitions. These equalities, of course, are valid because of specific laws of motion. While a pendulum clock performs a certain number of oscillations, a certain amount of liquid leaks out of a container, an airplane or a beam of light traverses a certain distance, etc. However, the feasibility of these definitions is also based upon laws of physiology and psychology. While the hands of a spring watch traverse a certain angle, the human heart performs a certain number of beats and the audience of a lecture is exhausted to a certain degree. From all these considerations, we learn that the operational definitions of "time interval" do not presuppose the previous existence of a "mental concept of time interval." We would rather say that the "mental" concept of time interval is just as well an "operational definition" as the physical one. The latter has been introduced only because the definition of subjective estimation has turned out not to be practical for some purposes. We could not stipulate that a lecture should last only until the audience is exhausted or bored to a certain degree and not even until all members of the audience have performed the same number of heart beats. These definitions of the length of a time interval would be just as practical as the definitions by mechanical clocks if we knew all the variables upon which exhaustion, boredom, and frequency of heart beat depend and could keep some of them at constant values in the same way as we keep temperature and pressure constant when we define "length." To be slightly flippant, we could say: If we knew laws by which we could compute the ability of lecturers to interest different audiences, we could use the degree of boredom produced in the audience as an operational definition of time distance.

It is clear from all this that every change in our knowledge of natural laws must produce a change in the operational definition of which we are making use. We learned, for example, that the contraction of bodies by motion brings about a change in the "operational definition" of length; this definition must now contain the speed of the body, the length of which is to be defined. In the same way, a new knowledge about how heat in a classroom affects the susceptibility of the audience brings about a change in the operational definition of a time interval by the degree of boredom aroused during this time. From all these considerations, we see that the development of "operational definitions" is very closely connected with our knowledge of physical laws.

5. Induction by Intuition and Induction by Enumeration

In the popular presentation of contemporary philosophy of science, we frequently meet the assertion that there are two different and even incompatible ways of establishing general laws of nature: by "induction" and by "intuition." The first way is to collect a series of observed events in which we recognize that some sequences of events repeat themselves again and again, like the periodic change of light and darkness in our everyday life experience, the trajectory of a projectile after it has been launched with a certain velocity, etc. In this context, we mean by "law of induction" the assertion that after such uniformities of sequences have been observed through many repetitions without exception or with few exceptions, this uniformity will go on forever, provided that the conditions in the surroundings are not changed. We believe that the change of day and night will never stop and that a projectile launched with a certain velocity will always traverse the same trajectory. We see here that the "law of induction" says basically the same thing as the "law of causality." By observing uniformities in nature, we are led by "induction" to the assertion of natural laws. This method of arriving at general laws has often been called the scientific way in the modern sense, the positivistic way in contrast to the Aristotelian way of deriving general laws from intelligible, self-evident principles.

The second way of finding general laws attempts to find these laws by what we may call "intuition" or "imagination" or, perhaps, just "guessing," and to test the results of this intuition by comparing the result with actual sense observations. As we learned previously (Section 3), Whewell regards this way as the procedure which has been actually used in the history of science for discovering new laws. This procedure leads us also from the observation of single facts to the statement of general laws because there is no guessing at a general law before a certain number of individual facts have been observed. Hence, this procedure is also called "inductive," and a distinction is made between two kinds: first, "induction by enumeration," the establishing of laws from the observance of a great many sequences of facts, and, second, "induction by intuition," or "imagination," the discovery of laws by the construction of "new concepts" on the basis of relatively few observations and the confirming of the law by a great number of observations.

If we agree with Whewell, we would say that, somehow, modern science proceeds in a way similar to that of ancient science (Section 1). It starts from relatively few actual observations and attempts to establish by "imagination" or "guessing" a simple law from which these observations can be derived. If this law is conspicuously simple, the scientist has confidence that a great many other observable facts can be derived from it. The work of recording a very great number of observations does not contribute much to the discovery of a law, but it is indispensable for its justification and confirmation. We accept a scientific law if it allows us to derive from a simple formula a great number of apparently unconnected observational facts. From the very fact that a certain sequence of observations appears again and again we cannot actually derive a scientific law. We are facing the same difficulty as in the theory of causality. No observational fact repeats itself completely; there are only special components of the complexity of observation that are repeated. The problem is always: What are the components which, by their frequent repetition, allow us the inference of a perpetual uniformity. If a projectile is launched, the repetition of the initial location does not allow any inference that the trajectory will be repeated too; but if the location and velocity

(speed and direction) are repeated, the trajectory as a whole will be
repeated.   In all such cases, we have to do with a physical law.
The sequence of days and nights allows the inference of a perpetual
repetition only because we regard this repetition as a consequence
of a physical law, the uniform rotation of the earth around its axis.
The type of induction in which we infer from a frequent repetition
a perpetual repetition is called "induction" in most of the popular
books and elementary classes.   Reichenbach calls it "induction by
enumeration."   He strongly stresses the point that by such an
"enumeration" no new law can actually be found.   He writes:[20]

> The scientist who discovers a theory is usually guided to his discovery
> by guesses; he cannot name a method by means of which he found the
> theory and can only say that it appeared plausible to him, that he
> had the right hunch, or that he saw intuitively which assumptions
> would fit the facts.

As we can learn from our previous argument (Sections 2 and 3),
John Stuart Mill attempted to show that new theories can be found
through "induction by enumeration"; Whewell, however, who
had thoroughly studied the origin of theories, was rather critical
toward these attempts, and ascribed new theories to "guesses,
hunches, and intuitions," if we quote Reichenbach's words.   Mill[21]
described the procedure of induction by sentences like the following:

> We observe that every occurrence of the combination ABC is followed
> by the combination abc, and every combination ABD by abd.   When
> we have observed this sequence very frequently we draw the inference
> that D is the "cause" of d and d is the "effect" of D.

Whewell,[22] however, comments on this "induction by enumeration" or by frequent repetition of the same sequence:

> Upon these methods, the obvious thing to remark is that they take for
> granted the very thing which is most difficult to discover, the reduction
> of the phenomena to formulae such as we have presented to us.

He refers to the "induction" that led to the discovery of some
individual physical laws as examples:

> If we look at the facts of the planetary paths, of falling bodies, of
> refracted rays, of motions, of chemical analysis . . . where are we
> to look for our ABC and a, b, c?   Nature does not present to us the
> cases in this form.

Whewell directs our attention to the fact that even after the theories have been discovered it is very difficult to point at the ABC and abc elements in the history of science.   "Who will carry," he writes, "these formulae through the history of the sciences, as they have really grown up?"   The prominent astronomer John Herschel[23] very strongly stressed the point that what matters primarily in scientific discoveries is the finding of a formula; if a simple formula is found that covers a wide field of observations, it does not contribute much to the belief in the validity of this formula if we add a great number of individual facts which can be derived from this formula and are true.   "No doubt," writes Herschel, "such inferences are highly instructive; but the difficulty in physics is to find such, not to perceive their form when found."

The belief in discovery through observation of a huge number of observable facts has frequently been regarded as the characteristic of "positivism" or rather of the "positivistic" approach to science. It seems to me that this opinion is based upon a very superficial study of positivism and its philosophy of science.   If we investigate the attitudes of scientists with "positivistic" leanings at the turn of the century (1900), we shall certainly find that the most powerful figure of that period was the physicist Ernst Mach.   In his book he develops his ideas about induction which were, in fact, very similar to Whewell's.   Mach writes:[24]

> The mental operation by which one achieves new concepts and which one denotes generally by the inadequate name of induction is not a simple but rather a very complicated process.   Above all, it is not a logical process although such processes can be inserted as intermediary and auxiliary links.   The principal effort that leads to the discovery of new knowledge is due to *abstraction* and *imagination*.

The fact that method cannot produce much in this matter was emphasized by Whewell himself.   In a line that is very similar to Whewell's, we find in Mach's book a strong emphasis upon the primary role of unifying and simplifying ideas in the discovery of new scientific laws.   Mach writes:

> Guided by one's interest for the whole of things again and again one directs one's attention beyond the facts, whether these facts may be straight sensations or belonging to the domain of representations. . . .

Then, one would, perhaps, in a happy moment, contemplate the simplifying and fertile thought.

The close connection between induction and intuition is stressed in a paper, *Induction and Intuition*, published by the Swedish philosopher Alf Nyman.[25]   He makes an attempt to connect the view presented by the Swedish philosopher Hans Larssen[26] in his book *Intuition*, with the view of scientists like Whewell, Herschel, and Mach.

Although the "induction by enumeration" is not of primary importance for the discovery of new theories, it would be very wrong to claim that the "induction by enumeration," the induction that is based upon recurrence of sequences, is of no importance at all in science.   "The same scientist," writes Reichenbach,[27] "who discovered his theory through guessing presents it to others only after he sees that his guess is justified by the facts."   Reichenbach made a great effort to show that the method of this justification is based upon the recurrence of sequences.   If a theory predicts such a recurrence, the acceptance of this theory is justified if a great many of these recurrences can be actually observed.   The more the derived recurrences are actually observed, the more probable is the validity of the theory.   Hans Reichenbach has perhaps tried harder than anyone else to compute the probability of a theory from the observed recurrences of the derived facts.   He distinguishes clearly between the *context of discovery* and the *context of justification*.   In the first, induction proceeds by the guessing of new concepts; in the second, by observing recurrent facts.   While the first process cannot be carried out by a scheme or a method, the second employs the method of "inductive inference."   If we start from the observed positions of planets on the sphere, we can find Newton's laws of motion by "intuition" or "guessing."   However, when we know Newton's laws, we can ask to what degree these known laws are confirmed and made plausible by the observed facts.   We must enumerate as many facts as possible which can be derived from the laws and tested by observation.   The greater the number of observations with positive results, the more "probable" these derived facts make our theory.   For this reason, the study of "inductive inference" belongs to the theory of probability.   It would be incorrect to believe in a close analogy between deduction and induction: The facts are simply derived from the theory by deduction, but the theory cannot be simply inferred from the facts by induction. "Observational facts," wrote Reichenbach, "can make a theory only probable, but will never make it absolutely certain." While conclusions, conditional on the validity of a known theory can be derived with certainty by the procedures of deductive logic, new theories are based upon the "principle of induction," which asserts that if, in a long series of events, a certain event recurs again and again with a certain frequency (e.g., the number one if we throw a die), approximately the same frequency will hold for the future. This is the simplest type of prediction by inductive inference, and is called, as we have pointed out, "induction by enumeration." Then, of course, the question arises whether every inductive inference can be reduced to "induction by enumeration." This was certainly the opinion of Mill. Reichenbach states bluntly that all forms of inductive inference are reducible to induction by enumeration and even that the possibility of such a reduction can be proved. The validity of this claim is not very obvious because we can easily point out single examples of inductive inference by enumeration that would lead to wrong results. For centuries, Europeans had known only white swans and it was natural to draw the inductive inference that "all swans are white." One day black swans were discovered in Australia; the inductive inference had led to a false conclusion. Should we now say that this principle is false or, at least, that it cannot be applied in every case? Reichenbach attempts to show that the principle of inductive inference was applied in an incorrect and oversimplified way, but that the principle itself is sound. He wrote:

> It is a matter of fact that other species of birds display a great variety of color among their individuals; so the logician should have objected to the inference by the argument that, if color varies among the individuals of those species, it may also vary among swans.

The inductive inference by enumeration would now be: We observe again and again that the individuals of a certain species have different colors and we infer that this should be the case in all species. The example shows that one induction can be altered by another induction. In fact, all inductive inferences are made not in isolation but within a network of many inductions.

This claim of reducibility to induction by enumeration leads certainly to a very simple hypothesis about how inductive inferences of very simple types can be built up. However, we are still very far from actually knowing "the method of inductive inferences" that would lead us from the observations of the planets to Newton's laws of motion.[28]

14

The Validation of Theories

1. Induction and Statistical Probability

Using Reichenbach's way of speaking, we put now the question: How by testing observationally the conclusions drawn from a given theory, can we find the "probability" of that theory, or, more precisely, the probability of that theory's validity. "Inductive inference" is the method by which this probability is computed or, at least, approximately computed. However, we must always remember that, in any valuable scientific discourse, we must employ only concepts that have "operational meaning." (Chapter 13, Section 4.) Hence, we must ascertain the operational meanings of the expressions "probability" and "inductive inference" before we can apply these terms in the language of science.

In the ordinary calculus of probability, as it has developed for the mathematical treatment of games of chance, the "probability of an event" is defined as the "relative frequency" of this event if we regard it as a member of a given long series of events. If we play dice and consider a long series of throws, say n throws, we can ask for the probability of the event that an ace (number one) appears at the top. If this happens m times among n throws, we call m/n the relative frequency of this event. If as n increases, the frequency tends toward a value p, we call this value the "probability" of the event. The probability of throwing an ace is of course $1/6 (p = 1/6)$. It is obvious that no probability can be defined unless one regards the event in question as a member of a series in which the frequency tends toward a limit. Such a series is called, as von Mises suggested, a "collective." If we consider the statement "the probability that a Mr. X. Y. will die next year is small," the statement has an operational meaning only if we regard the death of X. Y. as a member of a given collective; the value of this probability depends upon what collective we choose. If we regard X. Y. as a member of the collective that consists of all men on earth, his death is much more probable than if we regard him as belonging to the inhabitants of the United States, but one choice is as legitimate as the other.

The question arises whether in the sentence, "Newton's theory has a certain probability," the term "probability" has the same operational meaning as in the sentence, "The probability of throwing an ace on a die is 1/6." Reichenbach states bluntly that in the expression, "The validity of a certain theory has a certain probability," this word has exactly the same meaning as in the sentence, "The probability of throwing an ace on a die is 1/6." Hence one can, according to Reichenbach, assign to the validity of every scientific theory a numerical value which can be computed on the basis of the experimental confirmations of this theory by applying the methods of the usual calculus of probability. He[1] proposes two methods of computing the probability of a theory that actually correspond to two different operational definitions. In the first one, which he calls "probability of the first kind," he proposes to regard as the basic collective the ensemble of all observable facts that can logically be derived from the theory: The number of these facts may be n. Then those facts that are confirmed by actual observation or experiment can be singled out: Their number may be m. Then the ratio $p = m/n$ is the relative frequency of the confirmed results of the theory, and is to be regarded as the "probability" of the theory in the same sense as $p = 1/6$ is the probability of throwing an ace on a die. In the definition of what Reichenbach called the "probability of the second kind," the basic collective is the ensemble of a certain domain of observable facts (e.g., the motions of material bodies) which have been accounted for by an ensemble of theories. We denote by n the number of all facts in this domain that have actually been observed. An individual theory (e.g., Newton's laws of motion) allows the derivation of m facts among these n facts.

Then we define the fraction $p = m/n$ as the probability of Newton's theory of motion.

If we ask whether this definition given by Reichenbach for the probability of a theory or a hypothesis is a "correct definition," the answer depends upon what purpose this definition is to serve. From the scientific point of view, such a definition is "correct" if the term defined by it is useful in the formulation of scientific laws. (See Chapter 13, Section 4.) As we learned previously, an operational definition is useful only if there are several "operations" which ascribe one and the same value to a certain variable as, for example, a temporal distance can be defined by a pendulum clock and a spring watch. Hence, if the term "probability of a theory" is, on the one hand, defined by the operations described by Reichenbach, the value p obtained in this way must also tell us something about the willingness of scientists to accept the theory and call it a "valid theory." Richard von Mises,[2] who was instrumental in providing the logical foundations for the theory of probability, firmly denied that there was much connection between Reichenbach's "probability p" of a theory and the willingness of scientists to accept this theory. "It must be noticed," writes von Mises in his book *Positivism*, "that even in unprecise, colloquial talk physicists hardly ever use the expression that a theory has a greater or smaller probability." Actually, the reasons for which scientists accept a certain theory are very little connected with the "probability" of the theory. We could, if we used an exaggerated example of Reichenbach's method, assume that a theory consists in the direct enumeration of all observable facts in the domain in question. If all these facts are actually "observed," we could conclude, according to Reichenbach, that the theory has a probability of 100 per cent, or $p = 1$. The scientist would not, however, regard this enumeration as an acceptable theory, but rather as no theory at all. The theories which the scientist likes to accept have a simplifying and unifying character; they allow us to account for a great many facts by means of a few sentences that are used as hypotheses or axioms. Von Mises writes about the probability of theories:

> The physicist judges the usefulness, the possible acceptance or rejection of a theory by various criteria quite different from the ones above —to mention but one example: by the point of view of economy of thought.

Some authors have been inclined to say that theories "should" be judged according to their "probability on the basis of observational evidence." We shall see later, however (Chapter 15, Sections 2 and 3), that the evaluation of a criterion for the acceptance of a theory makes sense only if we indicate the purpose for which the theory is to serve. Let us, as an example, consider the probability of the hypothesis that "in throwing a die, an ace will appear on top." If we compute its probability according to the method of Reichenbach, or a similar method which is based upon the calculus of probability, we achieve as a result $p = 1/6$. This would mean that the probability for the validity of this hypothesis is 1/6 or about 16 per cent. According to the parlance that has actually been used in science, however, one would say on the basis of our experience in throwing dice that the hypothesis predicting that in every throw an ace would appear is simply wrong. Another example is given by a close collaborator of von Mises; Hilda Geiringer[3] writes:

> Let us assume that someone advances the hypothesis H that "every triangle has an obtuse angle." In order to test his assertion, we picked out a hundred triangles at random and measured them. The result may be that H is right in seventy cases and wrong in thirty cases. Then the scientist would obviously say that "H is wrong," and not that it is "valid with a probability of 70%."

There is, however, another objection to applying the ordinary calculation of probability. Obviously, the result of our measurements on the triangles depends very much upon the specific way in which we pick the triangles at random. This way determines the "collective" in which the triangle is embedded. A triangle can be characterized in different ways: The first one may consist in giving the length of the three sides, a, b, c; the second in giving one side a and two adjacent angles β, γ. If we select a series of triangles at random, we can do it by assuming that all values of a, b, c turn up with equal frequency. But we can also construct a random series by assuming that all values of a, β, γ turn up with equal frequency. Hence, we have to do with two "collectives" which are different from each other. The ratio of the triangles with obtuse angles to the total number of triangles will not be the same in both collectives. Hence, the "probability" of the hypothesis that "every triangle has an obtuse angle" is not unambiguously defined, but depends upon the arbitrary way in which we define the collective. For this reason, the probability of the hypothesis that every triangle has one obtuse angle cannot be defined by means of the ordinary calculus of probability. According to Hilda Geiringer, the scientist would say: "If a hypothesis *H* of the form '*B* is followed by *A*' is examined, and it turns out that in 10 out of 100 cases it is not in agreement with the observations, this hypothesis *H* is wrong, and is not valid with a probability of 90%."

2. Statistical and Logical Probability

Rudolf Carnap[4] made the attempt to define the "probability of a theory or a hypothesis" in a more general way which is not based upon the traditional calculus of probability. He starts from a material of sense-observations or measurements which he calls briefly the given empirical evidence (*e*). Then he assumes that we have found by imagination or guessing a hypothesis *h* from which statements about observations are derived. If we know *e* and *h*, we can ask: What is the probability that *h* is valid on the basis of the observational material *e*? The hypothesis *h* is found or "guessed" on the grounds of the empirical evidence *e* by "induction." Carnap's goal has been to set up a mathematical criterion for the degree in which *h* is "justified" by *e* (degree of confirmation). This degree is also interpreted as the "inductive probability" that *h* is justified on the basis of the evidence *e* or, in other words, the probability that the induction which leads from the evidence *e* to the hypothesis *h* is a valid induction. The meaning of the term "inductive probability" is not the "statistical probability" used in the common interpretation of probability in statements that occur in the statistical theories of physics and genetics. In the latter case, "probability" is used in the sense of "relative frequency." As we hinted in our presentation of Reichenbach's and von Mises' views, it seems to be very complicated and artificial to ascribe a "statistical probability" to the validity of scientific hypotheses. Carnap suggested the use of the term "statistical probability" in cases where we can reduce probability statements to statements about relative frequencies, but in other cases a new term "inductive probability." In this terminology, the statement, "the inductive probability of a hypothesis *h* on the basis of a certain evidence *e* is high" means the same as saying that "the evidence *e* confirms the hypothesis *h* in a high degree," or that "the degree of confirmation is high." The concept of "inductive probability" or "degree of confirmation" is, as Carnap introduces it, a purely logical concept. It is therefore also called "logical probability." The truth of a statement about the inductive probability of the hypothesis h on the basis of the evidence e does not depend upon the truth of e and h, just as in deductive logic the truth of the statement "e implies h" does not depend upon the truth of h and e.

Carnap[5] attempts to build up an "inductive logic" that is in many respects analogous to deductive logic. He gives the following example of this analogy. In deductive logic the observational evidence e may be: "All men are mortal, Socrates is a man." From this evidence we can conclude h: "If this is the case, Socrates is mortal." This conclusion can be drawn without knowing whether it is true that all men are mortal and Socrates is a man. We need only to know the logical structure of the evidence and the laws of conclusion (or logical implication). Then, an elementary statement of deductive logic says: "e implies h." An analogous example in inductive logic would start from the observational evidence e that "the number of inhabitants of Chicago is three million. Two million of them have black hair and b is an inhabitant of Chicago." By using the rules of inductive logic we would infer that the inductive probability of this hypothesis h that b has black hair on the grounds of the evidence e is equal to $2/3$. The truth of this inference does not depend upon whether it is true that Chicago has three million inhabitants, two million of which have black hair, nor upon whether it is true that b is an inhabitant of Chicago. In the same way, the validity of "e implies h" depends only upon the rules of implication, not upon the truth of the evidence e.

The easiest way to formulate Carnap's general definition is perhaps to start from this example concerning the inhabitants of Chicago. The given evidence e defines a range of men (b) who are inhabitants of Chicago. The hypothesis h defines a range of men (b) who have black hair. From the evidence, it follows that these two ranges (inhabitants of Chicago and black-haired men) have one range in common, which is defined by men who are inhabitants of Chicago *and* black haired. If s is a statement of the form "b has a certain property (pr)," the function $m(b)$ assigned to the property pr a positive number, the "measure" of the range, that contains all men (*b*) who have the property *pr*. Then *m*(*e*) is the range of all men *b* who are inhabitants of Chicago, while *m*(*h*) is the range of all men *b* who have black hair. The logical conjunction *h* · *e* states that a man *b* is an inhabitant of Chicago *and* also has black hair. Then *m*(*h* · *e*) is the range of all inhabitants of Chicago who have black hair. Thus, on the basis of the evidence *e*, it is obvious that $\frac{m(h \cdot e)}{m(e)} = \frac{2}{3}$, and it is understandable that Carnap defines his inductive probability *i* of the hypothesis *h* on the grounds of the evidence *e* by $i = \frac{m(h \cdot e)}{m(e)}$. While *m*(*s*) is a function of one sentence *s*, the inductive probability $i = \frac{m(s,r)}{m(s)}$ is a function of two sentences, the evidence *e* and the hypothesis *h*.

In the example from which we started the measure, *m*(*e*) is simply the number of inhabitants or the number of black-haired people. Generally, *m*(*e*) is the measure of all the observed facts for which our hypothesis *h* is to account. As we learned in our discussion of causality (Chapters 11 and 12), every result of our observations of a physical system can be described by ascribing to the state variables of the system specific values or, in other words, by describing the "state of the system." For our simple example, the "state" was described by the number of inhabitants and the "range of all possible" states was described by all possible numbers of inhabitants, *i.e.*, all positive integer numbers. The "range of all possible states of a physical system" is described by all possible value systems of the state variables. While in our simple example the "range" of the evidence is a certain range among positive integer numbers, the range of evidence for a general physical system is given as a certain range of the state variables. The evidence *e* is characterized by certain values that are ascribed to the state variables as the result of actual observations. The hypothesis *h* is characterized by certain values that are ascribed to the state variables as the result of logical deviation from a system of principles.

We can understand these general considerations easily by means of a simple example. The evidence may consist in observations of the locations of a mass-point on a plane. The "range" of a single observation then is the area of a small circle around a point because we must always consider that because of errors of observation a sin-

gle observation never yields a geometrical point but a small area
around a point. If we make a number N of observations, the sum
of the areas that correspond to N observations is the "range of the
evidence e." In this simple case, the "measure of the evidence"
$m(e)$ is the sum of all the circular areas that are obtained by N
observations. We can, for example, regard them as the positions
of a planet during its motion around the sun. As is well known,
Kepler inferred from this evidence the hypothesis that all these positions are situated on an elliptic orbit. We can now raise the question: What is the probability of the hypothesis on the basis of the
evidence e that is presented by our N observations? As the "measure" of the evidence is given by the sum of N circular areas, the
measure of the hypothesis h is given by the range of the positions
which are derived from this hypothesis. If we again make
allowance for the error of observation, the range consists of an
area between two elliptic curves. The measure $m(h)$ is equal to
the area between these ellipses. The circular areas which correspond to the observational evidence e may have some area in common, whether it is in the area $m(h)$ or not. In the area which they
have in common at every point the conjunction $h \cdot e$ is valid.
Therefore, the total area which they have in common is given
by $m(h \cdot e)$, the measure of $h \cdot e$. Then the inductive probability of Kepler's hypothesis is given according to Carnap by
$i = \dfrac{m(e \cdot h)}{m(e)}$. If the N small circular areas are situated in such a
way that the whole area between the ellipses is covered by them,
the areas $m(e)$ and $m(e \cdot h)$ are the same, and the inductive probability approaches the value one. The probability of the Kepler
hypothesis approaches certainty. Generally, the greater the common area of the experimental evidence and elliptic belt, the greater
is the inductive probability of the Kepler hypothesis.

We must, however, be aware of the fact that the Keplerian
hypothesis presents only a very simple example for the computing
of the inductive probability. The state variables which are used
for the formulation of the hypothesis are exactly the same as for the
experimental evidence: the coordinates of mass-points in a plane.
The statement becomes immensely more complicated, however, if
we ask, for example, what is the inductive probability of Newton's
laws of motion (law of inertia, etc.). The main difficulty consists

in the fact that from Newton's laws by themselves no positions of material bodies can be derived without assumptions about the system of forces acting upon the bodies and the structure of the bodies (elastic, plastic, rigid, etc.).   It is difficult to compute the inductive probability of the Newtonian laws from the evidence because this evidence is not only dependent upon these laws, but upon the wide variety of structural influences.   Hence, the number of state variables would be very large.   We have learned (Chapters 11 and 12) that the law of causality has a practical meaning only if the number of state variables is small.   In the same way, the law defining the inductive probability of a hypothesis in a certain system is without practical meaning if the number of state variables in this system becomes very large.   Basically speaking, a hypothesis like Newton's laws says nothing about what states are involved and does not allow a computation of the formula $i = \dfrac{m(h \cdot e)}{m(e)}$.   Carnap writes:

> There are many inductions in science which by their complexity make the application of inductive logic practically impossible.   For instance, we cannot expect to apply inductive logic to Einstein's general theory of relativity.

This would be, however, no serious objection to inductive logic. As we learned in our presentation of "causality," this law cannot be applied to situations of great complexity.   It is a question of "applied inductive logic" to find out whether the situations in which the computation of $i = \dfrac{m(h \cdot e)}{m(e)}$ is possible are of sufficient complexity to make them a sufficient approximation to practical situations or whether only practically irrelevant situations can be treated in this way.

Since, according to Carnap, the statements of inductive logic are purely logical, they say nothing about physical facts, or, in other words, they are not statements that are the results of observations. They are of the same type as the formal system of geometry, Euclidean or non-Euclidean,[6] before operational definitions (like straight lines or light rays) are introduced.   In order to make statements about inductive probability that can be checked by observations, we must add an operational definition of the term "inductive probability."   If we say that "a straight line means a light ray in a vacuum," or means "the edge of a sharpened knife," then the statement obtains an exact meaning only if we present the operations by which we produce a light ray or the edge of a knife. If we speak of the operational meaning of "inductive probability," we must state what actions are induced by statements in which the term "inductive probability" occurs. Carnap very strongly stresses the point that "inductive logic itself can make statements about inductive probability, but is not concerned with the practical application of its theorems, any more than pure geometry is concerned with the application of geometrical theorems for the purpose of navigation."[7] As a matter of fact, we know, to speak precisely, that even all statements about a triangle of steel or wood belong in this sense to applied geometry, or, to use a more general term, to physical geometry. We learned in discussing geometry and mechanics[8] that the operational definitions themselves always contain terms which are not symbols but words of our everyday language. Physical operations are formulated in a vocabulary which is not essentially different from the vocabulary which we need to describe our breakfast table. For this reason, the presentation of the way in which we apply geometry or mechanics in practice is not itself a part of geometry or mechanics. The systems that we call "geometry" or "mechanics" or "theory of relativity" are instruments which we use in order to make our lives more pleasant. Hence, their usefulness is essentially of the same type as that of any tool, whether it be a hammer, a yardstick, an airplane, or an atomic bomb. According to Carnap, the same thing holds for the system of inductive logic; this system allows only the computation of values of "inductive probability" which is achieved by a chain of deductive conclusions within the system. He writes:[9]

> The analysis of the application involves also certain assumptions and concepts of a psychological nature (e.g., concerning the measurement of preference and valuation). The problem and difficulties here involved belong to the methodology of a special brand of empirical science, the psychology of valuation, as a part of the theory of human behavior and that therefore they should not be regarded as difficulties of inductive logic.

Human decisions, according to Carnap, cannot be guided by the theory of inductive logic if we restrict the theory to "pure logic," to the rules of computation of probability by means of the formula

$i = \dfrac{m(e \cdot h)}{m(e)}.$ We must add to this theory explicit rules of action
which are indispensable if we wish to make of the theory a system
of advice on how to act in a certain situation.   The first rule formulated by Carnap says: "Assume that those events will occur which
have a high value of i (inductive probability) on evidence e, and act
as though you knew that these events were certain."   The second
rule says, in a more specific way: "Expect that event which has the
highest inductive probability and act as though you know that this
event is certain."   If then we add "rules of action" to the rules
which tell us how to compute inductive probabilities, we obtain a
theory which teaches us how to act in a given situation.

It is instructive to compare these rules of action with the theory
which we obtain if we do not start from "inductive probability,"
but from "statistical probability," which is used in the usual presentation of "probability calculus" and its use in science.   As we
mentioned previously, the definition of "statistical probability"
starts from an infinite random series in which each event has a certain "relative frequency," *e.g.*, the appearance of an ace in a series
of throws with an ordinary die.   Then the "relative frequency" is
called the "statistical probability" of this event.   In our case, the
probability p of our ace is obviously $p = 1/6$.   Von Mises showed
that from his definition we can deduce all the rules of the traditional
calculus of probability.[10]   By means of these rules, we can derive
other collectives from a given collective, and compute the relative
frequencies of events in each series.   It is clear that nothing can be
derived about "probability of individual events."   It is a meaningless operation to ask whether an individual throw may "probably"
yield an ace.   If we wish to derive advice on how to decide in a certain case by this method, we must add "rules of decisions" as we
do in "inductive logic."   For example, we must accept the rule that
we should act as if events of very high probability within a series
were practically certain events in individual cases.   The concepts
of "inductive" and "statistical probability" at first glance look
fundamentally different from each other.   Carnap writes:[11]

> An elementary statement of statistical probability is factual and
> empirical; it says something about the facts of nature and hence must
> be based upon empirical procedure.

From these statements which ascribe a particular value (*e.g.*, $p = 1/6$) to a certain event we must distinguish the theorems of the mathematical theory of probability. Writes Carnap: "They say something about connections between values of statistical probability." On the other hand, he writes:

> An elementary statement of inductive probability, *e.g.*, one which attributes to two given arguments (*e* and *h*) a particular number (*i*) as value of inductive probability is either logically true or logically false. . . . It is independent of the contingency of facts because it does not say anything about facts, although the two arguments (*e* and *h*) do, in general, refer to facts.[12]

However, if we apply both concepts of probability to one and the same concrete case, we soon notice that both concepts are closely related to each other; it is sometimes difficult even to distinguish them from each other well. We may start from the simple statement: "The probability of throwing an ace with this die is 1/6." This statement has often been interpreted as a typical example of statistical probability. It would say that in a long series of throws the relative frequency of an ace would be 1/6. However, Carnap pointed out that this statement can also be interpreted as a statement about inductive probability. For this purpose, we regard the statement about relative frequency as the evidence and look for the inductive probability of a hypothesis *h* on the basis of the evidence *e*. The evidence *e* is that the relative frequency of an ace is 1/6. Then, we investigate the hypothesis *h*, that the next throw of our die will yield an ace, and ask: What is the inductive probability of this hypothesis on the basis of the evidence *e*? We conclude from the definition of inductive probability $\left(i = \dfrac{m(e \cdot h)}{m(e)} \right)$ that in the example $i = 1/6$ or, in words, that the inductive probability that the next throw will yield an ace is 1/6. We attribute to an individual event (the next throw) a numerical value of probability. If we identified "probability" with "statistical probability," it would certainly be meaningless to ascribe the probability 1/6 to an individual throw. If, however, we attribute to this single event a numerical value of inductive probability, we do not assert that the statement can be checked by experiment. A statement of "inductive probability" is not a statement about an observable fact, but about the logical connection between given statements. In our example, it says that on the grounds of the observed relative frequency of an ace we compute the probability $i = 1/6$ of an individual throw. Carnap writes:

> The concept of inductive probability is applied also in cases in which the hypothesis h is a prediction concerning a particular event, *e.g.*, the prediction that it will rain tomorrow or that the next throw of this die will yield an ace.[13]

If we affirm the statement that "the inductive probability of an ace is 1/6 for a particular throw" and accept Carnap's rules of decision, we shall act in exactly the same way as if we knew from experience that the "statistical probability" in a long series of throws was 1/6. Neither a statement about "inductive probability" of an individual event, nor a statement of "statistical probability" within a long series directly provides a rule of action unless operational definitions or, in other words, rules of decision are added.

There has been during the last decades, since about 1920, a cleavage among scientists and philosophers concerning the correct "theory of probability." In his fundamental paper of 1919,[14] Richard von Mises advanced a set of principles from which all the calculus of probability could be derived. In this system, probability was defined as "statistical probability," and the author made a strong plea for the assertion that this is the only concept of probability which is compatible with an empiric or positivistic conception of science. Since Carnap has been regarded during all these decades as one of the strongest advocates of what we have called empiricism and positivism in science and philosophy, he has been accused of grave inconsistency in advocating a second concept of probability besides the statistical and empirical one. Certainly the main principle of empiricism, or even logical empiricism as Carnap understood it, is the principle of verifiability or confirmability. The strict adherents of the statistical conception of probability would say that no statement that a particular event will occur with a certain probability can be verified. Hence it is, according to the doctrines of logical empiricism, meaningless. In discussing the objections of empiricists and positivists, Carnap writes:

> They might say, for example: "How can the statement that the probability of rain tomorrow on the evidence of given meteorological obser

vations is 1/5 be verified?" We shall observe either rain or not rain tomorrow, but we shall not observe anything that can verify the value 1/5.

This objection, however, is based upon a misconception concerning the nature of the statements of inductive probability. This statement does not ascribe the value 1/5 to the inductive probability of tomorrow's rain, but rather to a certain logical relation between the prediction of rain and the meteorological report.[15]

Carnap makes the point that such a statement is a purely logical statement and is, therefore, not in need of verification by observation of tomorrow's weather. Carnap attempts to clarify this situation by comparison with deductive logic. He starts from the sentence *h* "there will be rain tomorrow," and *j* "there will be rain and wind tomorrow." Then, he says, one can certainly conclude from deductive logic that "*h* follows logically from *j*." Then even the strictest logical empiricist will not request that the statement must be confirmed by rain observation. According to Carnap:

The statement, "the inductive probability of *h* on the evidence *e* is 1/5" has the same general character as the former statement. . . . The difference between the two statements is merely this: While the first states a complete logical implication, the second states only, so to speak, a partial logical implication.[16]

While von Mises advocated the exclusive use of "statistical probability," Keynes[17] and Jeffreys[18] recommended a logical concept of probability which was in some respects similar to Carnap's 'inductive probability.'

3. Which Theory of Probability Is Valid?

Richard von Mises and Hans Reichenbach have strictly maintained that the definition of probability as "relative frequency" is the only satisfactory scientific basis of probability calculus, and moreover, of any discourse on probability. A great many scientists and philosophers have agreed with this view and have become "frequentists." On the other hand, many authors have pointed out their agreement with J. M. Keynes, H. Jeffreys, and other proponents of "logical probability." We must raise the question of what is really meant in maintaining that the "frequency theory of probability" or the "logical theory of probability" is the right one.

Does it make sense to maintain that the "frequency theory" is a valid theory in the sense that we say that the wave theory of light is valid?

Von Mises and Carnap would both agree to the statement that our common-sense conception of "probability" is a vague one and that we have to crystallize out of it clearly defined concepts in order to use "probability" in scientific discourse. Von Mises[19] maintained that the only concept which can be precisely defined and actually used in science is his concept of "relative frequency" or "statistical probability," while authors like Keynes have employed a different process of refinement and produced a concept of "logical probability." Rudolph Carnap[20] recommended a kind of compromise and suggested the use of a refining process that yields two end products: "statistical" and "inductive probability." From a scientific point of view, we can formulate a judgment about these concepts only by investigating the criteria for the validity of scientific theories. We must examine scientific theories in which the concept of probability occurs and find out which of these concepts is more useful. We may, for example, examine theories like the kinetic theory of gases, or of electrons, or any other theory of statistical physics.

As we have repeatedly pointed out, neither Carnap's definition of "inductive probability" nor von Mises' "statistical probability" will lead to any statement about individual observable phenomena unless "rules of action" or "operational definitions" are added to the abstract definitions. However, if these rules are added, both concepts of probability lead to one and the same statement about actions. If we present, for example, the kinetic theory of gases, the observable results do not depend on which concept of probability we employ in our presentation. If this is so, why is Carnap not satisfied with the use of the frequency theory of probability; why does he require the use of inductive and logical probability also in the presentation of science? Again and again he has pointed out that the concepts of science should be as close as possible to the concepts of common sense. When we speak of probability in our everyday language, we are accustomed to speak of the probability that it will rain tomorrow or that there will be a war this year. Since the frequency concept of probability would not permit us to speak in this way, Carnap recommends the introduction of two concepts of probability into science which are of very different logical status but agree in their application to the results of physical experiments and observations. He strongly emphasizes the fact that the sophisticated philosopher may refuse to speak of the probability of a particular event, but that the layman never will. He discusses[21] a simple example. He assumes that observations have yielded the information that a certain die is symmetrically built, that six thousand throws have been made with it under ordinary circumstances, and that a thousand among them have yielded an ace. If we call h the hypothesis that the next throw will yield an ace, "there will be," according to Carnap, "almost general agreement that the inductive probability of h on the evidence described will be (exactly or approximately) 1/6." Then he adds the instructive comment:

> It is true that there are a few theoreticians who would refuse to make any statement in terms of "probability" with respect to h because, according to their conception, a probability statement with respect to a single event is meaningless. . . . However, the man in the street and the practical scientist in the laboratory have no such scruples. If we give them the evidence e and ask them what is the probability of h, the overwhelming majority will not hesitate to give an answer, and the overwhelming majority of the answers will show good agreement with one another.[22]

It is very instructive to note that Carnap distinguishes between the "theoretician" on one hand and the "man in the street" or "practical scientist in the laboratory" on the other. The "practical scientist in the laboratory" is regarded by him as using the same type of everyday language as "the man in the street." Their speech determines whether a theory should be accepted. Their speech habits are in some cases regarded as more relevant than the criticism of "theoreticians." We shall realize that this point is relevant when we discuss (in Chapter 15, Section 2) the criteria for the acceptance of theories and the place of agreement with common sense among these criteria.

The more thoroughly we investigate these criteria, the more we note how difficult it is to distinguish in individual cases which of two alternative theories is closer to everyday language. It is, for example, not certain that acceptance of the frequency theory of probability actually excludes the use of statements about the probability of single events. Reichenbach, who has been a strict "frequentist," gave an interpretation of probability statements on single events which fits into the frequency theory. He starts with the remark that, strictly speaking, statements about causal connections in their common-sense formulation cannot be checked by any experiment or observation. If we say, for example, "If one turns on the faucet, the water *must* flow," there is no way of confirming by experiment that this is actually a "must." This point has been well known since David Hume.[23] Reichenbach, however, wrote:[24] "The man who believes that if he turns on the faucet, the water must flow, has developed a good habit insofar as his belief will lead him to correct statements about the totality of such events." The same thing holds also, according to Reichenbach, for a probability statement about a single event which cannot be confirmed by experiment either. He writes: "Similarly the man who believes that a probability of 75% applies to a single case has developed a good habit." If "probability" is "relative frequency," there is no probability of a single event in the strict sense of the definition, but if one develops the habit of speaking about probability of a single event "his belief will induce him to say that of a great number of similar cases, 75% will have the result referred to."

We note again that, if we judge science as a guide in action, it makes no difference whether we regard the probability of a single event as a statement of inductive probability or whether we keep to statistical probability and interpret statements about probability of single events as good habits, but do not regard them as confirmable reports about empirical facts. If we sum up all these considerations, we shall note that the question of whether the "frequency theory of probability" or the "logical theory of probability" is "right" cannot be answered by a simple "yes" or "no." As we shall soon learn (Chapter 15, Sections 2 and 3), the answer depends upon what good we want to attain by this theory. The acceptance of one of the "theories of probability" depends, for example, upon what importance we give to the agreement with common-sense language or what importance we give to the fitness of our theory to decide between rival scientific theories, such as the wave theory and the corpuscle theory of light. We can easily note, for example, that the difficulties which we mentioned in discussing Reichenbach's theory appear in a very similar way when we apply Carnap's theory. We remember that von Mises and Hilda Geiringer raised the objection that in Reichenbach's theory a theory was said to be "valid with a probability of 70%," if 70 per cent of the conclusions of the theory were confirmed by experiment. However, we know that, in the discourse actually used by scientists, a theory which is in disagreement with experience in 30 per cent of the actually performed experiments is called "false" or "wrong." We come to the same conclusion if we apply Carnap's "inductive probability."

An attempt to avoid these difficulties and to advance a radically different approach to the probability of a hypothesis or theory was made by Jacob Bronowski.[25] His purpose[26] was to formalize the criterion for the validity of a theory that was advocated by men like John Frederick Herschel and William Whewell. As we remember, these scientists and philosophers saw the main achievement of a theory in its unifying and simplifying power. The simpler a theory was in comparison with the complexity of the observable facts covered by the theory, the more probable the theory. If the theory consists in a complete enumeration of all observable facts, the "theory" would have a very high probability if we used the definition of "probability of a theory" advocated by Reichenbach and Carnap. These definitions are essentially based upon statistics of the observable facts that follow from a theory. A theory was regarded by those authors as highly probable if a great number of the derived facts would be actually confirmed by experiment and observation. However, von Mises and Bronowski rejected this type of application of the calculus of probability. The usefulness of a theory for actual scientific work cannot be judged exclusively by the agreement of its results with actual observations; there could be a "theory" which agrees with all observed facts but is a mere record of observations and no theory at all. If we have two theories which yield the same observable facts, the scientist prefers the theory which is more economical or just simpler. Bronowski compares the scientific theory with a code which serves us to describe the observable facts. We prefer that code which is more practical, more efficient. In order to improve the code, we try systematically, to quote Bronowski, "to break down the code into its constituent symbols and their laws of arrangement." The one hundred odd chemical elements form a code which allows us the description of the chemical phenomenon. If we break down these elements into three kinds of elementary particles (protons, neutrons, and electrons) and the forces acting between them, we have a code which describes, for example, the interaction between hydrogen and oxygen in a way from which can be derived much more information than from any theory in which "oxygen" and "hydrogen" themselves occur as primitive symbols. If we consider theories that are not obviously contradictory to the observable facts, Bronowski calls a theory the more probable, the more the code provided by the theory is broken down into constituent symbols and laws of arrangement. Every acceptance of a debatable theory is due to a compromise between Reichenbach's and Bronowski's criteria: agreement with facts and efficiency as a code (Chapter 15, Section 2).

15

Theories of High Generality

1. The Role of Causality in Twentieth-Century Science

A great many authors have summarized the results of comtemporary atomic physics in a simple slogan: physics until the end of the nineteenth century was based upon the principle of causality, this principle has been dropped by twentieth-century atomic physics. We can read this summary of modern physics in the writings of biologists, psychologists, sociologists, philosophers, lawyers, physicians, and especially in the sermons of clergymen and the speeches of politicians. Such a formulation is, conservatively speaking, a gross oversimplification. It would be unjust to deny, however, that these misunderstandings outside physics have their origin in the superficial presentations that physicists have frequently used in formulating generalizations of what is scientifically confirmable. If we keep in mind the critical concept of causality that has been developed in Chapters 11 and 12, we can form a judgment about the proper place of causality in modern atomic physics.

This discussion of causality may seem a little lengthy and perhaps even pedantic. As a matter of fact, if we have to do with sciences that have been established with little change over long periods of time, everyone knows without much talk about causality how such a science formulates its laws and applies them in predicting the future. But when radical changes in the conceptual scheme of a science occur, it is no longer obvious how the concept of "causal law" is to be formulated, or even whether it should be retained and allowed to survive in the next phase of this science. The logical analysis of geometry became important and interesting only after the non-Euclidean geometries had been advanced; and Mach's analysis of Newtonian mechanics did not manifest its real meaning until Einstein's theory of relativity was accepted by the physicists. For the same reasons, an elaborated analysis of causality did not become attractive and relevant until toward the end of the nineteenth century, when contemporary atomic physics was in the making; this brought about a radical change in the laws of motion as exemplified in the rise of quantum and wave mechanics in the twentieth century.

Perhaps the best thing to do to acquire an easy and exact understanding is to restate one of the basic experiments in atomic physics, and to reformulate it by using the terms "causal law," "statistical law," and "causality." We refer to the passing of electrons through two slits in a diaphragm and the production of scintillations on a screen parallel to the diaphragm. This experiment was thoroughly discussed in Chapters 8 and 9. The initial conditions (or the "cause") in this experiment consist in a swarm of electrons (generally, atomic objects) which are emitted by a source and move in a direction perpendicular to the diaphragm and the scintillation screen. The distance between the slits (a), the speed of the electrons (v), and the distance between source, diaphragm, and screen also belong, of course, to the initial conditions. If the swarm of particles is a dense one, we observe a definite pattern of fringes at intervals which can be calculated by a mathematical rule from the initial conditions of the experiment, specifically from a and v. We can certainly say that whenever these initial conditions are established, a definite pattern of fringes follows. This is certainly a "causal law." The "cause" determines the "effect" unambiguously if we mean by "effect" the pattern of fringes as a whole. This is a causal law in terms of observable facts. The situation changes if the swarm of particles is a thin one. Then the pattern on the screen is built up from individual scintillations that follow each other at long intervals. We cannot predict these single scintillations; we can only predict a statistical distribution which can be checked when we have a great number of hits.

The situation is not totally different from the case of aiming a mass-point at a target. We cannot predict the precise location of

hits, but only the statistical distribution. But there is one difference: By narrowing the dispersion in the initial conditions in the
aiming process, we can narrow down the dispersion of hits around
the center of the target. When "atomic objects" pass through the
slits in a diaphragm, however, we can never make the scintillations
occur at a certain point of the target (the screen). We are in
exactly the same situation as in the experiment of tossing coins;
however we manage the initial conditions, we can only make a statistical prediction about the "effect" of our shooting atomic particles through the diaphragm. We can say that although it looks
like aiming projectiles at a target, the outcome of this experiment
is more nearly similar to that of tossing coins. Briefly speaking, in
atomic physics observable phenomena follow "causal laws" only if
the flow of atomic objects is a very large one; then the "pattern"
on the screen can be unambiguously predicted. The mutual distance between fringes is a mathematical function of the initial conditions a and v. A similar situation also exists in the Compton
effect; the frequency of X-rays is altered by collision with a swarm
of electrons. This alteration can be precisely computed from the
initial conditions, but no precise location of an electron can be predicted. The frequency of the spectral lines emitted by a hydrogen
atom can, according to Bohr's spectral theory, also be predicted
exactly from the initial conditions; but the location of a single electron in its path around the nucleus cannot be computed. The frequencies in the hydrogen spectrum, as in the Compton effect,[1] are
properties of a pattern produced by a great number of "atomic
objects" like electrons.

All this adds up to the fact that, as far as directly observable phenomena are concerned, the situation in atomic physics is not essentially different from the situation in so-called "classical physics,"
for example, in Newtonian mechanics. From the observable initial conditions, results can be predicted with certainty if we have
to do with properties of a great number of objects; but we observe
unpredictable "fluctuations" if we observe phenomena of a small
density. To repeat again a familiar fact: Even in aiming projectiles
at a target, the hits cannot be predicted if we launch only a small
number. The objection has often been made that every single shot
could be exactly predicted if one knew the initial conditions exactly,
but for this purpose we need the omniscient Intelligence introduced

by Laplace, discussed in Chapter 12. If we humans are the observers of observable facts, precise predictions are only possible on the basis of a great number of cases.

If we wish to understand the difference between the roles of causality in twentieth-century atomic physics and in nineteenth-century physics, we must examine the axioms, the relations between symbols, the conceptual schemes, which form the bases of these two physical theories. We learned (in Chapter 12, Sections 2 and 3) that the fundamental equations in Newtonian mechanics give the increments in time (first derivatives) of the state variables, $\dfrac{du_k}{dt}$, as functions of the present values $u_1, \cdots u_n$ of these variables: $f_k(u_1, \cdots u_n)$. These $u_1, \cdots u_n$ are the components of coordinates and momenta of mass-points. In the field theory (Chapter 12, Section 4) the causal laws are of the form $\dfrac{\partial u}{\partial t} = F\left(x,y,z,t, \dfrac{\partial u}{\partial x}, \dfrac{\partial u}{\partial y}, \dfrac{\partial u}{\partial z}\right)$ where $u(x,y,z,t)$ gives the intensity of the field as a function of the location in space and time. By the values of u at the present $(t = 0)$, $\dfrac{\partial u}{\partial t}$, the increase of u per unit of time is determined and the future distribution of u in space can be calculated by mathematical operations. The principle of causality does not say for which variable u such a "causal law" is valid, but only that there are variables which have this property. We learned in Chapter 8 that in atomic physics the mathematical scheme for computing future phenomena consists in introducing the amplitude of the de Broglie waves. The initial conditions of an experiment can be formulated as the spatial distribution of these amplitudes. The mathematical scheme consists then in a differential equation that allows one to compute the future values of these amplitudes if the values at the present $(t = 0)$ are given. These amplitudes are known in technical literature on quantum theory or wave mechanics as Schroedinger[2] functions and are usually denoted by $\psi(t,x,y,z)$. Hence, they are also referred to as psi-functions. These functions obey a differential equation which has the form of a causal law:

$$-\frac{h}{2\pi i} \cdot \frac{\partial \psi}{\partial t} = -\frac{h^2}{8\pi^2 m^2}\left(\frac{\partial^2 \psi}{\partial x^2} + \frac{\partial^2 \psi}{\partial y^2} + \frac{\partial^2 \psi}{\partial z^2}\right) + V(x,y,z)\psi$$

where m is the mass of the particle moving in a field of force of the potential energy $V(x,y,z)$.

If we wish to use this law for the prediction of observable phenomena, we must add to the differential equation (the symbolic scheme) the operational definitions of the symbols. In the differential equation of the Schroedinger function ψ is in general a complex function of x, y, z, t. If we form the "norm" of ψ, $i.e.$, the product of ψ and its complex conjugate, we obtain a real function of the space coordinates that means the average frequency of "point-events" (that is, scintillations) in a unit-volume of the region around the point with the coordinates x, y, z. Therefore, by integrating the Schroedinger equation, we cannot predict single point-events at a definite location in space. All predictions of observable phenomena are statistical laws, but the principle of causality is somehow fulfilled, since there is a state variable ψ which obeys this law. There is, however, no causal law for observable phenomena because the operational definition of ψ does not link its value to single point-events, but to a statistical average computed from a great number of point-events.

If we ask the blunt question whether or not the principle of causality is valid in modern atomic physics, we cannot answer by a simple yes or no. We are in the same situation as if we were asked whether or not in non-Euclidean geometry (for example, Lobatchevski's geometry) the Euclidean theorem is valid—that "two straight lines that are at all points the same distance from each other are parallel." This question is meaningless because equidistant straight lines do not exist in non-Euclidean geometry. If all points are drawn that are at an equal distance from a given straight line, they cannot be connected by a straight line. In a similar way, the causal laws of Newtonian mechanics do not make sense in atomic physics. In traditional mechanics, the future values of coordinates and velocities (state variables) are determined and predictable by the present values. If we ask whether this law is preserved in atomic physics, the question has no meaning. As a matter of fact, position and velocity are not state variables. There is no state of a system that is described by the present values of positions and velocities; therefore, we cannot ask whether or not in atomic physics future positions and velocities are determined by the present ones. Such a description of a state does not exist.

However, we can, as we learned in Chapters 8 and 9, describe the approximate state of an atomic object by ascribing to it a position and a momentum each with a certain margin called indeterminacy. Instead of introducing one particle with a certain indeterminacy of coordinates, we can introduce a swarm of particles with a certain dispersion of coordinates. Then according to the present Section the swarm would have a certain dispersion of momenta. If we denote the dispersion of coordinates and momenta by D_q and D_p, we find from the relation of indeterminacy that $D_q \cdot D_p = h$. This relation can be derived from the Schroedinger equation, the causal law which the ψ-function obeys. We learned that the operational meaning of ψ is connected with the distribution of particles around a certain point x, y, z in space, and hence with the dispersion D_q. We shall not go into this derivation further because it would lead us too far into mathematical technicalities.

What is relevant for us is to stress the following point: If we try to approximate the "motion" of an atomic object by the motion of a swarm of real particles, we find that the dispersions in this swarm D_q and D_p cannot disappear simultaneously, since $D_q \cdot D_p = h$. We must also keep in mind that this relation is not connected with any kind of "philosophical" interpretation; it is derived from those principles of atomic physics which are used for the prediction of those observable facts that are regarded as effects of atomic events. Very often in popular presentations we read: An electron can never have a precise position, and therefore the future cannot be unambiguously determined by the present. What is actually meant by such a statement is that by approximating atomic objects by swarms of real particles the dispersions of coordinates and momenta in such a swarm cannot disappear simultaneously.

To summarize the role of the causal law in atomic physics, we can say: There are no laws by which we can predict from any observable initial conditions the precise future locations of point-events. In other words, there are no state variables the initial values of which we can keep within such a narrow margin that we can achieve a precise predictability of single point-events in the future. If we try to introduce real particles as an approximation, we notice that the dispersions of their coordinates and momenta cannot be narrowed down simultaneously in the initial state of our atomic object. We must understand, on the other hand, that there are variables (like ψ) in atomic physics that allow us to predict the future values if the present ones are given, but these variables are connected with observable point-events by operational definitions in such a way that the precise knowledge of ψ at a certain future instant of time provides us with only a statistical knowledge of future localized point-events.

All these considerations lead to the result that the question whether the law of causality has survived in twentieth-century atomic physics cannot be answered by a simple "yes" or "no". There is a gradual change—the determinism of Newton and Laplace that is based on a definition of state in which position and velocity can both be held within narrow margins must be replaced by Bohr's theory of complementarity in which the "indeterminacy" or dispersion of location implies a certain margin of momentum.

Bohr says very aptly:[3] "The viewpoint of complementarity may be regarded as a rational generalization of the very idea of causality."

2. The "Scientific" Criteria for the Acceptance of Theories

After the considerations (of Chapter 14) which belong to the field of deductive and inductive logic, we are going to wind up our argument by discussing the "acceptance" of theories as an activity of the scientist. We turn from the "logical" to the "pragmatic" component of our argument. This component becomes particularly important when we have to do with theories of high generality, like the theory of relativity, Bohr's theory of complementarity, the theory of spontaneous generation (evolution of organisms from inorganic matter), etc.

The distinction between the "logical" and the "pragmatic" components in the presentation of a science has been closely connected with the twentieth-century rise of new ideas about the logical structure of science.

In the Aristotelean and Scholastic tradition, the presentation of science had been based upon a scheme which consisted of two elements (a "dyadic" scheme): the real objective world and the picture of this world given by the scientist. Both were regarded as agreeing with each other like a photograph with its original. Truth was, to speak in terms of Thomistic philosophy, the agreement of man's intellect with the things of the real world. This view had been preserved until the end of the nineteenth century in various schools of philosophy. As a drastic example, one can quote the almost religious fervor with which, under the name "theory of reverberations," this "dyadic" scheme is applied to science in the official Soviet philosophy, which has followed the lead of Lenin's main book on philosophy.[4] It teaches that theory is to reflect reality.

However, at the end of the nineteenth century, it has been proposed by C. S. Peirce to introduce rather a triadic scheme[5] in the presentation of science. This scheme consists of the observed object, the working scientist, and, as the third element, the signs that the scientist invents in order to give his presentation.[6] This scheme has been adopted in the twentieth century by great movements in the philosophy of science. In particular, the followers of pragmatism, logical positivism, operationalism, and general semantics have adopted the triadic scheme. It has been clearly defined and elaborated in the *International Encyclopedia of Unified Science*[7] by Rudolf Carnap and Charles Morris. According to them, science investigates, first, the relations between physical objects and signs or symbols; the result is called the "semantical" component of science. The relations between symbols form the "logical" component. Moreover, we have to study, as a third component, the relations between the scientist and his signs, or, in other words, the relations of the social and psychological circumstances under which the scientist is working to his theories. The study of these relations yields the "pragmatic" component. In their usual work, scientists have considered mostly the logical and semantical components. They accept a theory if it is logically consistent and agrees with the observed facts. However, if we have to do with theories of very high generality, we notice that they are not uniquely determined by these criteria. We also have to consider the pragmatical component, or, in other words, the impact of psychological and social factors upon the systems of signs which have been built up by the scientist as a part of the physical and psychological world. This will lead us into what is called today the "behavioral sciences."

It is generally understood among scientists that from the purely scientific angle, a system of propositions is an acceptable theory if, and only if, the system is logically correct, and its conclusions are in agreement with observable facts. Since, certainly, not all conclusions can be checked by experiment, we should rather say that the theory is acceptable if no conclusion is in disagreement with experiment, provided the number of tests is sufficiently great. It makes no difference for "science in the modern sense" which type of concepts and which type of relations between concepts occur in the proposition of the theory, provided no conclusion drawn from the theory is in disagreement with observations. Of course, we must always consider that the propositions of the theory consist not only of the relations between the basic concepts (or the basic symbols), but also of the "operational definitions" which connect those statements about basic symbols with statements about observable facts. According to these criteria, a theory (relation between symbols and operational definitions of the symbols) is confirmed if it is in agreement with observable conclusions which have been tested by actual observations. But if a theory has been "confirmed" in the sense which we have described above, it cannot be concluded that it is "valid," but only that it *may* be valid. According to what criteria have the scientists made a choice between several theories which may be valid?

In general, scientists would say that among several theories that are set up to account for a certain domain of observed facts, one will stand out as the best and will be accepted generally. If we follow Reichenbach's advice (Chapter 14, Section 1), we would say that we should accept the "most probable" theory. This means, according to the statistical theory of probability, that the theory should be accepted which shows "more" agreement with observed facts than the other theories. However, this agreement cannot be the only criterion of acceptance. If this were so, the best theory would be the mere description of facts; but this would be no theory at all. As we have mentioned repeatedly, the actual advance of science has always been engineered by a criterion of economy and simplicity. The criteria of Reichenbach and Carnap, which are based, like John Stuart Mill's inductive logic, upon agreement with observations, have to be complemented by the criterion of economy and simplicity which was advanced in the history of science by men like William Ockham, Isaac Newton, and Ernst Mach. In our twentieth century, the importance of criteria other than mere agreement with observation was stressed by von Mises and Bronowski (Chapter 13).

Most contemporary scientists would claim that among all theories that can account for the same observed facts the "simplest" theories are chosen, but the question arises of how to define the degree of simplicity. If we restrict ourselves to the concept of "mathematical simplicity," everyone would agree that an algebraic equation of the first degree is simpler than an equation of the second or third degree. The Copernican theory led to concentric circles as a geometrical description of the planetary motion which corresponded to an analytical description by single trigonometric functions. This description was certainly "simpler" than the description by the Ptolemaic theory, which made geometric use of "loops" that were analytically represented by series of trigonometric functions (Fourier series). During the long dispute between the wave theory and the corpuscular theory of light, one of the reasons for preferring the corpuscular was an argument of "simplicity." This theory led mathematically to the differential equations for the motion of a particle which are formulated by Newton's laws of motion. These laws lead to ordinary differential equations of the second order. The wave theory, however, led to the wave equation, which was a partial differential equation of the second order that had to be solved under restriction by boundary conditions. This was in the beginning of the nineteenth century a mathematical problem much less simple than the solution of ordinary differential equations. Hence, mathematical simplicity could be invoked in favor of the corpuscular theory. This difference in simplicity became, of course, less and less obvious the more the theory of partial differential equations developed. Hence, it is clear that how we judge the mathematical simplicity of a theory depends on the state of science at a certain period. There have been periods in which a certain theory was regarded as simple if it avoided the use of infinitesimal calculus and restricted itself to "elementary mathematics."

There is, of course, the question *why* "simple theories" should be preferred. Some scientists say that they prefer them because "simple" formulae allow easier and quicker computation of the result; they are "economical" because they save time and effort. Other authors, however, say that simple theories are more "elegant," more "beautiful"; they prefer simple theories for "aesthetic" reasons. However, we know from the history of the fine arts that a certain aesthetic preference is the result of a certain way of life, or a certain culture or social pattern. The same thing is true if we judge the "beauty" of a mathematical formula. A great many scientists of mathematical background are enthusiastic about Einstein's theory of gravitation because its formulae are of extreme mathematical simplicity and beauty. However, among the experimental physicists and observational astronomers, we find many who would say that these formulae are extremely complicated and that it is hardly worth while to introduce such complicated formulae in order to derive very few and even debatable facts.

If we investigate which theories have actually been preferred because of their simplicity, we find that the decisive reason for acceptance has been neither economic nor aesthetic, but rather what has often been called "dynamic." This means that the theory was preferred that proved to make science more "dynamic," *i.e.*, more fit to expand into unknown territory. This can be made clear by using an example that we have invoked frequently in this book: the struggle between the Copernican and the Ptolemaic systems. In the period between Copernicus and Newton a great many reasons had been invoked on behalf of one or the other system. Eventually, however, Newton advanced his theory of motion, which accounted excellently for all motions of celestial bodies (*e.g.*, comets), while Copernicus as well as Ptolemy had accounted for only the motions in our planetary system. Even in this restricted domain, they neglected the "perturbations" that are due to the interaction between the planets. However, Newton's laws originated in generalizations of the Copernican theory, and we can hardly imagine how they could have been formulated if he had started with the Ptolemaic system. In this respect and in many others, the Copernican theory was the more "dynamic" one or, in other words, had the greater heuristic value. We can say that the Copernican theory was mathematically "simpler" and also more dynamic than the Ptolemaic theory.

We find, by investigating decisions between theories that have actually been made, that it seems to be a general rule that mathematically simple theories are also dynamic, are fit to be generalized into theories that cover a wide range of facts. We have previously presented two examples: Maxwell's theory of the electromagnetic field and Einstein's theory of gravitation. They show clearly how a mathematical simplification of observed facts can lead to the advancement of very general theories of which these facts are only very special consequences. It has been made clear by now that the requirements for the acceptance of a theory by scientists in the modern sense are "agreement with observation" and "simplicity." There is, of course, a question that has not been touched at all by stating these two requirements. Which of them is the more important one? This seems at first glance to be an idle question, but as a matter of fact there are many cases in which we are faced with the question: If we have to choose between a theory that is in agreement with the facts but is very complicated, and a theory that is much simpler but does not agree as well in all details with the facts, which theory are we to choose? If we ask a scientist, he will probably answer that the decisive point is the agreement with the observed facts and that "simplicity" is of secondary importance. If, however, we give some thought to the question, we shall realize that such an answer would be a fallacy. The value of a theory consists, obviously, in its property of being simpler than a mere record of observations. There is certainly no theory that is in complete agreement with all our observations. If we require such a complete agreement, we can certainly achieve it by merely recording the observations. However, nobody would regard this record as an acceptable theory, although it is in complete agreement with observations. What makes a theory is the quality of being simpler and shorter than the record of observations. Hence, the acceptance of a theory is always the result of a compromise between the requirement of "agreement with facts" and of "simplicity."

If we look at the reasons why theories have been actually accepted, however, we soon notice that agreement with facts and simplicity are not the only qualities which a scientific theory is expected to possess. When we remember, for example, Francis Bacon's attitude toward the Copernican theory,[8] we note that he prefers the geocentric (Ptolemaic) theory because it is more in agreement with common sense. We have discussed this requirement on several occasions, and must recognize that there are actually three requirements that have been admitted by scientists: agreement with observations, simplicity, and agreement with common-sense experience. We should certainly point out that what is regarded as "simplicity" and "common sense" is a matter of the social background of a theory. There is, therefore, a certain justification for restricting "purely scientific" criteria to the agreement with facts. Then we should regard "simplicity" and "agreement with common sense"

as sociological criteria. Since, however, scientists have actually
accepted them in most cases, it is difficult to draw a clear dividing
line between strictly scientific and sociological criteria, if we enter
into the "pragmatic" part of the argument.

3. The Role of "Extrascientific" Reasons

When, in the seventeenth century, the chain between science and
philosophy had been broken,[9] "scientific truth" seemed to be based
only upon the criterion of agreement with facts. As we learned in
the previous section, this is not literally true. "Simplicity" and
"agreement with common sense" have always played their role and
brought some sociological and psychological reasons into the decisions of the scientists. If we keep in mind that reasons of this type
have always played some role in science, we will not wonder that the
philosophical end of the chain between "facts" and "principles"
has never completely disappeared. In other words, science has
never been completely restricted to technological use. Therefore,
criteria which are different from the "scientific" ones in the narrower sense have always intervened. If we again invoke an old
example, the Copernican theory, we can easily see that a great
many scientists and philosophers who have admitted that this system is "mathematically simple" and "in agreement with the facts"
have rejected it for reasons of a completely different kind.

It is easily seen from well-known examples that fitness to support
desirable conduct on the part of citizens or, briefly, to support moral
behavior, has served through the ages as a reason for the acceptance
of a theory. In antiquity, the physics of Aristotle and Plato seemed
to be fitter, in this respect, than the physics of Epicurus. According to the first, the celestial bodies were made of a nobler material
than our earth, while according to the "materialistic" doctrine of
Epicurus, all these bodies consisted of the same elements. This latter doctrine, however, made it more difficult to teach the existence
of a difference between material and spiritual beings. Since a great
many educators and statesmen have been convinced that the belief
in this difference is important for the education of good citizens, the
Epicurean doctrine was rejected by powerful groups. An instructive example is Plato, who in the description of "good government"
included the requirement that the followers of Epicurean philosophy
should be silenced.

A second historical example is, of course, the struggle against the Copernican and Galilean doctrines on the grounds that they made it more difficult and complicated to uphold Aristotelian physics. The law of inertia, for example, made it somehow difficult to assume, as did Aristotle and St. Thomas, that no motion is possible unless the body is moved by another body.   This assumption was used in important proofs for the existence of a prime mover; which was, in turn, important in the demonstrations of the existence of God.   We can easily quote similar examples from contemporary science.   The most obvious one can be taken from the influence of governments that regard materialistic philosophy as the support of a desirable moral conduct, and favor scientific theories which give support to materialism.   A familiar example is the fight against Einstein's theory of relativity in Soviet Russia.

Scientists and scientifically minded people in general have often been inclined to say that these "nonscientific" influences upon the acceptance of scientific theories are something which "should not" happen; but since they do happen, it is necessary to understand their status within a logical analysis of science.   We have learned by a great many examples that the general principles of science are not unambiguously determined by the observed facts.   If we add requirements of simplicity and agreement with common sense, the determination becomes narrower, but it does not become unique. We can still require their fitness to support desirable moral and political doctrines.   All these requirements together enter into the determination of a scientific theory.   The firm conviction of most scientists that a theory should be accepted "on scientific grounds" only, forms a philosophy which they absorbed as young students in the period when they started to acquire knowledge about the world; this philosophy claims that a "true" theory gives us a "picture of the physical reality," and that this theory can be found on the basis of observed facts.   If a theory built up exclusively because of its agreement with observed facts told the "truth" about the world, it would be foolish to assume seriously that the acceptance of a scientific theory should be influenced by reasons of simplicity and agreement with common sense, let alone moral, religious, or political grounds.   However, we have learned that "agreement with observed facts" never singles out one individual theory.   There is never only one theory that is in complete agreement with all observed facts, but several theories that are in partial agreement. We have to select the final theory by a compromise. The final theory has to be in fair agreement with observed facts and must also be fairly simple. If we consider this point, it is obvious that such a "final" theory cannot be "The Truth."

However, this metaphysical concept of a true theory as a "replica of physical reality" is not prevalent in the scientific philosophy of today. A theory is now rather regarded as an instrument that serves some definite purpose. It must be helpful in predicting future observable facts on the basis of facts that have been observed in the past and in the present. The theory should also be helpful in the contribution of devices which can save us time and labor. The scientific theory is, in a way, a tool that produces other tools according to a practical scheme. Scientific theories are also accepted, however, because they give us a simple and beautiful world picture and support a philosophy which, in turn, supports a desirable way of life.[10]

The question of which theory "should" be accepted can only be answered if we know whether predictions of facts, construction of devices, beauty, simplicity, or fitness to support moral and political aims "should" be preferred. We can understand this situation if we compare it with the question about the choice of an airplane. In the same way as we enjoy the beauty and elegance of an airplane, we enjoy also the beauty and "elegance" of the theory that makes the construction of the plane possible. If we speak of an individual airplane, it makes no sense to ask whether it is "true" in the sense of being "perfect." We can only ask whether it is "good" or "perfect" for a certain purpose. If we set "speed" as our purpose, the perfect airplane will be different from the one which is perfect for the purpose of "endurance." The criterion of perfection will be different again if we choose "safety" or "fun" or "convenience for reading or sleeping" as our purpose. It is impossible to construct an airplane which fulfills all these purposes in a maximal way; we must achieve some compromise. In order to determine the kind of compromise which "should" be achieved, we must decide which is more important: speed or safety, fun or endurance? The answer to this question can certainly not be derived from physical or engineering science. From the view of "science proper,"[11] the purpose is arbitrary; science can only teach us how to construct a plane if we are given in advance the speed and the degree of safety that should be achieved. However, the desirable relation between speed and safety is dependent upon moral, political, and even religious opinions. The compromise depends upon debatable predilections. The policy-making authorities from the point of view of "science proper" are free to make, according to their predilections, the choice of the plane to be put into production. However, if we look at the situation from the point of view of a unified science which includes both physical and social science, we shall understand how the compromise between speed and safety, between fun and endurance, is determined by social and psychological conditions. If we express ourselves in an oversimplified and rather flippant way, we would say that the compromise is determined by the conditioned reflexes of the policy makers. The conditioning may be achieved, for example, by writing letters to Senators. If we keep to the pragmatic philosophy of science, we would say that the acceptance of a scientific theory is not essentially different from the acceptance of an airplane.

We may ask why a certain scientific theory, *e.g.*, the Copernican theory of planetary motion or Einstein's theory of relativity, has been accepted as true or perfect. According to the previous considerations, this question can only be answered if we first answer the question: What purpose is the theory to serve? Is it only the purely technical purpose of predicting observable facts? Or is it the purpose of obtaining a simple and elegant theory which will allow us to derive a great many facts from simple principles? We have to prefer the theory that fits our purpose. For some groups, the main purpose of scientific theories may be to serve as a support in teaching a desirable way of life, or in discouraging an undesirable one. These groups would accept theories which may give a rough picture of observed facts, provided we can get from them a picture of the world in which man plays the role that they regard as desirable.

If we wish to speak in a brief and rather perfunctory way, we may distinguish between two main purposes of a theory: use for the construction of devices (technological purposes), and use for direct guidance of human conduct. The actual acceptance of theories has always been a compromise between the technological and the sociological value of the theory. Human conduct has been directly influenced by the latter because specific religious and political views were encouraged, while technological influence upon human conduct has been rather indirect—technological changes bring about social changes which manifest themselves in changes in human conduct. Everybody knows of the "industrial revolution" in nineteenth-century England and the accompanying changes in human behavior. Probably the rise of atomic technology in our twentieth century will produce analogous changes in human life.

A great many scientists and educators believe that a conflict between the technological and sociological goals of scientific theories existed at some "dark" periods of history and exists still in "dark" countries, but has largely disappeared with the advance of science, particularly of free science. According to this opinion, it can now be unambiguously determined, by using the "method of science," which theory is the valid one. This opinion is certainly in error if we consider theories and premises of very high generality. In twentieth-century physics we note, for example, that some particular formulation of the principles of quantum theory[12] is accepted or rejected according to whether or not it is believed that the introduction of determinism to physics gives comfort to desirable ethical commitments. Many educators, and even politicians, have been firmly convinced that "free will" is not compatible with Newtonian physics, but very much so with quantum theory. They have been convinced also that it is desirable that the citizen should believe in free will, and they have exerted a certain influence in favor of the indeterministic formulation of subatomic physics. What they have in mind is certainly a sociological purpose of science, whatever the technological purposes may be.

This double role of scientific theories becomes even more obvious in biology. If we investigate the attitude of biologists toward very general questions, we may take as an example the question of whether or not living organisms have developed from inanimate matter. Here we find the conflict between the technological and the sociological purpose of theories in full bloom. Some prominent biologists say that "spontaneous generation" is highly probable (e.g., George Wald,[13] G. G. Simpson[14]), while others claim that according to their computations this probability is almost zero.[15] When we proceed in a strictly scientific way, according to the methods of mathematical physics, we can easily see that we cannot find any reliable value for this probability.   One group believes that the biological theories should uphold human dignity because otherwise moral behavior could not be founded upon science.   This dignity would be impaired if man is descended not only from apes, but even from earth and stone.   Another group, however, believes that the assumption of spontaneous generation would uphold the belief in the unity of nature as a whole and, on this basis, support a moral human behavior.

From all these considerations, we see that the validity of a scientific theory cannot be judged unless we ascribe a certain purpose to that theory.   The achievement of that purpose depends upon the degree to which the different criteria for the acceptance of a theory are satisfied, agreement with observed facts, simplicity and elegance, agreement with common sense, fitness to support desirable human conduct, etc.   Hence, the validity of a theory cannot be judged by "scientific" criteria in the narrower sense: agreement with observations and logical consistency.   After application of all these criteria, there remains often a choice among several theories.   However, if we mean by "science" not only physical science, but also the sciences of human behavior (psychology and sociology), then we can decide which among several physical theories achieves a certain human purpose in the best way.

To sum up: The problem of deciding between different theories of the physical sciences cannot be solved within these sciences if we have to do with theories of high generality.

New lines of research arise for the scientist who wants to achieve a real understanding of his science.   We are guided into the wide field which embraces science as a part of human behavior in general. We may speak of a "sociology of science" or of the "humanistic background of science" if we want to give these new fields a frame of reference in our traditional parlance.

The philosophy of science presented in this book touches this pragmatic aspect in the chapters on the metaphysical interpretation of science (Chapters 7 and 10).

Very frequently science has served by means of its metaphysical interpretations as a direct guide to human conduct.   Science in its technical interpretations has supported mechanical or electrical or nuclear engineering; by its metaphysical interpretations it has served what has been called, occasionally, "human engineering."   If we want to express ourselves as soberly as possible, we may say that "philosophy of science" leads eventually to a research in the "pragmatics of science," which envisages a coherent system containing the physical and biological as well as the sciences of human behavior.

Footnotes

Footnotes for Chapter 1

1. Edgar Allan Poe (1809–1849), American author of poems and tales. He has often been regarded as the man who introduced the "detective story" and "science fiction" into literature.

2. Walt Whitman (1819–1892), American poet. In 1855 he published his main collection of poems, *Leaves of Grass*. He said that the "mother-idea" of his poems is democracy, and that "democracy carries far beyond politics . . . even into philosophy and theology."

3. Charles Sanders Peirce (1839–1914), American philosopher, logician, and scientist. In a paper published in January 1878 in the *Popular Science Monthly*, he presented the principles of pragmatism, and can probably be regarded as the originator of this philosophical school.

4. *The Monist* (1891), a magazine devoted to the philosophy of science.

5. Auguste Comte (1798–1857). His chief book, *Cours de Philosophie Positive*, started in 1830 and finished in 1842, was published in Paris.

6. Auguste Comte, *Positive Philosophy*, freely translated and condensed by Harriet Martineau (New York: E. Blanchard, 1858), Vol. II.

7. Suzanne K. Langer, *Philosophy in a New Key* (Cambridge: Harvard University Press, 1942). Reprinted as a Mentor Book in 1948. The author is a living American philosopher and logician.

8. *Ibid.*, p. 15 in the Mentor Book edition.

9. St. Thomas Aquinas, *Summa Theologica*, Part I, Question XVI, *On Truth*, First Article. The Great Book Foundation published in a paper-bound edition a small part of the *Summa Theologica*, under the title, *St. Thomas, On Truth and Falsity, On Human Knowledge*.

10. A living American philosopher. He has developed original ideas on the philosophy of science, in particular about the interrelation between the philosophy of science and the political philosophy of a certain group.

11. F. S. C. Northrop, *The Meeting of East and West* (New York: The Macmillan Company, 1946), Ch. 10.

12. Aristotle (384–322 B.C.), Greek philosopher. Not long after his death, died the famous conqueror and statesman, Alexander the Great, who had been Aristotle's pupil, and the famous orator Demosthenes, who was Alexander's bitter enemy among the Greeks. Aristotle is, for all time, one of the leading figures in science and philosophy. His role in science has often been misunderstood and minimized. For example, Will Durant writes in his book, *The Story of Philosophy* in 1926, which is

perhaps the best popular presentation of philosophy: "For lack of a telescope Aristotle's astronomy . . . is a tissue of childish romance."

13. Aristotle, *Physics*, from *The Works of Aristotle*, edited by W. D. Ross (London: Oxford University Press, 1908–1952).

14. *Ibid.*

15. Isaac Newton lived and worked around 1700 (1642–1727). He advanced the Newtonian laws of motion, which have been the basis for the understanding of all changes and events in the physical world. There were no radical modifications of these laws until the first decades of the twentieth century.

16. Gregor Mendel (1822–1884) was an Austrian monk who advanced the fundamental laws of genetics (theory of heredity). By these laws we can predict, for example, the results of the crossing of a yellow, round pea with a green, wrinkled pea. This theory (Mendelism) was published in 1866.

17. "Scholastic philosophy" is the philosophy that developed in the Middle Ages when ancient Greek thought, in particular the doctrines of Plato and Aristotle, were modified in order to be made compatible with Christian Revelation. The evolution of this "great tradition" started with St. Augustine (354–430), reached its peak with St. Thomas Aquinas in the thirteenth century, and attained a more skeptical and flexible state in the fourteenth and fifteenth centuries. This "late scholasticism" paved the way for the age of science that started in the sixteenth century.

18. The name "positivism" was originally given to the "positive philosophy" of Auguste Comte (see footnote 5). In a more modern and broad sense, positivism is presented in the book, *Positivism, An Essay in Human Understanding*, by Richard von Mises, translated by Jeremy Bernstein and Rodger Newton (Cambridge: Harvard University Press, 1951). The German original, *Kleines Handbuch des Positivism*, was published in 1938.

19. The doctrine of pragmatism was introduced into the theory of knowledge by Charles S. Peirce in 1878.

20. Hans Reichenbach (1893–1953) was trained as a physicist. He has become one of the most competent and prominent men who have built up the philosophy of science in the first half of the present century. The theory of relativity, quantum theory, and the theory of probability owe to him basic books on their logico-empirical and philosophic foundations. His latest book, *The Rise of Scientific Philosophy* (Berkeley: University of California Press, 1951), contains a brief and almost popular survey of the present state in and the historical background of these fields.

21. Plato (427–347 B.C.) was one of the greatest Greek philosophers. His famous theory of ideas starts from the assumption that behind the surface phenomena that are perceived by our senses there *are* generalizations, regularities, and senses of direction (values). While the surface phenomena are perceived by our senses, the ideas (generalizations, regularities, and values) are conceived by reason and thought. According to Plato's theory, these "ideas" are more "permanent" and more "real" than the particular things that are perceived by the senses.

22. Reichenbach, *op. cit.*, Ch. 2, p. 20.

23. St. Thomas Aquinas lived in the thirteenth century in Italy (1225–1274). On the basis of Aristotelian philosophy, he developed a coherent system which was in accordance with Christian theology. He has become the representative philosopher of the Roman Catholic Church. In contrast to later scholastic philosophers, he attempted to derive the existence and the properties of God by the light of reason, without resorting to divine revelation. He is the greatest representative of "rationalism" in medieval thought.

24. St. Thomas Aquinas, *Summa Theologica*, translated by Fathers of the English Dominican Province (New York: Benziger Brothers, 1947), Part I, Sect. XVI, Second Article, Obj. 1.

25. *Ibid.*, Part I, Question I, *Knowledge of the Divine Persons*, First Article, Reply, Obj. 2. The purpose for which St. Thomas made this distinction was to draw a line between proofs of the existence of God on the one hand, and proofs of the existence of the Trinity on the other. The existence of God, according to St. Thomas, can be derived by human reason, by a logical chain from self-evident principles. But belief in the Trinity can only be shown to have plausible consequences, while its "existence" cannot be proved by reason, but only by Divine revelation.

26. *Ibid.*

27. In the dialogue, *Laws, The Dialogues of Plato*, Book XII, p. 967, translated by Benjamin Jowett (New York: Charles Scribner's Sons, 1871), Plato declares that if the celestial bodies "had been things without soul, and had no mind, they could never have moved with such wonderful numerical exaction." Thus, everyone who teaches the material nature of the sun and stars gives rise to impiety and atheism. In X, 907, 908 of the same dialogue, Plato proposes harsh punishment for impiety.

28. Plato, *Republic*, VII, 527–530.

29. Pierre Duhem (1861–1916), French physicist who became one of the most prominent historians and philosophers of science.

30. Pierre Duhem, *Système du Monde* (Paris: Hermann et fils, 1913), Part I, Ch. II, Sect. XIII, p. 100*ff.*

Footnotes for Chapter 2

1. See Chapter 1, footnote 25.

2. Ohm's law, "the intensity of the electric current in a circuit is proportionate to the resistance in the wire," was advanced in 1827 by the German physicist George S. Ohm (1787–1854).

3. The law says that two mechanical masses exert upon each other a force which is inversely proportionate to the square of their distance. The law was advanced by Sir Isaac Newton in 1685.

4. The differential equations of the electromagnetic field are due to the British physicist, James Clerk Maxwell (1831–1879).

5. See Chapter 1, footnote 16.

6. The corpuscular (also called ballistic) theory assumed that a beam of light consists of small material particles which move according to the laws of mechanics. The wave (or undulatory) theory assumed that the propagation of light is to be regarded as the propagation of waves in a continuous medium, similar to the propagation of sound waves in the air or water waves in the ocean.

7. The best way to get acquainted with Aristotle's original doctrine is to read his book, *On the Heavens* (De Caelo), *The Works of Aristotle*, edited by W. D. Ross (London: Oxford University Press, 1908–1952).

8. Hermann Helmholtz, a prominent German physicist, physiologist, and psychologist, wrote in his famous paper *On the Conservation of Energy* (1847): "The task of physical science is finally to reduce all phenomena of nature to forces of attraction and repulsion. . . . Only if this problem is solved are we sure that nature is conceivable."

9. Alfred North Whitehead (1861–1947), *Science and the Modern World* (New

York: The Macmillan Company, 1925), Ch. I. Used by permission of the publishers. English mathematician and philosopher. In 1924 he crossed the Atlantic to become professor of philosophy at Harvard University.

10. *Ibid.*

11. William James (1842–1910), American psychologist and philosopher; leader of the movement known as pragmatism.

12. William James, Pragmatism Lectures delivered at the Lowell Institute in Boston, 1907. Published in New York and London (1907). In Lecture I, James exemplifies "these two types of mental make-up" by the following contrasts: "rationalistic" (going by principle) and "empiricist" (going by facts); idealistic and materialistic; dogmatic and skeptical; "free-willist" and fatalist.

13. Gustavo Giovannoni, *The Legacy of Rome,* edited by Cyril Bailey (London: Oxford University Press, 1923), p. 433. The author is at the School of Applied Engineering in Rome.

14. Aristotle, *Politics,* from *The Works of Aristotle,* edited by W. D. Ross (London: Oxford University Press, 1908–1952), Book I, 5.

15. *Ibid.*

16. Plutarch (46–120 A.D.), Greek biographer and popular philosopher. He wrote forty-six parallel lives, biographies in pairs, taking a Greek and a Roman together. *Plutarch's Lives,* translated by John Langhorn and William Langhorn (New York: Harper and Brothers, 1846). A selected edition (including "Pericles") appeared as a Mentor Book.

17. Pericles (490–429 B.C.), famous Athenian statesman. He is largely responsible for the splendor of Attic art in his time. He was the patron of Phidias and other great artists.

18. Phidias (about 500–434 B.C.) has been regarded as the greatest of Greek sculptors.

19. Anacreon, Greek lyric poet, born about 560 B.C.

20. Marcellus, Roman general who captured Syracuse in 212 B.C. His biography was written by Plutarch (see footnote 16).

21. Archimedes (287–212 B.C.), Greek mathematician and inventor, born in Syracuse, Sicily.

22. See footnote 20.

23. Plutarch, in his biography of Marcellus, *op. cit.*

24. Whitehead, *op. cit.*

25. *Ibid.*

26. Eccentrics and epicycles were first introduced into Greek astronomy by Apollonius and more precisely elaborated by Hipparchus (around 130 B.C.) and Ptolemy (130 A.D.).

27. Roger Bacon (about 1214–1298) was a contemporary of St. Thomas Aquinas.

28. Roger Bacon, *Opus Magnus,* edited by J. H. Bridges (London: Oxford University Press, 1897), Vol. 2, pp. 169–170.

29. See Introduction, footnote 3.

30. James Bryant Conant, "Scientific Discoveries May Be Disregarded," *Science and Common Sense* (New Haven: Yale University Press, 1951), Sect. 7.

31. François Jean Dominique Arago (1786–1853), French physicist. In 1850, he

presented the idea of a "crucial experiment" that was to decide between the corpuscular and the undulatory theories of light.

32. Albert Einstein, *Annalen der Physik*, 17 (1905).

33. Pierre Duhem, *The Aim and Structure of Physical Theory*, translated by P. Wiener (Princeton: Princeton University Press, 1954), Part II, Ch. V, Sect. 3.

34. Einstein, *op. cit.*

35. Francis Bacon, Baron Verulam (1561–1626), English philosopher and statesman.

36. Francis Bacon, *Descriptio Globi Intellectualis* (written probably 1612). See *The Philosophical Works of Francis Bacon*, edited by Ellis and Spedding (London, 1857).

37. *E.g.*, in the book *Facts and Fiction in Modern Science* (1944) by H. V. Gill, written from the point of view of Thomistic philosophy.

38. William Whewell, in his *History of the Inductive Sciences* (London, 1847), writes in Vol. II, Book VI, Ch. III, Sect. 3 about the reception of the Newtonian theory abroad that "even those whose mathematical attainments most fitted them to appreciate its proofs were prevented by some peculiarity of view from adopting it as a system, like Leibniz, Bernouilli, Huyghens, who all clung to one modification or other of the system of vortices." The chief reason for their attitude was their aversion to the law of inertia. This reason is elaborated in Chapter 7, Section 1. Leibniz presents his points in his *Letters to Samuel Clarke*, 1715–1716, *On Newton's Mathematical Principles of Philosophy* in Leibniz, *Selections*, edited by Philip Wiener (New York: Charles Scribner's Sons, 1951) p. 216*ff*.

39. Isaac Newton in a reply to Leibniz, published in the *Memoirs of Literature* (1712), XVIII.

40. See Chapter 3, Section 10; Chapter 4, Sections 6 and 7; Chapter 13, Section 4.

41. By "semantical rules" relations between symbols are connected with statements which have a meaning in our common-sense language.

42. Scientists who keep to the "stubborn facts" and distrust broad generalizations will regard the theory of relativity or the Darwinian theory in biology as "unscientific." Since every human being has his "tender spots" (see footnote 12), these scientists will obtain satisfaction by keeping to the generalizations which they have imbibed in their childhood.

43. Rene Descartes, *Principles of Philosophy*, original edition in Latin, 1644, French translation, 1647.

44. See Chapter 7, Sections 1 and 4.

45. *Ibid.*

46. See footnote 21.

47. Ernst Mach (1838–1916), Austrian physicist, psychologist, and philosopher. The discussion of Archimedes' theory of the lever is taken from Mach's *Science of Mechanics* (1883).

48. Thales (about 600 B.C.) regarded water as the primary stuff, but Anaximenes (about 550 B.C.) chose air, and Heraclitus (about 500 B.C.) fire.

49. The role of induction in science is elaborated in Chapters 13 and 14.

50. If we investigate precisely how new general principles of science may be found, it becomes clear that a principle like the law of inertia or the principle of relativity cannot be invented by any formal method (deductive or inductive), but only by

using a certain amount of inventive power, also called "imagination," or, occasionally, "intuition." This was strongly stressed by Einstein in his Spencer Lecture (see footnote 54).

51. Herbert Dingle is a prominent British astrophysicist and philosopher of science. He has organized a program of study in history and philosophy of science at the University College in London. This course has become an example for a good many other institutions.

52. Herbert Dingle, "The Nature of Scientific Philosophy," *Proceedings of the Royal Society of Edinburgh* (1949), *62*, Part IV, p. 409.

53. *Ibid.*

54. Albert Einstein, *On the Methods of Theoretical Physics*, Herbert Spencer Lecture, given at Oxford 1933, reprinted in *The World as I See It*, translated by Alan Harris (Toronto: George McLeod, Ltd., 1934).

55. Herbert Dingle, *op. cit.*, p. 403.

56. Edouard Le Roy (1870–   ), French philosopher, "Science et Philosophie," *Revue de Metaphysique et du Monde* (1899), I, 375*ff*.

57. Philipp Frank, "Metaphysical Interpretations of Science," Section 4, "Science and Common Sense," *The British Journal for the Philosophy of Science*, Vol. I.

Footnotes for Chapter 3

1. In his paper "The Architecture of Theories," *The Monist* (1891).

2. Plato, *Republic*, Book VI, *The Dialogues of Plato*, translated by Benjamin Jowett (New York: Charles Scribner's Sons, 1871), 525B*ff*. There is a translation of the *Republic* in the Mentor edition. Plato stressed the point that the study of geometry and mathematics is the indispensable preliminary for the approach to philosophy.

3. Peirce, *op. cit.*

4. *Ibid.*

5. René Descartes, *Discourse on Method*, translated by John Veitch (Chicago: Henry Regnery Company, 1949), Part I, pp. 17, 18.

6. *Ibid.*

7. *Ibid.*

8. *Ibid.*

9. Blaise Pascal (1623–1662), French scientist and philosopher. His interpretation of science attempted to point out that the whole personality is involved in this activity. "The heart," he wrote, "has its reasons which reason does not know."

10. Blaise Pascal, "The Difference Between the Mathematical and the Intuitive Mind," *Pensées (Thoughts)*, translated by W. S. Trotter (New York: Modern Library, 1941), Sect. 1, p. 1.

11. According to A. C. Ewing, *Idealism: A Critical Survey* (London: Methuen & Co., 1933), Kant is called an "idealist" because he treated physical objects in a characteristically idealist fashion, reducing them to elements in human experience and leaving to the realist only the unknowable thing-in-itself.

12. He has been called a "rationalist" because he believed that true general statements about the material world can be found by the power of reason without sense experience. In this respect, his views were similar to the views of scholastic philosophers of the type of St. Thomas Aquinas.

13. A judgment is "synthetical" if it cannot be proved by mere logic. It is "*a priori*" if its truth can be demonstrated without sense observations. According to Kant, mathematical judgments (such as $7 + 5 = 12$) are "synthetical" *and* "*a priori*."

14. Immanuel Kant, *Prolegomena to Any Future Metaphysics*, edited in English by Paul Carus (Chicago: Open Court Publishing Company, 1902), Sect. 4.

15. Wooster Bemann and David Smith, *New Plane and Solid Geometry* (Boston: Ginn and Company, 1899).

16. Descartes presented his views on the foundations of geometry first in his *Rules for the Direction of the Mind*, published in 1701 after his death, Rule IV, Rule XIV. In 1644 he published his *Principles of Philosophy*. In Principles 197 through 200 he interprets all phenomena of nature by shape, magnitude, and motion.

17. John Stuart Mill (1806–1873), British philosopher and economist.

18. John Stuart Mill, *A System of Logic* (1843), Book II, Ch. V.

19. Kant, *op. cit.*

20. Euclid, a Greek mathematician, was younger than Plato and older than Archimedes. His main work, *The Elements*, in which he presented his deductive system of geometry was written about 325 B.C.

21. Non-Euclidean geometry was advanced by Nikolai I. Lobatchevski and Wolfgang Bolyai, who were in turn stimulated by Karl Friedrich Gauss.

22. Nikolai Ivanovich Lobatchevski (1793–1856), Russian mathematician.

23. Wolfgang Bolyai (1775–1856), Hungarian mathematician.

24. Rudolph Carnap, living American logician and philosopher, *Formalization of Logic* (Cambridge: Harvard University Press, 1943), Introduction.

25. William Kingdon Clifford (1845–1879), English mathematician and philosopher. His most important book, *The Common Sense of the Exact Sciences*, was first published in 1875. A new edition was published in New York in 1946 by Alfred A. Knopf, Inc.

26. David Hilbert (1862–1943), German mathematician. His *Foundations of Geometry* (Leipzig, 1899), inaugurated modern axiomatic method.

27. Henri Poincaré (1854–1912), French mathematician, astronomer, and philosopher.

28. Francis Bacon, *Novum Organum*, I, 14 and I, 11.

29. Louis Rougier, living French philosopher, *La Philosophie Geometrique de Henri Poincaré* (Paris: F. Alcan, 1920).

30. Bernhard Riemann (1826–1866), German mathematician, *On the Hypothesis upon Which Geometry Is Founded* (1867).

31. Hermann Helmholtz (1821–1894), German physicist, physiologist, and philosopher.

32. Emile Borel (1871–   ), French mathematician and statesman.

33. Giuseppe Veronese (1854–1917), Italian mathematician. He published *Elements of Geometry* (Verona, 1897).

34. Frederigo Enriques and Umberto Amaldi, *Elementi di Geometria* (Bologna: Zanichelli, 1905). Enriques was a mathematician who was also a prominent author in the history and philosophy of science.

35. Albert Einstein (1879–1955). His address "Geometry and Experience" was first given to the Prussian Academy (Berlin) in 1921. An English translation appeared in *Sidelights on Relativity*, translated by G. B. Jeffery and W. Perrett (New York: E. P. Dutton & Co., Inc., 1921), and (London: Methuen and Co., Ltd.).

36. Moritz Pasch (1843–1931), German mathematician.

37. See footnote 27.

38. Einstein, *op. cit.*

39. P. W. Bridgman, *Logic of Modern Physics* (New York: The Macmillan Company, 1927).

40. Rougier, *op. cit.*

41. Hermann Helmholtz, *Popular Lectures on Non-scientific Subjects*, translated by E. Atkinson (London: Longmans, Green & Company, 1873).

Footnotes for Chapter 4

1. Herbert Butterfield, *The Origins of Modern Science, 1300–1800* (London: George Bell & Sons, Ltd., 1950).

2. See Chapter 1, footnote 23.

3. Dante Alighieri (1265–1321), the greatest of Italian poets. In his main work, *The Divine Comedy*, he describes his travels through Hell, Purgatory, and Paradise. Whereas he is guided through the two lower realms by the poet, Virgil, he is received at the border of Paradise by Beatrice, a girl whom he had met when they were both in their ninth year. His love for her remained always in the realm of imagination; it is described in his *Vita Nuova (New Life)*. Dante was deeply interested in scholastic philosophy, and had studied the *Summa Theologica* of St. Thomas Aquinas.

4. Werner Jaeger, *Aristotle*, translated by Richard Robinson (Oxford: Clarendon Press, 1950).

5. St. Thomas Aquinas, *Summa Contra Gentiles*, translated by Antoni Pegis (Garden City: Image Books, 1955), Vol. I, Ch. XIII.

6. Moses Maimonides (1135–1204), Jewish scholastic philosopher. His main book is *The Guide of the Perplexed* (1194), translated by Leon Roth (London: Hutchinson's Home University Library, 1948).

7. Aristotle, *On the Movements of Animals*, from *The Works of Aristotle*, translated by W. D. Ross (London: Oxford University Press, 1908–1952), Vol. V. In order to understand well the organismic theory of motion, it is particularly instructive to study this small book because it starts from the motion of organisms in order to gain an understanding of the motions of celestial bodies.

8. Aristotle, *Physics, ibid.*, Vol. II, Book VIII.

9. See Chapter 2, footnote 39.

10. David Hume (1711–1776), British philosopher, historian, and political economist. The passages on "volition" and "causation" are from his *An Inquiry Concerning Human Understanding* (Chicago: Open Court Publishing Company, 1949). This book was originally published under the title, *Philosophical Essays*.

11. Auguste Comte, *Positive Philosophy*, translated by Harriett Martineau (London: George Bell & Sons, Ltd., 1896), Book III (Chemistry), Ch. I.

12. Niels Bohr (1885–   ), Danish physicist. He advanced his "model of the atom" in 1917. This work, together with Einstein's and Planck's ideas on quantum theory, opened a new chapter in the history of science. Bohr's work is elaborately discussed in Chapters 8, 9, and 10.

13. Nicolaus Copernicus (1473–1543). His main book, *Revolutions of the Celestial*

Bodies, was finished about 1530, but published after his death. He made use of "organismic" arguments, *e.g.*, it is a "dignified" assumption that the most brilliant body, the sun, is in the center of the Universe, just as the largest lamp is placed in the center of the room.

14. Galilei, Galileo (1564–1642). In his book, *Dialogues Concerning Two New Sciences*, the law of inertia was formulated in the introduction to the fourth Dialogue in the following way: "Imagine any particle projected along a horizontal plane without friction. Then we know . . . that this particle will move along this same plane with a motion which is uniform and perpetual, provided the plane has no limits."

15. In Newton's *Mathematical Principles of Natural Philosophy* (1867), the law of inertia is formulated as the first law of motion. Every body continues in its state of rest, or of uniform motion in a straight line, except insofar as it is compelled by impressed force to change its state.

16. Giordano Bruno (1548–1600), Italian philosopher. He rejected Aristotelian astronomy for that of Copernicus, which allowed for the possibility of innumerable worlds. He was inclined toward pantheism and was burned at the stake in 1600.

17. In addition to the first law (quoted in footnote 15), Newton formulated the second law: The rate of change of momentum is proportional to the impressed force.

18. See Chapter 3, footnote 27.

19. Ludwig Wittgenstein (1899–1950), Austrian philosopher. He spent a great part of his life in England and was, from 1939–1947, professor of philosophy at the University of Cambridge. His *Tractatus Logico-Philosophicus* (1921) was published in a bilingual edition (German and English) (New York: Harcourt, Brace & Company, Inc., 1933). It has greatly influenced the "Vienna Circle."

20. Lucretius (98–55 B.C.), Roman poet and philosopher. His greatest didactic epic, *On the Nature of Things*, presents the philosophy of Epicurus in verses of great beauty, translated by W. H. D. Rouse (Cambridge: Harvard University Press, 1941).

21. Ernst Mach, *Mechanics and Its Evolution*, titled in its English translation, *The Science of Mechanics* (Chicago: Open Court Publishing Company, 1893). His criticism of Newton's principles is developed in Chapter 2, Sections 6 and 7. The first German edition appeared in 1883.

22. Sir Joseph John Thomson (1856–1940), British physicist. He published *Recent Researches in Electricity and Magnetism* in 1893. The mass of electrically charged particles is derived in this book.

23. See footnote 3.

24. Sir Francis Bacon, *Advancement of Learning* (1605) (London: Macmillan & Co., Ltd., 1917).

25. Aristotle, *Physics, op. cit.*

26. David Gregory (1661–1708), "David Gregory, Isaac Newton, and Their Circle," *Extracts from D. Gregory's Handbook* (London: Oxford University Press, 1937).

27. Isaac Newton, *Mathematical Principles of Natural Philosophy*.

28. Lucretius, *op. cit.*

29. Ernst Mach, *op. cit.*

Footnotes for Chapter 5

1. Aristotle, *Physics*, from *The Works of Aristotle*, translated by W. D. Ross (London: Oxford University Press, 1908–1952), Vol. II.

2. St. Augustine (354–430) was born a pagan and baptized in 387. His *Confessions*, an autobiography stressing his intellectual and emotional development, was published about 400 A.D. It has been translated from the Latin original into all literary languages, and has become a classic in theology and psychology, widely read all over the world. The edition quoted was translated by Edward B. Pusey (Mount Vernon, N. Y.: Peter Paul Press).

3. Philipp Frank, *Einstein, His Life and Times* (New York: Alfred A. Knopf, Inc., 1947), Ch. VIII, Sect. 5, p. 178.

4. Chapter 2, Section 6.

5. In his *History of Modern Philosophy* (London: Macmillan & Co., Ltd., 1900), Harold Hoeffiling writes: "Space is for Newton not an empty form, but the organ by means of which God works as omnipresent in the world and, at the same time, immediately perceives the conditions of things. It is an 'unlimited and homogeneous sensorium.'"

6. Leon Foucault (1819–1868), French physicist, *Comptes Rendus de l'Academie* (1850), Vol. 30.

7. The deflection of the starlight by annual aberration was discovered in 1725 and explained in 1727 by James Bradley (1693–1762).

8. In 1879, which was, by a singular coincidence, the year of Einstein's birth.

9. Albert Abraham Michelson (1832–1931), American physicist, born in Germany, was graduated from the United States Naval Academy in 1873.

10. Heinrich Hertz, (1857–1894), German physicist.

11. See Chapter 4, footnote 22.

12. Hendrik Antoon Lorentz (1853–1928), Dutch physicist.

13. Vladimir Ilyich Ulyanov Lenin (1870–1924), *Materialism and Empirocriticism: Critical Observations on a Reactionary Philosophy* (1909).

14. Albert Einstein (1879–1955), German physicist who lived and worked from 1933 until his death in the United States.

15. A typical example of the attempts to draw a clear dividing line between "natural science" and "philosophy of nature" is the book, *Philosophy of Nature* by Jaques Maritain (New York: Philosophical Library, 1951).

16. See Chapter 4, footnote 22.

17. See footnote 12.

18. Max Abraham (1875–1922), German physicist.

Footnotes for Chapter 6

1. See Chapter 5, Sections 6 and 7.

2. *Ibid.*

3. Ernst Mach, *The History and Root of the Principle of the Conservation of Work* (Prague, 1872). Later this analyisis is elaborated in his book, *Mechanics and Its Evolution* (first German edition, 1883). English translation, *The Science of Mechanics* (Chicago: Open Court Publishing Company, 1893).

4. Isaac Newton, *Opticks*, Book III, Part I.

5. Max Planck (1858–1947), German physicist. He was the first (1900) to advance the hypothesis that there is a minimum quantity of energy, the "quantum hypothesis." See *Annalen der Physik* (1901), Vol. 4, p. 553.

6. Planck published his argument in a paper "On Mach's Theory of Physical Knowledge," *Physikalische Zeitschrift* (1910), Vol. 11, p. 1186*ff*.

7. "On the Influence of Gravitation upon the Propagation of Light," *Annalen der Physik* (1911), Vol. 35, p. 898*ff*.

8. In 1915, Einstein published in the *Berichte der Preussischer Academie* his definitive general theory of relativity. He recognized clearly that Mach's forces emanating from the rotating system of fixed stars are actually gravitational forces.

9. Gilbert Ryle, *The Conception of the Mind* (New York: Barnes & Noble, Inc., 1949).

10. See Chapter 1, Section 6.

11. Joseph Louis Lagrange (1736–1813), French mathematician. His fundamental treatise, *Analytical Mechanics*, was produced in Berlin between 1776 and 1796.

12. Hermann Minkowski (1864–1909), German mathematician.

Footnotes for Chapter 7

1. Aristotle, *Physics*, from *The Works of Aristotle*, translated by W. D. Ross (London: Oxford University Press, 1908–1952), Vol. II.

2. St. Thomas Aquinas, *Summa Theologica*, translated by Fathers of the English Dominican Province (New York: Benziger Brothers, 1947), Part I, First Article, Reply, Obj. 2.

3. See Chapter 3, footnote 11.

4. In his book *Metaphysische Anfangs Gründe der Naturwissenschaft* (1787) (*Metaphysical Principles of Natural Science*), Part III, "Metaphysical Principles of Mechanics."

5. See Chapters 9 and 10.

6. Kant, *op. cit.*

7. James Clerk Maxwell, *Matter and Motion* (London, 1877), Ch. III, Article 27.

8. See Chapters 4 and 5.

9. Ernst Mach, *The Science of Mechanics* (Chicago: Open Court Publishing Company, 1893).

10. Herbert Spencer, *Synthetic Philosophy*, Vol. II, *First Principles*, Part II, Ch. IV.

11. *Ibid.*, pp. 352–354.

12. See Chapter 5, Section 8.

13. See Chapters 5 and 6.

14. Edmond Ware Sinnott, *Two Roads to Truth: A Basis for Unity under the Great Tradition* (New York: Viking Press, Inc., 1953).

15. Pitirim Sorokin, *Social and Cultural Dynamics* (New York: American Book Company, 1933).

16. Henri Bergson, *Time and Free Will* (London, 1910). (The French original appeared in 1888.)

17. In his article on *Relativity* in the *Encyclopaedia Britannica*.

18. Lincoln Barnett, *The Universe of Dr. Einstein* (London: George J. McLeod, Ltd., 1948). This excellent book has been reprinted as a "pocket edition."

19. See Chapter 5, Section 8.

20. Mary Baker Eddy (1821–1910), founder of Christian Science, *Science and Health* (Boston: Armstrong, 1897).

21. Herbert Wildon Carr, *A Theory of Monades, Outlines of the Philosophy of the Principle of Relativity* (Leiden, 1922).

22. See Chapter 5.

23. In its article on materialism in the 1936 edition.

24. In the first, *i.e.*, pre-war, edition.

25. In an article, "The Theory of Relativity as a Source of Philosophical Idealism," *Under the Banner of Marxism* (1938).

26. See Chapters 9 and 12.

27. *Ibid.*

28. Delivered in a camp of the National Socialist (Nazi) Student Association, 1936.

29. In "The New Physics and Dialectical Materialism," *Under the Banner of Marxism* (1938).

30. As presented in Chapters 5 and 6.

31. Hermann Minkowski (1864–1909), German mathematician.

32. In *Science and the Modern World* (New York: The Macmillan Company, 1925), Ch. VIII. Used by permission of the publishers.

33. See Chapter 5, Section 8.

34. In his Herbert Spencer lecture, given at Oxford in 1933, printed in *On the Method of Theoretical Physics* (Oxford: Clarendon Press, 1933).

35. The role of operational theory in Einstein's Relativity is elaborately discussed in Chapter 5, Section 6.

36. In an article in the British journal, *Nature* (1938).

37. See Chapter 5, Section 5.

38. Ludwig Klages (1872–   ), German psychologist and philosopher. His most widely known philosophical book was *Der Geist als Widersacher der Seele* (*The Mind as the Adversary of the Soul*) (1929).

Footnotes for Chapter 8

1. Isaac Newton, *Opticks*, First Book, Part I, Definition I. First edition published in 1704, but written in 1675.

2. *Ibid.*, Second Book, Part III, Proposition XII, and Definition.

3. *Ibid.*, Queries 27, 28, 29.

4. *Ibid.*, Query 29.

5. Christian Huyghens (1629–1695), Dutch Physicist. His wave theory of light was first published in 1690 in his *Traité de la Lumière*, written in 1678. Translated by Silvanus P. Thompson (London: Macmillan & Co., Ltd., 1912).

6. Newton, *op. cit.*, Query 28.

7. Quoted from Foucault, see footnote 8.

8. *Recueil des Travaux Scientifiques de Léon Foucault*, Volume I (Paris: Gauthier-Villars, 1878).

9. Thomas Young (1773–1829), British scientist. He was one of the first successful workers on the deciphering of Egyptian hieroglyphic inscriptions. In his

Bakerian Lecture in 1801 on *The Theory of Light and Colours*, he developed the theory that light consists in undulations of a medium, rare and elastic in high degree, that pervades the universe.

10. Jean Fresnel (1788–1827), French physicist. He gave to Young's hypothesis its final mathematical form.

11. Pierre Duhem, *La Théorie Physique; Son Objet et Sa Structure* (Paris, 1906). Translated by Philip P. Wiener (Princeton: Princeton University Press, 1954).

12. In 1864 Maxwell derived from his equations of the electromagnetic field that given an oscillating circuit, electromagnetic waves will be dissipated into the surrounding space, as a candle sends out light energy.

13. Maxwell's prediction (footnote 12) was not confirmed until twenty years later by Heinrich Hertz.

14. Philipp Lenard (1866–1947), German physicist. By his study of the photoelectric effect [*Annalen der Physik*, Vol. 8 (1902)] he stimulated Einstein to advance the hypothesis of "photons" (light quanta). After the First World War, Lenard adopted an ultranationalistic and "Nazi" world view. He gave to physical laws philosophical interpretations which were to support his political, moral, and religious creed. Lenard is an outstanding example of how a great physicist can believe in these interpretations if they serve his political goals.

15. Einstein advanced his interpretation of Lenard's experiments in *Annalen der Physik*, Vol. 17 (1905).

16. See footnote 14.

17. See footnote 15.

18. Bohr's model of an atom was advanced first in three papers in *Philosophical Magazine*, Vol. 26 (1913).

19. Louis de Broglie, *Annalen der Physik*, Vol. 3 (1925).

Footnotes for Chapter 9

1. The "Uncertainty Principle" was advanced by Werner Heisenberg, *Zeitschrift für Physik*, Vol. 43 (1927). See also his *Principles of Quantum Theory*, translated by Carl Eckart and Frank C. Hoyt (Chicago: University of Chicago Press, 1930).

2. Chapter 8, Section 5.

3. Bohr's argument presented in this chapter is in the main taken from his paper "Discussions with Einstein on Epistemological Problems in Atomic Physics," in Schilpp, Paul A., editor, *Albert Einstein: Philosopher-Scientist* (New York: The Library of Living Philosophers, Inc., 1949, 1951), Vol. VII, pp. 209, 210.

4. Niels Bohr, *Nature*, Vol. 121 (1928).

5. As we saw in Section 1.

6. See footnote 3.

7. *Ibid.*

8. In his book, *The Nature of Thermodynamics* (Cambridge: Harvard University Press, 1941), Bridgman speaks about the verbal compulsion to use such sentences as "energy has entered" as if "energy" were a moving thing.

9. Hans Reichenbach (1891–1951), *Philosophic Foundations of Quantum Theory* (Berkeley: University of California Press, 1944).

10. Alfred Landé, German physicist now in the United States. Published *Quantum Theory* (New Haven: Yale University Press, 1955).

11. The Warsaw Meeting (1938) was arranged by the International Institute of Intellectual Cooperation of the League of Nations. Its proceedings were initially published in Paris in 1938. English translation, *New Theories of Physics* (New York: Columbia University Press, 1938).

Footnotes for Chapter 10

1. In his play *Too True to be Good.*

2. See Chapter 4, Sections 2 and 3.

3. Bernard Bavink, "The Natural Sciences in the Third Reich," *Unsere Welt,* Vol. 25 (1933), 225.

4. Jan Christian Smuts (1870–1950), South African statesman with great interest in philosophy, *Holism and Evolution* (New York: The Macmillan Co., 1926).

5. This view was presented in all solemnity by General Smuts in his opening address at the celebration of the centenary of the British Association for the Advancement of Science in 1931, printed in *Nature* (1931), p. 521*ff.*

6. Sir James Jeans, *The Mysterious Universe* (New York: The Macmillan Co., 1930), New Revised Edition, 1948, p. 186.

7. Aloys Wenzel (1887–   ), German philosopher, *Metaphysics of Contemporary Physics* (Leipzig: Felix Meiner, 1935).

8. Bernard Bavink, *Science and God,* translated by H. Stafford Hatfield (New York: The Century Co., 1933).

9. See Chapters 8 and 9.

10. London (1933) Ch. VI, 2, and Ch. VI, 4.

11. Advanced by Werner Heisenberg in 1927.

12. See Chapter 7, Section 2.

13. See footnote 6, Sir James Jeans, *Physics and Philosophy* (Cambridge: The Cambridge University Press, 1943), Ch. VII, p. 216.

14. Erwin D. Canham, "The Twilight of Materialism," *The Christian Science Monitor* (February 11, 1950).

15. See Chapters 8 and 9.

16. See Chapter 9, Section 3.

17. Alfred Landé, *Quantum Mechanics* (New York: The Macmillan Co., 1951).

18. Henry Margenau, *The Nature of Physical Reality: A Philosophy of Modern Physics* (New York: McGraw-Hill Book Co., Inc., 1950).

19. See Chapter 15, Section 1.

20. William Henry Werkmeister, *The Basis and Structure of Knowledge* (New York: Harper and Brothers, 1948).

21. Niels Bohr, "Discussions with Einstein on Epistemological Problems in Atomic Physics," in Schilpp, Paul A., editor, *Albert Einstein: Philosopher-Scientist* (New York: The Library of Living Philosophers, Inc., 1949, 1951), Vol. VII, pp. 209, 210.

22. Margenau, *op. cit.*

23. See Chapter 8, Sections 4 and 5.

24. Sir Arthur Eddington, *The Nature of the Physical World* (New York: The Macmillan Co., 1928), p. 350.

25. Sir Arthur Eddington, *The Philosophy of Physical Science* (Cambridge: The Cambridge University Press, 1949).

26. Bishop Fulton J. Sheen, *Philosophy of Religion, the Impact of Modern Knowledge on Religion* (New York: Appleton-Century-Crofts, Inc., 1948), p. 148.

27. Nalimi Kanta Brahma, *Causality and Science* (London: George Allen & Unwin, Ltd., 1939).

28. Auguste Valensin, "Du Libre Arbitre," *Etudes Philosophiques* (Paris: Presses Universitaires, 1953), p. 16*ff.*

29. Benedict Spinoza (1632–1677), Jewish-Dutch philosopher. His *Ethics* was finished in 1666, but published later. Part II, prop. 48. Translated by R. H. M. Elwes (London: G. Bell & Sons, 1883–1884).

30. Reprinted from *New Introductory Lectures on Psychoanalysis*, by Sigmund Freud, translated by W. J. H. Sprott, by permission of W. W. Norton & Company, Inc., and The Hogarth Press, Ltd. Copyright, 1933, by Sigmund Freud.

31. Thomas Merton, *Seeds of Contemplation* (Our Lady of Gethsemane Monastery, 1949), reprinted in paper cover by Dell Company.

Footnotes for Chapter 11

1. Bertrand Russell, in his paper "On the Notion of Cause," in the book, *Mysticism and Logic* (New York; W. W. Norton & Company, Inc., 1929).

2. Hans Kelsen (1881–   ), Austrian lawyer and philosopher of law now living in the United States. He published *Society and Nature*, a sociological inquiry (Chicago: University of Chicago Press, 1943).

3. Russell, *op. cit.*

4. Pierre Simon Laplace (1749–1827). The first edition of his theory of probability, *Théorie Analytique des Probabilités*, appeared in 1812. His *Essai Philosophique* (1814) is a more popular exposition of the same subject. In 1796 he published *Exposition du Système du Monde*, a popular exposition of his celestial mechanics and astronomical history. *Oeuvres Completes* (Paris: Gauthier-Villars, 1878–1912).

5. Napoleon I (1769–1821), Emperor of France. This episode is discussed in *Naturalism and Agnosticism*, by James Ward (London: Macmillan & Co., Ltd., 1906), Lecture I.

6. Philipp Frank, "Causality and Experience" ("Kausalgesetz und Erfahrung") *Annalen der Naturphilosophie*, vol. 6 (1907). English translation in Philipp Frank, *Modern Science and Its Philosophy* (Cambridge: Harvard University Press, 1950).

7. Philipp Frank, *The Law of Causality and Its Limits* (*Das Kausalgesetz und Seine Grenzen* (Vienna: Julius Springer, 1932).

8. *Ibid.*

Footnotes for Chapter 12

1. Philipp Frank, *Modern Science and Its Philosophy* (Cambridge: Harvard University Press, 1949), Ch. I.

2. Philipp Frank, *Das Kausalgesetz und Seine Grenzen* (*The Law of Causality and Its Limits*) (Vienna· Julius Springer, 1932).

3. David Hume, *Enquiry Concerning Human Understanding*, Section VII, Part II.

4. Immanuel Kant, *Prolegomena to Any Future Metaphysics*, Edited into English by Paul Carus (Chicago: Open Court Publishing Company, 1902), p. 46.

5. Heinrich Rickert, *Die Grenzen der Naturwissenschaftlichen Begriffsbildung*, (*Limitations of the Conceptions of Natural Science*), 2 volumes (1896–1902).

6. See Chapter 3, Section 5.

7. The word "ergodic" is a composition of the Greek work *ergos* (work) and *hodos* (path). The "ergodic theorem" refers to a surface of constant energy (ergodic surface) upon which the path of the system is situated.

Footnotes for Chapter 13

1. John Stuart Mill (1806–1873), British philosopher and economist. His *System of Logic* appeared in 1843.

2. William Whewell (1794–1866), British philosopher and historian of science. He published his *History of Inductive Sciences* (1837), and *Philosophy of Inductive Sciences* (1840).

3. See Chapter 1, Section 7, and Chapter 2, Section 2.

4. Francis Bacon, Baron Verulam (1561–1628). His book, *Novum Organum Scientiorum* (*New Method of Scientific Discovery*) appeared in 1620.

5. In his *Novum Organum*, Aphorism XIX, introduction and notes by Thomas Fowler (Oxford: The Clarendon Press, 1889).

6. *Ibid.*, Aphorism XX.

7. See Chapters 1, 3, 4, 5, 6, 7.

8. Curt John Ducasse (1881–   ), French philosopher, teaching at Brown University. His book, *Philosophy as a Science*, appeared in 1941.

9. In his paper "On Whewell's Philosophy," *Philosophical Review*, LX (1951).

10. The most adequate impression can be obtained by reading Whewell's small book, *Of Induction. With Especial Reference to Mr. J. S. Mill's System of Logic* (London: 1849).

11. Johann Kepler (1571–1630), German astronomer and mathematician. Kepler's law, which is discussed by Whewell and Mill in this section, says: "Every planet describes an ellipse, the sun being in one focus."

12. See Chapter 3.

13. See Chapter 7, footnote 7.

14. In his paper, "On Faraday's Lines of Force" (1855).

15. See Chapters 5 and 6.

16. See Chapter 3, Section 5.

17. See Chapter 5, Section 10.

18. In his book, *The Nature of Thermodynamics* (Cambridge: Harvard University Press, 1943).

19. *Ibid.*, Ch. I, p. 67.

20. Hans Reichenbach, *The Rise of Scientific Philosophy* (Berkeley: University of California Press, 1951), p. 230.

21. See Introduction, footnote 10.

22. *Ibid.*

23. John Herschel, *Discourse on the Study of Natural Philosophy* (1831).

24. In his book, *Erkenntnis und Irrtum* (*Knowledge and Error*), Third Edition, p. 318*ff*.

25. In the Swedish Journal, *Theoria*, Vol. XIX (1953).

26. Hans Larsson, *Intuition* (Stockholm: A. Bonnier, 1920).

27. Reichenbach, *op. cit.*

28. A new logical analysis of induction was advanced in the small book by Nelson Goodman, *Fact, Fiction, and Forecast* (Cambridge: Harvard University Press, 1955).

Footnotes for Chapter 14

1. Hans Reichenbach, *Erkenntis*, Vol. V., p. 277*ff*.

2. Richard von Mises, *Positivism, An Essay in Human Understanding*, translated by Jeremy Bernstein and Roger Newton (Cambridge: Harvard University Press, 1951), p. 173.

3. In a paper, "On the Probability of Hypotheses," *Journal of Unified Science*, Vol. VIII (1938), p. 151*ff*.

4. Rudolph Carnap, "Inductive and Deductive Logic," *Logical Foundations of Probability* (Chicago: University of Chicago Press, 1950), p. 200.   Copyright, 1950, by the University of Chicago Press.

5. *Ibid.*, "The Usefulness of Inductive Logic" Ch. IV, p. 349.

6. See Chapter 3, Section 6.

7. Carnap, *op. cit.*, Ch. 4, p. 253*ff*.

8. See Chapters 3 and 4.

9. Carnap, *op. cit.*, Ch. 4, p. 254.

10. Richard von Mises, "Grundlagen der Wahrscheinlichkeitsrechnung," *Mathematische Zeitschrift*, Vol. 5 (1919).

11. Carnap, *op. cit.*, Ch. 2, Sect. 10, p. 32.

12. *Ibid.*

13. *Ibid.*, p. 30.

14. Richard von Mises, "Grundlagen der Wahrscheinlichkeitsrechnung," *Mathematische Zeitschrift*, Vol. 5, (1919).

15. Carnap, *op. cit.*

16. *Ibid.*

17. Sir John Maynard Keynes (1883–1946), English economist.   The great role of statistics in economics led him into mathematics, where he was especially interested in the philosophical foundations of probability.   His *Treatise on Probability* appeared in 1921.

18. Harold Jeffreys (1891–   ), British astronomer.   His first paper on probability appeared in the *Philosophical Magazine* (1919 and 1920).   He wrote *Scientific Inference* (Cambridge: Cambridge University Press, 1931), and *Theory of Probability* (Oxford: Oxford University Press, 1939).

19. "The 'subjective' or 'logical' theory of probability, preferred by school philosophy, seeks in vain for a basis of probability measurement that would be different from the frequency of the occurrence of the event in question."   Richard von Mises, *Positivism, An Essay on Human Understanding*, p. 166.

20. Carnap, *op. cit.*

21. *Ibid.*, Ch. 4, p. 235.

22. *Ibid.*

23. David Hume, *Enquiries Concerning Human Understanding,* Sect. IV, Part I.

24. Hans Reichenbach, *The Rise of Scientific Philosophy,* (Berkeley: University of California Press, 1951), Ch. XIV, p. 239.

25. Jacob Bronowski (1908–   ), British scientist, philosopher, poet, and literary critic. His book, *The Common Sense of Science* (Cambridge: Harvard University Press, 1951) presents a lively and lucid approach to contemporary philosophy of science in a broad, historical context.

26. "The Logic of Experiment," *Nature,* Vol. 171 (1953).

Footnotes for Chapter 15

1. Arthur Compton, American physicist. In 1922 he discovered the Compton effect, *Physical Review,* Vol. 21 (1923), p. 715*ff.*

2. Erwin Schroedinger, Austrian physicist. His first paper on wave mechanics appeared in *Annalen der Physik,* Vol. 79 (1926). Translated by James F. Shearer and W. M. Deans, *Collected Papers on Wave Mechanics* (London and Glasgow: Blackie & Son, 1928).

3. In his *Discussions with Einstein.* See Chapter 9, footnote 3.

4. Vladimir Lenin, *Materialism and Empirio-Criticism* (New York: International Publishers, 1927).

5. Charles Sanders Peirce suggested in a fragment written in 1897 that the logical structure of science cannot be presented in dyadic relations because it is based upon irreducible "triadic" relations. *Collected Papers of C. S. Peirce* (Cambridge: Harvard University Press, 1932), Vol. II, Ch. 2.

6. According to Peirce, for example, "a sign is something which stands to somebody for something in some respect or capacity," *Ibid.,* Vol. II, Ch. 2, Sect. 228.

7. Rudolf Carnap, *Foundations of Logic and Mathematics,* and Charles Morris, *Foundations of the Theory of Signs,* both found in the *International Encyclopedia of Unified Science* (Chicago: University of Chicago Press, 1955), Vol. I, Part I.

8. See Chapter 2, Section 9.

9. See Chapter 2, Section 1.

10. Philipp Frank, "The Reasons for the Acceptance of Scientific Theories," *Scientific Monthly,* Vol. 79 (September, 1954).

11. See Chapter 2, Section 8.

12. See Chapters 8 and 9.

13. George Wald (1906–   ), American biologist. See *Scientific Monthly,* Vol. 79, (1954).

14. George Gaylord Simpson (1902–   ), American vertebrate paleontologist. *The Meaning of Evolution* (New Haven: Yale University Press, 1949). Later reprinted in an inexpensive Mentorbook edition.

15. Lecomte du Nouy, *Human Destiny* (New York and London: Longmans, Green & Co., 1947).

Index

S